PROTEOME ANALYSIS
INTERPRETING THE GENOME

PROTEOME ANALYSIS
INTERPRETING THE GENOME

Edited by

David W. Speicher

The Wistar Institute
3601 Spruce Street
Philadelphia, PA 19104
USA

2004

ELSEVIER

Amsterdam – Boston – Heidelberg – London – New York – Oxford
Paris – San Francisco – Singapore – Sydney – Tokyo

ELSEVIER B.V.
Sara Burgerhartstraat 25
P.O. Box 211, 1000 AE Amsterdam
The Netherlands

ELSEVIER Inc.
525 B Street, Suite 1900
San Diego, CA 92101-4495
USA

ELSEVIER Ltd
The Boulevard, Langford Lane
Kidlington, Oxford OX5 1GB
UK

ELSEVIER Ltd
84 Theobalds Road
London WC1X 8RR
UK

First edition 2004

Library of Congress Cataloging in Publication Data
A catalog record is available from the Library of Congress.

British Library Cataloguing in Publication Data
A catalogue record is available from the British Library.

ISBN: 0 444 51024 9

♾ The paper used in this publication meets the requirements of ANSI/NISO Z39.48-1992 (Permanence of Paper).
Printed in Hungary.

Preface

The new discipline of proteomic can trace its birth to the mid-1990's when the terms "proteome" and "proteomics" were first introduced by Wilkins and Williams. Although some investigators initially considered "proteomics" to simply be a new name for conventional protein biochemistry, it rapidly became clear that this was a misperception. Proteomics represents an exciting new way to pursue biological and biomedical science at an unprecedented pace. Proteomics takes a broad, comprehensive, systematic approach to understanding biology that is generally unbiased and not dependent upon existing knowledge. It is primarily discovery-based rather than "hypothesis limited". That is, conventional one protein-at-a-time reductionist research is based upon and limited by existing knowledge. A reductionist scientist usually chooses to study a protein because prior research indicates that it is, or may be, important in a biological or disease process of interest. However, there could easily be 10 or more proteins that are more critical to the biological or disease process being studied, which have not yet been identified as associated with this process. Proteomics provides the means for identifying some or all of these more critical protein targets. Hence proteomics should have a dramatic impact on biological, biomedical and pharmaceutical research for the foreseeable future.

In comparison to genomics, proteomics is a far more complex and challenging problem. Genomes, particularly mammalian genomes, are certainly quite complex and, for example, simply assigning all open reading frames (ORFs) and splice variants of the human genome remains a daunting, incomplete task. However, even these complex genomes are finite and largely static over the lifetime of an organism. In contrast, proteomes are constantly changing in response to external stimuli and during development. The proteomes of higher eukaryotes are especially complex because alternative gene splicing and posttranslational modifications produce up to 20 to 100 times more unique protein components with distinct biological activities than the number of ORFs in the genome. The simplest form of a complex organism, a single cell type in culture, will express different subsets of this vast array of biologically distinct protein components under different conditions, and the moment the cells are exposed to a new condition, a different proteome will result. Hence, there are an essentially infinite number of proteomes that even such a relatively simple biological system can express. Despite this nearly infinite complexity, current proteomic tools provide the capacity to discover many novel individual proteins and groups of proteins (biosignatures) that are associated with normal biological processes as well as polygenetic disorders such as cancers and cardiovascular disease. These targets and biosignatures have substantial potential for development of novel therapies and diagnostics. Furthermore, proteomics together with modern genetic methods such as nucleic acid arrays and siRNA approaches are starting to generate complex datasets that should begin to provide a systems biology view of cells, organisms and complex diseases.

This volume is a compilation of the current status of proteomics written by an international panel of experts in the field. The major components of proteomics from basic discovery using a range of alternative analytical methods, to discovery validation, and use of proteome methods for clinical applications are discussed. State-of-the-art protein profiling methods described include high resolution two-dimensional gels, two-dimensional differential in-gel electrophoresis, LC-MS and LC-MS/MS using accurate mass tags, and protein identifications of proteins from gels. These chapters reflect the rapid evolution of key technologies, including "old" methods such as 2-D gels, over this discipline's first decade. However, further refinement of existing methods and continued introduction of new approaches are still clearly needed. Further advances are critical because no existing single analytical method or combination of methods can profile and quantitatively compare the majority of proteins in complex samples such as cells, tissues or biological fluids from higher eukaryotes. To address these limitations, most proteomics experts have turned to additional methods of sample separation. Some of the most promising approaches to obtain more global analysis of complex proteomes are to prefractionate proteomes into a series of well resolved fractions or to reproducibly isolate a single sub-proteome for detailed study. These approaches are discussed in a chapter describing electrophoretic prefractionation methods for global proteome analysis, and several chapters characterizing selected sub-proteomes based on specific posttranslational modifications including the phospho-proteome, the glyco-proteome, and nitrated proteins.

These conventional proteome analysis chapters are complemented by discussion of emerging technologies and approaches including affinity-based biosensor proteomics as well as the use of protein microarrays, microfluidics and nanotechnology. Strategies for improving throughput by automation are also discussed. Other chapters address the application of current proteome techniques to clinical problems and the availability of protein expression library resources for proteome studies.

Finally, I would like to express sincere appreciation to all the authors for their invaluable scientific contributions in a field where the rapid pace makes overcommitment and lack of time the norm. I would also like to acknowledge the invaluable administrative assistance of Ms. Emilie Gross with all aspects of editing and preparation of this volume for publication.

David W. Speicher

Contents

List of Contributors

Natalie G. Ahn
Department of Chemistry and Biochemistry, University of Colorado, Boulder, CO 80309, USA
and
Howard Hughes Medical Institute, University of Colorado, Boulder, CO 80309, USA

David G. Camp II
Macromolecular Structure and Dynamics, Environmental Molecular Sciences Laboratory, Battelle, Pacific Northwest National Laboratory, P.O. Box 999/MSIN K8-98, Richland, WA 99352, USA

Bruno Catimel
The Ludwig Institute for Cancer Research, Melbourne Tumour Biology Branch, P.O. Box 2008, The Royal Melbourne Hospital, Parkville, Vic. 3050, Australia

Angelika Görg
Technische Universität München, Fachgebiet Proteomik, D-85350, Freising-Weihenstephan, Germany

Patrick R. Griffin
Department of Molecular Profiling Proteomics, Merck Research Laboratories, Rahway, NJ 07065, USA

Sam M. Hanash
Department of Pediatrics, University of Michigan, 1150 W. Medical Center Drive, MSRBI, Room A520, Ann Arbor, MI 48109, USA

William A. Hanlon
Department of Molecular Profiling Proteomics, Merck Research Laboratories, Rahway, NJ 07065, USA

Peter James
Protein Technology, Wallenberg Laboratory II, Lund University, P.O. Box 7031, SE-220 07 Lund, Sweden

Judith Jebanathirajah
Department of Biochemistry and Molecular Biology, University of Southern Denmark, Camusvej 55, DK-5230 Odense M, Denmark

Ole Nørregaard Jensen
Department of Biochemistry and Molecular Biology, University of Southern Denmark, DK-5230 Odense M, Denmark

Joshua Labaer
Institute of Proteomics, Harvard Medical School, Department of Biological Chemistry and Molecular Pharmacology, 250 Longwood Avenue, Boston, MA 02115, USA

Thomas Laurell
Department of Electrical Measurements, Lund Institute of Technology, Lund University, P.O. Box 118, SE-221 00 Lund, Sweden

KiBeom Lee
The Wistar Institute, 3601 Spruce Street, Philadelphia, PA 19104, USA

Georgy Marko-Varga
Department of Analytical Chemistry, Lund University, P.O. Box 124, SE-221 00 Lund, Sweden

Gerald Marsischky
Institute of Proteomics, Harvard Medical School, Department of Biological Chemistry and Molecular Pharmacology, 250 Longwood Avenue, Boston, MA 02115, USA

Edouard Nice
The CRC for Cellular Growth Factors, Melbourne, Australia
and
The Ludwig Institute for Cancer Research, Melbourne Tumour Biology Branch, P.O. Box 2008, The Royal Melbourne Hospital, Parkville, Vic. 3050, Australia

Johan Nilsson
Department of Electrical Measurements, Lund Institute of Technology, Lund University, P.O. Box 118, SE-221 00 Lund, Sweden

Katheryn A. Resing
Department of Chemistry and Biochemistry, University of Colorado, Boulder, CO 80309, USA

Peter Roepstorff
Department of Biochemistry and Molecular Biology, University of Southern Denmark, Camusvej 55, DK-5230 Odense M, Denmark

Richard D. Smith
Macromolecular Structure and Dynamics, Environmental Molecular Sciences Laboratory, Battelle, Pacific Northwest National Laboratory, P.O. Box 999/MSIN K8-98, Richland, WA 99352, USA

David W. Speicher
The Wistar Institute, 3601 Spruce Street, Philadelphia, PA 19104, USA

Walter Weiss
Technische Universität München, Fachgebiet Proteomik, D-85350, Freising-Weihenstephan, Germany

Xun Zuo
The Wistar Institute, 3601 Spruce Street, Philadelphia, PA 19104, USA

Proteome Analysis. Interpreting the Genome.
D.W. Speicher (editor)

Chapter 1

Overview of proteome analysis

DAVID W. SPEICHER*

The Wistar Institute, 3601 Spruce Street, Philadelphia, PA 19104, USA

1. Introduction

Proteomics emerged as an exciting new discipline in the mid-1990s and has continued to grow rapidly since its inception. This new approach to biology and biomedical research was enabled by the emerging availability of complete sequences of genomes and the development of high sensitivity mass spectrometry techniques and instruments for analyzing proteins and peptides. The term

* Tel.: +1-215-898-3972; fax: +1-215-898-0664. E-mail: speicher@wistar.upenn.edu (D.W. Speicher).

'PROTEOME' was first introduced in the mid-1990s by Wilkins and Williams to indicate the entire 'PROTEin' complement expressed by a 'genOME' of a cell, tissue, or entire organism [1]. The genome of a cell or organism is finite and static over its lifetime, but in contrast, proteomes are highly dynamic and constantly changing in response to external stimuli and during development. Especially for complex organisms such as higher eukaryotes, the potential number of unique proteomes is essentially infinite.

Some investigators consider proteomics to be a new name for conventional protein biochemistry. However, this viewpoint overlooks at least two critical distinguishing features of proteomics as a new discipline. First, in contrast to conventional studies that focus on detailed characterization of one or at most a few genes or proteins, proteomics takes a broader, more comprehensive and systematic approach to understand biology. Second, proteomics is primarily discovery-based rather than hypothesis-driven. As a result it is not constrained by prior knowledge, and proteomics can provide exciting new opportunities to dramatically advance knowledge by identifying new drug targets, lead to development of superior diagnostics and therapeutics, correlate biological pathways and molecular mechanisms with disease, and enable a systems view of biology.

Conventional hypothesis-driven research has identified some important targets and key pathways for complex polygenetic diseases such as many cancers and cardiovascular diseases, which are being studied by conventional means. But, proteomics provides the capacity to discover many novel additional disease targets that may be even more important than known proteins and genes (Chapter 9). More importantly, proteomics as well as modern genetic methods such as nucleic acid arrays are beginning to provide complex datasets that will provide a systems biology view of cells, organisms, and complex diseases when adequate supporting bioinformatic tools are developed.

The best definition of proteomics is "any large-scale protein-based systematic analysis of the entire proteome or a defined sub-proteome from a cell, tissue, or entire organism." Proteomics is the next logical step after genome sequencing, but analysis of proteomes is much more complicated and challenging than sequencing genomes, especially in higher eukaryotes. This is due to the constantly changing nature of proteomes, the greater complexity of proteomes compared with the corresponding genome, and the limited capacities of available analytical platforms for profiling proteins.

2. Scope of the proteomics problem

Many prokaryotes have relatively simple genomes with several thousand or fewer genes, and the maximum number of proteins that can be produced is approximately equal to the number of genes because there is minimal posttranslational processing in these organisms. The simplest eukaryote, yeast, is somewhat more complex with over 6000 genes and moderate amounts of posttranslational processing. But a typical yeast proteome probably contains less than 6000 unique protein components because not all proteins are expressed at any given time. Therefore, various protein profiling methods are capable of detecting and analyzing a major portion of most prokaryotic and yeast proteomes.

In contrast, the proteomes of higher eukaryotes are far more challenging with complexities that substantially exceed the capacities of all existing analytical protein profiling methods. For example, although the human genome is now estimated to contain only about 35,000 genes, the total number of unique protein components encoded by this genome is of the order of several million. This is due to alternative splicing of many mRNAs and extensive, variable posttranslational modifications of most proteins [2]. There are currently no solid estimates of the numbers of unique protein components present in an 'average' single human cell type, but it seems likely that at least 20,000–50,000 or more protein forms will be present in most types of human cells. Of course the total number of proteins present in a human tissue or tumor will be even greater and may easily exceed 100,000 components.

A second important attribute of most proteomes including those of some simpler organisms is the wide range in protein abundance levels that are usually encountered. Single cell organisms or individual cell types often contain many low abundance proteins that are present at about 100 copies per cell, while the most abundant proteins may be present at 100,000,000 copies per cell for a dynamic range of about 10^6 [3]. Tissue specimens will have even greater dynamic ranges because they contain mixtures of cell types that have unique but overlapping proteomes. Current gel as well as non-gel protein profiling methods typically have detection dynamic ranges that are at best about 10^3–10^4. Furthermore, biological fluids usually have much wider dynamic ranges than cells and tissues. Human plasma or serum has a few major proteins that comprise more than 90% of the total protein in the sample. Albumin, the most abundant protein is present at about 40 mg/ml while cytokines and other low abundance proteins are present at a few pg/ml for a dynamic range of at least 10^{10} [3,4].

The greater complexity and dynamic range of most proteomes compared with available analytical tools have both driven improvements in analysis methods and interest in reliable sample fractionation methods. Important early considerations in designing a proteomics experiment are to select a good experimental system, a reproducible sample preparation method, and any optional fractionation method such that introduction of experimentally induced losses of specific proteins will be minimized. This is critical because experimentally induced protein changes such as proteolysis, aggregation, precipitation, and poor sample recoveries are frequently variable and can produce protein changes that can be difficult or impossible to distinguish from authentic changes associated with the biological question being investigated. Experimental manipulations reflecting a biological question as well as those caused by poorly controlled experimental parameters will usually produce many detected quantitative changes. It is therefore essential that all experimental parameters be carefully controlled. For example, when working with cell lines the passage number as well as all environmental conditions should be as consistent as possible for the duration of the experiment, including extent of confluency, pH, nutrient depletion, pH changes, etc. Sample harvesting and processing methods should be very reproducible and designed to minimize or avoid proteolysis, oxidation, dephosphorylation, etc. Sorting out interesting discoveries from false signals will be more time consuming than optimizing experimental design and sample preparation before beginning the experiment, i.e. a common adage that applies here is, 'garbage in-garbage out'.

When dealing with higher eukaryotes, tissue specimens generally provide a more accurate picture of physiological conditions and proteome states than cells in culture, but the multiple cell types present in tissues increase the number of proteins present and widen the dynamic range. In addition, if the portions of cell types change between compared samples, large numbers of protein changes will be detected that merely reflect the change in proportions of cell types. Some common methods for reducing tissue heterogeneity include: laser capture microdissection, FACS sorting and magnetic bead capture [5–7]. However, limitations of these methods include potential perturbation of the proteome of interest during the frequently required lengthy sample processing, very limited sample sizes, and changes in the proteome caused by dissociation of tissues into individual cells for FACS sorting or magnetic bead capture.

3. Global versus targeted proteomics

The *scope* of proteomics experiments can be either global or targeted, regardless of the *type* of study. Types of proteome studies include: (i) quantitative comparisons of protein levels in multiple samples (protein profiling); (ii) analysis of protein–protein interactions including binary interactions and isolation/analysis of macromolecular complexes; and (iii) protein compositional studies, i.e. define all detectable proteins in a sample of interest without quantitation. Each of these types of studies could either have a global or targeted scope.

Ideally, global proteomics studies would be capable of examining all of the proteins in a cell or tissue, such as comprehensive comparison of protein levels in a cell line in the presence versus absence of a drug or transfected gene, or comparison of a normal and diseased form of a tissue. These unbiased studies have the potential to discover new critically important proteins and cellular pathways associated with the experimental parameter being investigated. Such proteome wide studies are quite practical for simple single cell organisms, but become dramatically more challenging for higher eukaryotes as discussed earlier. However, even though current capacities of proteomic technologies allow separation and detection of only a modest portion of the protein components present in complex proteomes, existing methods are nonetheless capable of producing potential new drug targets, diagnostic markers, and novel insights into disease mechanisms. In addition, proteome analysis methodologies, particularly mass spectrometry capacities are continuing to develop at a rapid rate. Hence the capacities of proteome analytical tools will probably continue to develop rapidly in the foreseeable future.

One strategy for dealing with proteomes that are substantially more complex than the capacity of available analysis tools is to divide the sample into a small number of well-resolved fractions that can be separately analyzed in parallel (Chapter 4). Another approach is to use a targeted proteomics strategy where a well-defined subset of an entire proteome or sub-proteome is characterized. Examples of targeted proteomics studies include analysis of: glycoproteomes (Chapter 5), phosphoproteomes (Chapter 6), organelles or other subcellular fractions, cellular machines such as ribosomes, the nuclear pore complex, etc. [8, 9]. Specific classes of proteins such as cysteine proteases can also be fished out of cell or tissue extracts using appropriate affinity methods [10]. Two of the biggest challenges in targeted proteomics are to develop highly reproducible isolation methods and to distinguish between specific and non-specific associations. An advantage of most targeted approaches is the closer match between sample complexity and analysis method capacity when dealing with complex proteomes from higher eukaryotes. Also, in many cases targeted proteomics by its nature addresses more focused biological questions than global studies, which facilitates data analysis and interpretation.

4. Top down protein profiling methods

Protein profiling, the quantitative comparisons of proteins in two or more samples, is the most common type of proteome study. Analogous to most nucleotide array experiments, the goal is to define the changes in levels of all proteins that define different physiological, experimental, or disease states.

Protein profiling methods can be classified as either top down or bottom up approaches. Top down methods separate *intact proteins* using one or more separation modes, and quantitative changes in levels of individual proteins are measured at the intact protein level. Proteins of interest, which are usually those exhibiting the largest changes in abundance, are then typically identified by fragmenting the protein with trypsin followed by MS/MS analysis (Chapter 7). With bottom up methods, two samples to be compared are differentially labeled with an appropriate stable isotope label, the two unfractionated protein extracts are mixed, and this mixture is digested with trypsin prior to any separation. The *tryptic peptides* are then separated by one or more separation methods prior to introduction into a mass spectrometer capable of MS/MS analysis. Protein identities are determined from the MS/MS spectra and relative abundances of proteins are determined by ratios of signals for the light and heavy forms of identified labeled peptides.

4.1. Two-dimensional gels

The most commonly used and oldest top down protein profiling technology is 2D gel electrophoresis (Chapter 2) where proteins are focused to their isoelectric points (p*I*s) under denaturing, reducing conditions in a first dimension followed by size separation in an SDS polyacrylamide gel. This basic method has been used for nearly 30 years, but only sporadic attempts have been made to improve the reproducibility, throughput and associated limitations. Probably the most significant early advance was the introduction in the 1980s of immobilized pH gradient gels (IPGs), which utilized covalently bound acrylamide derivatives with buffering groups to establish and maintain pH gradients [11]. The commercial availability of IPG gels with different pH ranges and gradient shapes substantially improved the reproducibility and convenience of 2DE compared with the previous method using soluble carrier ampholytes to establish pH gradients [12–14].

Until the emergence of proteomics as a major new discipline, gel product manufacturers invested little effort in improving the 2D gel technology. Although no company has yet invested a concerted effort in optimizing 2DE, the growing interest in proteomics has recently stimulated production of a wider range of options for gels and separations apparatus including partially or fully automated systems. One notable advance is the current availability of high-voltage IEF units with efficient cooling systems and programmable control of run conditions, which allows unattended gel rehydration and focusing in an overnight run [15]. Another important recent advance was the development of high-sensitivity non-covalent fluorescent stains for general proteins, glycoproteins, and phosphoproteins [16–18]. A major advantage of fluorescent stains compared with conventional chromogenic stains such as Coomassie blue and silver stains is a wider linear detection range.

Compared with alternative protein profiling methods, 2DE has several major strengths and weaknesses. The 2DE method can detect many but not all posttranslational modifications, particularly those that involve addition or loss of one or more charges to amino acid side chains. Similarly, detection of proteolytic processing is usually quite facile, and most alternatively spliced forms of a protein can be easily distinguished by shifts in either p*I* or size or both. In addition, 2DE is not limited to simultaneous comparisons of only two or three samples, and new samples can be compared to previously analyzed samples that may no longer be available. However, 2DE also has several major weaknesses, including: (i) a limited capacity for the total amount of an extract that can be analyzed directly, which limits detection to the most abundant proteins in the sample; (ii) very large, very small, and membrane proteins are poorly recovered; and (iii) although IPG gels have improved reproducibility compared with older soluble ampholyte-based systems, substantial gel-to-gel variation still persists.

4.2. Two-dimensional differential gel electrophoresis (2D DIGE)

The 2D DIGE method ([19] and Chapter 3) is a variation of 2DE that involves covalently labeling two or three different protein extracts with distinct fluorophores, which are cyanine dyes (Amersham Biosciences) that modify lysine residues. These dyes maintain a positive charge on the modified lysine and they have similar masses so that the migration position on 2D gels of a given protein is the same for different fluorescent derivatives, which have distinguishable fluorescence spectra. The primary advantage of 2D DIGE compared with conventional 2DE is that gel-to-gel variation is eliminated because two or three differently labeled samples can be mixed prior to electrophoresis. Another advantage is that the number of parallel and replicate gels required to obtain reliable results is greatly reduced.

Mild fluorescent dye coupling reaction conditions are used so that only a few lysines are modified per protein molecule, which minimizes migration shifts in the SDS-PAGE dimension due to the increased mass. As long as the extent of reaction is the same between samples to be compared, the mass shift will be uniform and the p*I* should be essentially the same as the unreacted protein. The fluorophores have distinguishable spectra, and therefore, the individual fluorescently labeled protein samples are combined and separated by conventional 2DE on a single gel. The proteins from each sample can then be visualized separately by imaging, using appropriate excitation and emission wavelengths for the different dyes. The differences between the quantities of the individual proteins from each sample can then be determined using specialized 2D image analysis software designed for this system.

One limitation of this method is that excision of spots of interest for identification by MS can be problematic. Due to the incomplete labeling of lysines, a major portion of most proteins may not be labeled and will migrate slightly more rapidly than the larger, fluorescently labeled form of the protein that is used for detection and quantitative comparisons. Even if a spot cutting robot is used that can detect fluorescent spots, the amount of the protein that is fluorescently labeled may be too low for facile protein identification using LC–MS/MS after in-gel digestion (Chapter 7). The position of the bulk amount of unlabeled protein may be estimated as being shifted about one spot diameter down (lower mass), but this strategy can lead to excision of a protein other than the one of interest [20]. Alternatively, one can attempt to correlate the 2D DIGE gel pattern with a colloidal Coomassie pattern from a parallel preparative 2D gel [21]. But this method is not entirely reliable because different stains show protein-to-protein variations in staining intensity and colloidal Coomassie is less sensitive. A second limitation of the 2D DIGE method is that many low abundance proteins are not detected because the partial lysine labeling method is usually less sensitive than a good silver stain. The recent development of fluorescent tags for cysteines instead of lysines may minimize these limitations and make the 2D DIGE method more robust.

4.3. Non-2D gel separation methods

The limitations of 2D gels discussed above have stimulated searches for both alternative top down protein separation methods and bottom up methods that fragment complex mixtures and separate peptides (see later). A number of different protein separation methods have been developed that use either two or more chromatographic separation modes or a preparative IEF method coupled with one or more chromatographic modes [22–25]. The final readout in most separation schemes is to use reverse phase chromatography as the final step with an ESI mass spectrometer as the detector. All these methods face common challenges including: (i) achieving good separation in the first dimension with minimal cross-contamination of proteins between adjacent fractions; (ii) preserving this resolution as fractions are collected and introduced into the second dimension; (iii) obtaining a reasonable throughput level; and (iv) detection of a large number of proteins at high sensitivity.

Despite some limitations, 2D gels remain as the gold standard to which any competing new method should be compared. A viable alternative separation method should provide clear advantages over 2D gels such as: detect more proteins, require less analysis time, detect lower abundance proteins, or have other advantages. At present, it is not apparent that any non-2D gel protein separation method provides major advantages over 2D gels in terms of the number of proteins detected, analysis time, or sensitivity. However, future advances in automation and chromatographic capacities may produce a clearly superior protein profile analysis technology. While non-2D gel protein separation methods may not currently be clearly superior protein profiling methods, they do have a number of attractive features. First, the use of a mass spectrometer as a detector enables very accurate mass measurements of intact proteins, and therefore posttranslational modifications and artifactual modifications, including those involving very small mass changes, can usually be detected. Second, the amount of sample used for fractionation can usually be readily scaled up, which increases the feasibility of detecting low abundance proteins. Third, most methods in this group involve collection of a portion of each fraction, and when interesting changes are observed in a fraction, the collected sample can be used for further fractionation or fragmentation to identify the protein of interest.

4.4. Protein and antibody arrays

Protein arrays, particularly antibody or antibody mimic arrays (Chapter 10), are exciting new alternatives for protein profiling studies that show great promise if key technical hurdles can be overcome. Among existing protein profiling methods, antibody arrays are most likely to ultimately offer the highest sensitivity and throughput for quantitatively comparing large numbers of proteins in complex protein extracts [26–28].

Protein arrays contain specific receptor or capture proteins (usually antibodies) that are immobilized or printed on a surface such as a glass slide or in a microtiter plate. Test samples are applied to the array, target proteins are specifically bound to the immobilized receptor, and the interaction is detected by one of the number of alternative methods including: (i) a second specific antibody; (ii) fluorescent labeling of the protein mixture before applying it to the array; or (iii) direct physical detection of the added mass using atomic force microscopy, a biosensor, or mass spectrometry. Protein arrays hold great promise including the potential for very high throughput, high sensitivity, and capacity to screen thousands of targets. At the same time, there are many challenges that must be solved before protein arrays become routine, robust protein profiling tools. Issues that need to be optimized include: (i) printed proteins must be maintained in a native, functional state; (ii) non-specific interactions must be avoided or corrected; and (iii) different capture reagents on an array must have similar binding affinities that are compatible with ligand concentration ranges in the sample, i.e. a high affinity capture reagent with a 1 nM K_d will not be able to efficiently trap a ligand present in the test sample at a 1 pM level. But perhaps the biggest hurdle to overcome is simply the monumental task of developing hundreds to thousands of capture reagents to desired target proteins that have appropriate binding affinities and specificities with similar linear detection ranges.

A critical component of protein arrays is the method used to detect protein–protein interactions. The most common detection methods for antibody arrays are based on standard ELISA (enzyme-linked immunosorbent assay) methods. For example, a cytokine microarray in microtiter plates used immobilized antibodies to capture the cytokines, soluble biotinylated detection antibodies that recognized different epitopes on the cytokines, and detection using a streptavidin–horseradish peroxidase conjugate with a chemiluminescent substrate to achieve a detection limit of 1 pg/ml [29]. An alterative detection method utilizes labeling of two samples to be compared with different fluorescent dyes analogous to the 2D DIGE approach, and this approach only requires one specific antibody for each protein target. The different intensities of the distinct fluorescent signals at each spot on the array reflect the relative abundances of the target protein in the two samples [30]. An attractive alternative detection method for antibody arrays is the rolling circle amplification method where captured antigens are coupled to antibody conjugated to an oligonucleotide that hybridizes with a circular DNA molecule. DNA polymerase produces a long single stranded DNA molecule with many tandem repeats of the circle DNA sequence which can be fluorescently labeled by direct incorporation of labeled nucleotides or by hybridization with a complementary fluorescently labeled probe [31].

Biosensor protein arrays ([32] and Chapter 10) use very different detection methods compared with those described above. These devices utilize capture reagents such as antibodies, which are linked to an optically or electronically active surface that is a mass sensitive detector. Protein–protein association produces an electrical signal proportional to the mass bound to the sensor surface. Real time monitoring allows determination of association and dissociation rates as well as quantitation of the total ligand bound. A number of different types of instruments are either commercially available or in production, including biosensor arrays that utilize grating coupled surface plasmon resonance instead of prism-based surface plasmon resonance [33].

5. Bottom up protein profiling methods

Bottom up protein profiling methods initially fragment complex protein mixtures, typically using trypsin, and the resulting peptide mixture is separated by multiple tandem chromatographic methods prior to introduction into a mass spectrometer. The most commonly used final analysis step is to interface a nano-capillary reverse phase HPLC column with a mass spectrometer capable of performing MS/MS measurements (LC–MS/MS). If qualitative protein profiling is desired, i.e. simply identifying as many proteins as possible in a sample, an unlabeled sample can be used. Protein identities are determined by using an appropriate program such as SEQUEST or MASCOT to compare MS/MS spectra to sequence databases [34,35]. If a quantitative comparison of two samples is desired, one sample must be labeled with a stable isotope either by metabolic incorporation or through chemical derivatization.

5.1. Multi-dimensional protein identification technology

A number of different research groups have developed multiple step chromatographic methods for separating the complex peptide mixtures resulting from trypsin digestion of complex protein mixtures. This general strategy has been given the colorful name 'MudPIT' for multi-dimensional protein identification technology by Yates and co-workers [36]. Using a high sensitivity nanocapillary ion exchange/reverse phase separation of tryptic peptides prior to MS/MS analysis, they demonstrated identification of ~ 1500 proteins from a yeast cell extract. The MudPIT method can be used for qualitative analysis of relatively simple mixtures such as macromolecular complexes or complete proteomes such as the yeast or higher eukaryote proteomes. The most common method of converting this analysis method to quantitative profile comparisons is by introduction of stable isotope labeling into one of two samples to be compared using either metabolic labeling or chemical modification [37–39].

5.2. Isotope-coded affinity tags

A recently developed, attractive method for quantitative comparisons of two proteomes is the isotope-coded affinity tag (ICAT) method. The ICAT reagent has a cysteine-specific reactive group at one end of the molecule and a biotin affinity tag at the opposite end. A hydrocarbon linker connecting these two functional groups can be synthesized with all hydrogens (light ICAT) or with eight deuteriums (heavy ICAT) [40]. One sample is reacted with the light reagent and the second sample is reacted with the heavy reagent using identical end point labeling conditions. After a solution trypsin digestion, the extremely complex tryptic peptide mixture is simplified by affinity purifying the cysteine containing derivatized peptides on an avidin affinity resin. The eluted peptides are then analyzed using LC–MS/MS for simpler samples or LC/LC–MS/MS for more complex samples. The ratios of MS signals for the light and heavy ICAT-labeled forms of the same peptide are compared to determine the relative abundance of the parent protein in the respective initial samples, and MS/MS data is used to identify the protein as described earlier.

The ICAT method has a number of advantages and disadvantages compared with 2D gels or alternative bottom up methods. One advantage compared with other bottom up methods is that the ICAT method can be used with any type of sample including those where metabolic incorporation of stable isotopes would not be feasible. In addition, the ability to fish out only cysteine-containing peptides has the advantage of simplifying the peptide mixture but has the disadvantage that proteins without cysteines are missed, as are the features of proteins such as posttranslational modifications or alternative splice forms that do not occur on cysteine containing peptides. Compared with 2DE, the amount of sample that can be used with the ICAT method is less restricted, the process is easier to automate, and specific classes of proteins such as very acidic, very basic, and membrane proteins, which are difficult to detect by 2DE may be more readily detected. However, conversely many protein changes that can often be readily detected on

2D gels may be difficult or impossible to detect using the ICAT method, including: proteolytic processing, changes in the splicedform of a protein that is expressed, and changes in posttranslation modifications including phosphorylation and glycosylation.

Several technical difficulties encountered with the original ICAT reagent have been resolved with a second generation cleavage version (Applied Biosystems, Foster City, CA, USA) that contains ^{13}C atoms rather than deuterium. The elimination of deuteriums reduces the possibility of chromatographic separation of the heavy and light forms of each peptide, which was often observed with the deuterium labeled reagent, cleavage of the linker improves yields from the affinity column, and removal of the bulky biotin improves mass analysis [41–44].

Like alternative LC/LC–MS/MS methods, ICAT methods can be automated and analysis speed as well as the numbers of proteins detected should continue to improve as mass spectrometry methods continue to improve. However, it is important to have realistic expectations and currently LC/LC–MS/MS methods require many hours, typically up to several days or more, of mass spectrometer instrument time and computer time, and in only a few cases have more proteins been quantitatively profiled than the 1000–1500 proteins that can be separated in a single high resolution 2D gel.

5.3. Accurate mass tag based protein profiling

An emerging alternative to above methods, which require repetitive identifications of proteins by correlating MS/MS spectra with sequence databases, is the utilization of mass spectrometers with very high mass accuracy to define accurate mass tags (Chapter 9). Use of elution times from high-resolution capillary LC columns combined with accurate masses obtained on Fourier transform ion cyclotron resonance mass spectrometers eliminates the need for routine, repetitive MS/MS measurements [45]. This can increase throughput, decrease computer analysis time, and simplify data interpretation. In this strategy, the protein origins of peptides are initially identified using MS/MS, and subsequent studies utilize only accurate mass tags and elution times to identify the proteins. This approach is obviously currently restricted to those laboratories that have mass spectrometers capable of the required high mass accuracy. But it is reasonable to expect that the rapid advances in mass spectrometry capacities over the past decade are likely to continue and this could make very high mass accuracy instruments relatively commonplace in the near future. Similar to the MudPIT and ICAT methods, comparisons of proteins abundances can be accomplished by incorporating stable isotopes into one of two samples to be compared [46].

6. Automation, miniaturization, and future prospects

Throughput and sensitivity of most protein analysis methods are being enhanced by increased automation (Chapters 10 and 12). As noted above, 2DE has been difficult to automate, but progress is being made even in this challenging area, including programmable IEF units for automated overnight IPG strip rehydration and focusing and even partially or fully automated 2D electrophoresis units. Even greater progress is being made in post-gel handling steps including use of robots for spot excision, sample alkylation, in-gel trypsin digestion, post-digestion cleanup/concentration, and sample spotting onto MALDI mass spectrometer targets or into injection vials for LC–MS analysis. Similarly, the non-2D gel protein profiling methods can be automated using nanoscale compatible autosamplers and sophisticated HPLC pumping systems and automated switching valves for unattended multi-dimensional separations. As automation methods become more robust, they can be expected to increase throughput and reproducibility, particularly between different laboratories.

In parallel with increased emphasis on automation, another high priority is to scale-down analysis methods to micro- and nanolevels (Chapter 13). Biological samples are frequently available in very limited amounts including tissue biopsies, laser microdissected specimens, etc. In addition, the amount of protein in low abundance excised 2D gel spots or capillary chromatographic fractions can present challenges for protease digestion and sample handling. Hence, new analytical protocols capable of processing nanoliter to picoliter volumes and low femtomole to attomole or lower amounts of proteins or peptides are being developed. Microfluidics and efficient processing of minute sample volumes without adsorptive losses and with improved reaction kinetics should help drive further improvements on proteome analysis sensitivity.

An interesting question is where is this new field of proteomics headed over the next 5–10 years. It is now clear that completion of the human genome sequence has given birth to an exciting new field that is essentially infinite in size and complexity and interpreting the genome will be the real task in understanding biology and diseases. In the short time since proteomics has emerged as a discipline, it has already begun to yield new diagnostic and therapeutic targets of diseases and more comprehensive insights into biological processes. As proteomics analysis platforms continue to mature, it is apparent that much more rapid and more impressive advances can be expected in the very near future. The next decade should be an incredibly exciting time for proteomics and discovery driven biomedical research.

A related question is which proteome technology(ies) will be used 5–10 years from now? As discussed earlier, proteomics has both rejuvenated an older technology, 2DE, and stimulated rapid progress in newer methodologies, particularly mass spectrometry, protein arrays, automation, and nanotechnology.

The single technology with the greatest potential as a high throughput, high sensitivity method for profiling a wide range of proteins is antibody or antibody mimic arrays. However, there are numerous technological problems that must be resolved before antibody arrays can become a routine, robust protein profiling tool capable of evaluating large numbers of proteins.

The great complexity of higher eukaryotic proteomes, their constantly changing nature, and most importantly, the wide diversity of protein properties and behavior suggest that no single proteome analysis method will be able to effectively address all proteome analysis problems within the near future. Each major existing technology is particularly well suited at detecting some types of proteins and certain types of changes, but each method is also equally poor at detecting other types of changes. Hence, it seems most likely that all the major current analytical technologies including 1D and 2D gels, multi-dimensional chromatography coupled with MS instruments, and antibody arrays will continue to play important and complementary roles in proteome analysis studies in the near future.

7. Summary

Proteomics is an exciting new approach to biological and biomedical research, which has rapidly grown into a major commercial and research enterprise with strong prospects for dramatically advancing our knowledge of basic biological and disease processes. The most common type of proteomics experiments involves quantitative protein profile comparisons of two or more experimental states of cells, tissues, or organisms by either ‘top down’ (separation and analysis of intact proteins) or ‘bottom up’ (separation and analysis of peptides) methods. These comparisons can be either global in scope or targeted to a specific sub-proteome such as phosphoproteins, glycoproteins, a specific organelle or cellular machine, or a class of proteins such as cysteine proteases.

Ideally global profiling experiments should be capable of quantitatively comparing the majority of proteins present in a proteome of interest. Several alternative protein profiling technologies including 2DE or LC/LC–MS/MS methods can analyze a major portion of the proteins expressed by prokaryotes or yeast. But no single current technology platform can reproducibly separate and quantitatively compare more than about a few thousand proteins, and therefore only a modest portion of complex proteomes such as cell extracts, tissue extracts, or biological fluids from higher eukaryotes can typically be analyzed. One strategy for increasing the comprehensiveness of analysis of complex proteomes is to subdivide the proteome either by targeting a specific sub-proteome or by using a high-resolution prefractionation method such as one of several alternative solution based IEF separation methods.

MS methods and instruments continue to improve at an impressive rate, which affects most protein profiling methods because MS is an integral part of most proteome methods, usually as the final readout of protein identifications. Automation, miniaturization, and refinement of protein arrays, particularly antibody arrays, are key areas of progress that should further improve the repertoire of tools available for proteome studies. While it is possible that one technology may emerge as a clearly superior approach, the great diversity of protein properties suggests that all existing major technologies including 1D and 2D gels, multi-dimensional chromatography methods coupled with MS instruments, antibody arrays and related methods will continue to play important and complementary roles in proteome analysis studies.

Acknowledgements

The author gratefully acknowledges the administrative assistance of Emilie Gross. This work was supported by NIH grants CA94360, CA77048, and CA92725 as well as an NCI cancer center grant (CA10815), and the Commonwealth Universal Research Enhancement Program, Pennsylvania Department of Health.

References

1. Wilkins, M.R., Pasquali, C., Appel, R.D., Ou, K., Golaz, O., Sanchez, J.C., Yan, J.X., Gooley, A.A., Hughes, G., Humphery-Smith, I., Williams, K.L. and Hochstrasser, D.F., From proteins to proteomes: large scale protein identification by two-dimensional electrophoresis and amino acid analysis. *Biotechnology*, **14**, 61–65 (1996).
2. Godley, A.A. and Packer, N.H., The importance of protein co- and post-translational modifications in proteome projects, In: Wilkins, M.R., Williams, K.L., Appel, R.D. and Hochestrasser, D.F. (Eds.), *Proteome Research: New Frontiers in Functional Genomics*. Springer, Berlin, 1997, pp. 65–91.
3. Corthals, G.L., Wasinger, V.C., Hochstrasser, D.F. and Sanchez, J., The dynamic range of protein expression: a challenge for proteome research. *Electrophoresis*, **21**, 1104–1115 (2000).
4. Herbert, B.R., Sanchez, J.C. and Bini, L., Two-dimensional electrophoresis: the state of the art and future directions, In: Wilkins, M.R., Williams, K.L., Appel, R.D. and Hochstrasser, D.F. (Eds.), *Proteome Research: New Frontiers in Functional Genomics*. Springer, Berlin, 1997, pp. 13–33.
5. Barisoni, L. and Star, R.A., Laser-capture microdissection. *Methods Mol. Med.*, **86**, 237–255 (2003).
6. Wulfkuhle, J.D., Paweletz, C.P., Ssteeg, P.S., Petricoin, E.F. III and Liotta, L., Proteomic approaches to the diagnosis, treatment, and monitoring of cancer. *Adv. Exp. Med. Biol.*, **532**, 59–68 (2003).
7. Siegel, D.L., Selecting antibodies to cell-surface antigens using magnetic sorting techniques. *Methods Mol. Biol.*, **178**, 219–226 (2002).
8. Huber, L.A., Pfaller, K. and Victor, I., Organelle proteomics: implications for subcellular fractionation in proteomics. *Circ. Res.*, **92**, 962–968 (2003).

9. Dreger, M., Proteome analysis at the level of subcellular structures. *Eur. J. Biochem.*, **270**, 589–599 (2003).
10. Jeffery, D.A. and Bogyo, M., Chemical proteomics and its application to drug discovery. *Curr. Opin. Biotechnol.*, **14**, 87–95 (2003).
11. Görg, A., Postel, W. and Günther, S., The current state of two-dimensional electrophoresis with immobilized pH gradients. *Electrophoresis*, **9**, 531–546 (1988).
12. Görg, A., Obermaier, C., Boguth, G., Harder, A., Scheibe, B., Wildgruber, R. and Weiss, W., The current state of two-dimensional electrophoresis with immobilized pH gradients. *Electrophoresis*, **21**, 1037–1053 (2000).
13. Celis, J.E. and Gromov, P., 2D protein electrophoresis: can it be perfected? *Curr. Opin. Biotechnol.*, **10**, 16–21 (1999).
14. Rabilloud, T., Two-dimensional gel electrophoresis in proteomics: old, old fashioned, but it still climbs up the mountains. *Proteomics*, **2**, 3–10 (2002).
15. Choe, L.H. and Lee, K.H., A comparison of three commercially available isoelectric focusing units for proteome analysis: the multiphor, the IPGphor and the protean IEF cell. *Electrophoresis*, **21**, 993–1000 (2000).
16. Schulenberg, B., Beechem, J.M. and Patton, W.F., Mapping glycosylation changes related to cancer using the Multiplexed Proteomics technology: a protein differential display approach. *J. Chromatogr. B*, **793**, 127–139 (2003).
17. Steinberg, T.H., Agnew, B.J., Gee, K.R., Leung, W.Y., Goodman, T., Schulenberg, B., Hendrickson, J., Beechem, J.M., Haugland, R.P. and Patton, W.F., Global quantitative phosphoprotein analysis using Multiplexed Proteomics technology. *Proteomics*, **3**, 1128–1144 (2003).
18. Patton, W.F., Detection technologies in proteome analysis. *J. Chromatogr. B*, **771**, 3–31 (2002).
19. Unlu, M., Morgan, M.E. and Minden, J.S., Difference gel electrophoresis: a single gel method for detecting changes in protein extracts. *Electrophoresis*, **18**, 2071–2077 (1997).
20. Patton, W.F., A thousand points of light: the application of fluorescence detection technologies to two-dimensional gel electrophoresis and proteomics. *Electrophoresis*, **21**, 1123–1144 (2000).
21. Tonge, R., Shaw, J., Middleton, B., Rowlinson, R., Rayner, S., Young, J., Pognan, F., Hawkins, E., Currie, I. and Davison, M., Validation and development of fluorescence two-dimensional differential gel electrophoresis proteomics technology. *Proteomics*, **1**, 377–396 (2001).
22. O'Neil, K.A., Miller, F.R., Barder, T.J. and Lubman, D.M., Profiling the progression of cancer: separation of microsomal proteins in MCF10 breast epithelial cell lines using nonporous chromatophoresis. *Proteomics*, **3**, 1256–1269 (2003).
23. Godovac-Zimmermann, J. and Brown, L.R., Proteomics approaches to elucidation of signal transduction pathways. *Curr. Opin. Mol. Ther.*, **5**, 241–249 (2003).
24. Ge, Y., Lawhorn, B.G., ElNaggar, M., Strauss, E., Park, J.H., Begley, T.P. and McLafferty, F.W., Top down characterization of larger proteins (45 kDa) by electron capture dissociation mass spectrometry. *J. Am. Chem. Soc.*, **124**, 672–678 (2002).
25. Lubman, D.M., Kachman, M.T., Wang, H., Gong, S., Yan, F., Hamler, R.L., O'Neil, K.A., Zhu, K., Buchanan, N.S. and Barder, T.J., Two-dimensional liquid separations — mass mapping of proteins from human cancer cell lysates. *J. Chromatogr. B*, **782**, 183–196 (2002).
26. Lal, S.P., Christopherson, R.I. and dos Remedios, C.G., Antibody arrays: an embryonic but rapidly growing technology. *Drug Discov. Today*, **7**(18 Suppl.), S143–S149 (2002).
27. Sreekumar, A. and Chinnaiyan, A.M., Protein microarrays: a powerful tool to study cancer. *Curr. Opin. Mol. Ther.*, **4**, 587–593 (2002).

28. Valle, R.P. and Jendoubi, M., Antibody-based technologies for target discovery. *Curr. Opin. Drug Discov. Dev.*, **6**, 197–203 (2003).
29. Moody, M.D., Van Arsdell, S.W., Murphy, K.P., Orencole, S.F. and Burns, C., Array-based ELISAs for high-throughput analysis of human cytokines. *Biotechniques*, **31**, 186–194 (2001).
30. Haab, B.B., Dunham, M.J. and Brown, P.O., Protein microarrays for highly parallel detection and quantitation of specific proteins and antibodies in complex solutions. *Genome Biol.*, **2**, 0004.1–0004.13 (2001).
31. Schweitzer, B., Roberts, S., Grimwade, B., Shao, W., Wang, M., Fu, Q., Shu, Q., Laroche, I., Zhou, Z., Tchernev, V.T., Christiansen, J., Velleca, M. and Kingsmore, S.F., Multiplexed protein profiling on microarrays by rolling-circle amplification. *Nat. Biotechnol.*, **20**, 359–365 (2002).
32. Jenkins, R.E. and Pennington, S.R., Arrays for protein expression profiling: towards a viable alternative to two-dimensional gel electrophoresis? *Proteomics*, **1**, 13–29 (2001).
33. Baird, C.L. and Muszka, D.G., Current and emerging commercial optical biosensors. *J. Mol. Recogn.*, **14**, 261–268 (2001).
34. Link, A.J., Eng, J., Schieltz, D.M., Carmack, E., Mize, G.J., Morris, D.R., Garvik, B.M. and Yates, J.R. III, Direct analysis of protein complexes using mass spectrometry. *Nat. Biotechnol.*, **17**, 676–682 (1999).
35. Perkins, D.N., Pappin, D.J., Creasy, D.M. and Cottrell, J.S., Probability-based protein identification by searching sequence databases using mass spectrometry data. *Electrophoresis*, **20**, 3551–3567 (1999).
36. Washburn, M.P., Wolters, D. and Yates, J.R. III, Large-scale analysis of the yeast proteome by multidimensional protein identification technology. *Nat. Biotechnol.*, **19**, 242–247 (2001).
37. Washburn, M.P., Ulaszek, R., Deciu, C., Schieltz, D.M. and Yates, J.R. III, Analysis of quantitative proteome data generated via multidimensional protein identification technology. *Anal. Chem.*, **74**, 1650–1657 (2002).
38. Peng, J., Elias, J.E., Thoreen, C.C., Licklider, L.J. and Gygi, S.P., Evaluation of multidimensional chromatography coupled with tandem mass spectrometry (LC/LC–MS/MS) for large-scale protein analysis: the yeast proteome. *J. Proteome Res.*, **2**, 43–50 (2003).
39. Kubota, K., Wakabayashi, K. and Matsuoka, T., Proteome analysis of secreted proteins during osteoclast differentiation using two different methods: two-dimensional electrophoresis and isotope-coded affinity tags analysis with two-dimensional chromatography. *Proteomics*, **3**, 616–626 (2003).
40. Gygi, S.P., Rist, B., Gerber, S.A., Turecek, F., Gelb, M.H. and Aebersold, R., Quantitative analysis of complex protein mixtures using isotope-coded affinity tags. *Nat. Biotechnol.*, **17**, 994–999 (1999).
41. Smolka, M., Zhou, H. and Aebersold, R., Quantitative protein profiling using two-dimensional gel electrophoresis, isotope-coded affinity tag labeling, and mass spectrometry. *Mol. Cell. Proteomics*, **1**, 19–29 (2002).
42. Hansen, K.C., Schmitt-Ulms, G., Chalkley, R.J., Hirsch, J., Baldwin, M.A. and Burlingame, A.L., Mass spectrometric analysis of protein mixtures at low levels using cleavable ^{13}C-ICAT and multi-dimensional chromatography. *Mol. Cell. Proteomics*, **2**, 299–314 (2003).
43. Griffin, T.J., Lock, C.M., Li, X.J., Patel, A., Chervetsova, I., Lee, H., Wright, M.E., Ranish, J.A., Chen, S.S. and Aebersold, R., Abundance ratio-dependent proteomic analysis by mass spectrometry. *Anal. Chem.*, **75**, 867–874 (2003).
44. Ranish, J.A., Yi, E.C., Leslie, D.M., Purvine, S.O., Goodlett, D.R., Eng, J. and Aebersold, R., The study of macromolecular complexes by quantitative proteomics. *Nat. Genetics*, **33**, 349–355 (2003).

45. Conrads, T.P., Anderson, G.A., Veenstra, T.D., Pasa-Tolic, L. and Smith, R.D., Utility of accurate mass tags for proteome-wide protein identification. *Anal. Chem.*, **72**, 3349–3354 (2000).
46. Smith, R.D., Anderson, G.A., Lipton, M.S., Pasa-Tolic, L., Shen, Y., Conrads, T.P., Veenstra, T.D. and Udseth, H.R., An accurate mass tag strategy for quantitative and high throughput proteome measurements. *Proteomics*, **2**, 513–523 (2002).

Chapter 2

Protein profile comparisons of microorganisms, cells and tissues using 2D gels

ANGELIKA GÖRG* and WALTER WEISS[1]

Technische Universität München, Fachgebiet Proteomik, D-85350 Freising-Weihenstephan, Germany

*Corresponding author. Tel.: +49-8161-714265; fax: +49-8161-714264. *E-mail addresses:* angelika.gorg@wzw.tum.de (A. Görg), walter.weiss@wzw.tum.de (W. Weiss).
[1] Tel.: +49-8161-714260.

1. Introduction

Although promising progress has been made in the development of alternative protein separation techniques in proteomics, such as isotope-coded affinity tag methodology, or two-dimensional liquid chromatography–tandem mass spectrometry [1–6], there is still no generally applicable method that can replace two-dimensional polyacrylamide gel electrophoresis (2D PAGE) [7,8] in its ability to separate highly complex protein mixtures derived from microorganisms, whole cells or tissues. 2D PAGE separates proteins according to two independent parameters, i.e. isoelectric point (p*I*) in the first dimension and molecular mass (M_r) in the second dimension by coupling isoelectric focusing (IEF) and sodium dodecyl sulfate polyacrylamide gel electrophoresis (SDS-PAGE) (reviewed by Dunn [9]). Proteins separated on 2D gels are visualized by either staining with Coomassie blue dye, silver stains, fluorescent dyes, immunochemical detection, or by radiolabelling, and quantified using densitometers, fluoro- and/or phosphorimagers. Theoretically, 2D PAGE is capable of resolving up to 10,000 proteins simultaneously, with approximately 1500–2000 proteins being routine, and detect and quantify protein amounts of less than 1 ng per spot.

Once proteins have been separated, visualized and quantified, protein spots of interest (e.g. up- or down-regulated proteins) are identified, usually by mass spectrometry. Due to the fact that for almost 20 years no method was available for fast and sensitive identification of protein spots on 2D gels, 2D PAGE methodology was far ahead of its time. However, the situation has changed dramatically by recent advances in mass spectrometry methods for rapid protein analysis and the establishment of whole genome sequences and protein databases of a large number of organisms. For routine analysis, protein spots are excised from the 2D gel, digested into fragments by specific proteases such as trypsin and then identified with matrix-assisted laser ionisation/desorption time-of-flight (MALDI-TOF) and/or electrospray ionisation (ESI) mass spectrometry (MS) and database mining (reviewed by Godovac-Zimmermann and Brown [10]).

In this chapter, a brief overview of the current status of 2D PAGE technology for proteome analysis of microorganisms, cells and tissues, with special emphasis on 2D PAGE with immobilized pH gradients (IPGs), is given. Due to its superior resolving power, reproducibility and simplicity compared to carrier ampholyte (CA) based 2D methodology, 2D PAGE with IPGs (IPG-Dalt) (Görg et al. [8,11]), in combination with modern mass spectrometry technologies, has become the core technology of proteome analysis. Recent developments and modifications with respect to sample preparation and sample application, IEF running conditions, the use of different pH gradients up to pH 12, 'zoom' gels, extended separation distances, as well as protein detection and automation will be critically discussed in this chapter. Although the limitations of the 2D PAGE approach are well known, e.g. poor solubility of membrane proteins, limited dynamic range and difficulties in displaying and identifying low-abundance proteins, 2D PAGE will remain a powerful and versatile tool, and at least in the immediate future, the most commonly used technique in proteome analysis [12–16].

2. Challenges of protein separation methods for proteome analysis

While genomics has been a tremendous challenge, the analysis of the corresponding proteomes may be technically even more challenging because of the fact that the number of proteins expressed at a given time in a cellular system is at least in the thousands and also due to the concomitant need for reproducible high-resolution separation of these complex protein mixtures.

The major challenges for successful proteome analysis are: (i) the ability to analyse very alkaline, hydrophobic and/or low or high M_r proteins with high resolution under steady-state conditions; (ii) the ability to detect minor components in the presence of large quantities of housekeeping proteins; (iii) methods for protein quantitation that are sensitive, rapid, simple, reliable and inexpensive; (iv) simplification and automation of protein separation procedures and the ability to perform high-throughput analysis.

Although there is currently no adequate alternative to 2D electrophoresis in proteome analysis where thousands of proteins have to be resolved simultaneously, 2D PAGE still suffers from several shortcomings.

(A) The number of proteins that can be resolved on a single 2D gel is limited. There is also no standard 2D PAGE system which separates all kinds of proteins equally well, i.e. low and high M_r proteins, hydrophilic as well as hydrophobic proteins and/or highly basic proteins. Moreover, on standard 2D electrophoresis gels (especially those which cover a wide pH range, e.g. pH 3–10), many spots contain overlapping, different proteins due to limited resolution. Common approaches to overcome this problem are various sample pre-treatment strategies, e.g. pre-fractionation of cells into their organelles and running of consecutive gels, sequential extraction procedures with buffers of increasing solubilizing power to obtain fractions based on hydrophobicity and/or running of 'zoom' gels, i.e. IPG strips which cover a series of narrow, overlapping ranges of p*I* for improved resolution.

(B) Hydrophobic membrane proteins, which are the most interesting targets for drug discovery, are not readily solubilized in solvents used in first dimension IEF, particularly due to the fact that the presence of SDS (which otherwise would be an excellent solubilizing agent) is not compatible with IEF. Thus, most IEF sample buffers include high concentrations of chaotropes, such as urea, and 'mild', however, less efficient detergents, such as NP-40 or CHAPS. More recently, progress has been made in overcoming these limitations by the development of more effective chaotropes and detergents such as thiourea and sulfobetaines or by applying organic solvents for solubilization of hydrophobic proteins.

(C) Another major challenge is related to the absence of any protein amplification system analogous to the PCR method for amplifying genes. Low-abundance proteins, which often do the most 'interesting' jobs in a cell, may be present in only a few dozen copies per cell; at the same time, the number of high-abundance 'housekeeping' proteins is in hundreds of thousands or even millions of copies per cell. At present, these low-abundance proteins cannot be displayed without ultrasensitive protein detection methods or without enrichment by sample pre-fractionation.

(D) Moreover, 2D PAGE is not strictly quantitative. Most staining techniques, e.g. silver stains, exhibit a rather limited dynamic range only. Consequently, the intensity of a highly expressed spot is not necessarily linearly correlated to that of a less expressed spot, a fact which makes quantitation of differentially expressed proteins difficult.

(E) Finally, the 2D PAGE procedure is rather time consuming and is thus not well suited as a high-throughput technology, since large formate gels typically require up to 2 days to complete. 2D technology also requires skilled staff, manual dexterity and precision [14].

During the past 10 years, IPG-Dalt has constantly been refined to accomplish at least several of these goals, e.g. by the development of basic IPGs up to pH 12 for the analysis of very alkaline proteins or the introduction of overlapping narrow IPGs to stretch the first dimension for higher resolution and the analysis of minor components, as well as the development of ready-made IPG strips and automated procedures [11,17].

Limitations remain in the field of analysis of hydrophobic and/or membrane proteins, as well as the lack of sensitive and reliable techniques for protein quantitation, although the advent of sensitive fluorescent dyes has considerably improved the latter situation. Steps were also taken to develop radioactive as well as non-radioactive dual label techniques for the visualization of differentially expressed proteins.

In the following sections, the most important challenges of 2D electrophoresis for proteomics will be discussed in more detail.

2.1. Highly alkaline proteins

Theoretical 2D profiles calculated from sequenced genomes [18,19] indicate not only that the majority of proteins of a total cell lysate possess p*I*s between pH 4 and 9, but also that a considerable number of proteins with p*I* values up to pH 12 can be present. In classical carrier ampholyte generated 2D PAGE, these proteins can only be separated by non-equilibrium pH gradient gel electrophoresis (NEPHGE) [20], however, at the expense of resolution and reproducibility.

In contrast, well-resolved, highly reproducible 2D patterns of these proteins are obtained using IPG-IEF. This can be accomplished using either narrow-range pH 10–12 or pH 9–12 IPGs, or non-linear, wide-range IPGs flattened at the basic end.

Highly alkaline proteins such as ribosomal and nuclear proteins with closely related p*I*s between 10.5 and 11.8 can be separated using *narrow-range* pH 10–12 or pH 9–12 IPGs [21,22] (Fig. 1). In order to obtain highly reproducible 2D patterns, different optimisation steps with respect to running conditions, pH gradient engineering and gel composition were necessary, such as sample cup-loading near the anode, IEF under a protective layer of silicone oil, the substitution of *N,N*-dimethylacrylamide (DMAA) for acrylamide and the addition of isopropanol to the IPG rehydration solution in order to reduce the reverse electroendosmotic flow and counteract water transport to the cathode, which gives rise to highly streaky 2D profiles. However, with the advent of the IPGphor (Amersham Biosciences), this protocol has been considerably simplified due to

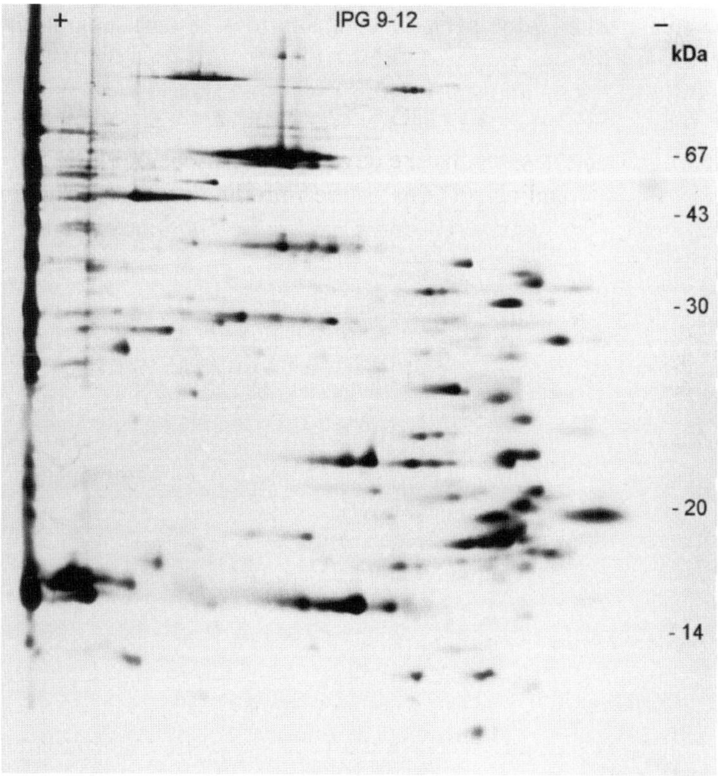

Fig. 1. Analytical IPG-Dalt of ribosomal proteins from mouse liver. First dimension: narrow-range alkaline IPG 9–12. Sample application by cup-loading near the anode. Sample load: 50 μg of protein. Second dimension: vertical SDS-PAGE (13%T). Silver stain (Reprinted with permission from: Görg, A., et al., *Electrophoresis*, **19**, 1516–1519 (1998)).

the ability to apply rather high voltages (up to 8000 V) with concomitant short running times to reach the steady state [23] (see Section 3.3.1.5).

In order to obtain an overview of the 'total' cellular or tissue proteome, *wide IPGs* covering the range pH 3–12, 4–12 and 6–12 (Fig. 2) have been generated [17,24]. Excellent 2D patterns of cell and tissue extracts and TCA/acetone-precipitated proteins for the visualisation of basic proteins with p*I* values exceeding pH 10 that are usually absent in lysis buffer extracts [17,23,24], as well as the Triton X-100 insoluble cell fraction of *Mycoplasma pneumoniae* [25] have been obtained. Moreover, a pH 4–12 IPG with a flattened region between pH 9 and 12 was found to give excellent separation of very basic proteins such as ribosomal proteins [24]. These wide gradients can be run under standard conditions without isopropanol, as the strong water transport to the cathode (reverse electroendosmosis) characteristic of narrow IPGs exceeding pH 10 is negligible. In addition, these results have been further improved by applying rather high voltages (up to 8000 V), probably due to shorter running times [23].

In summary, high-quality 2D patterns of very alkaline proteins can be obtained with relative ease when wide pH gradients, short (< 18 cm) separation distances

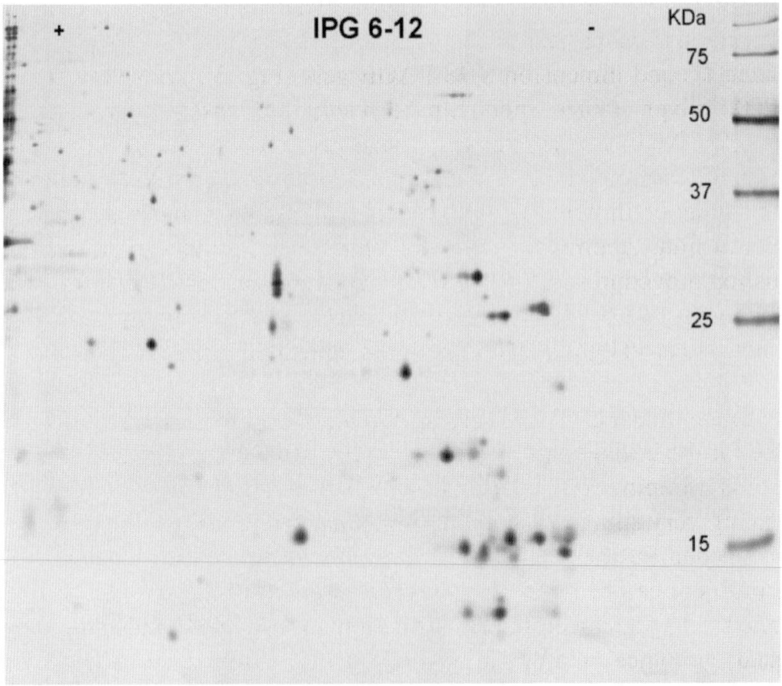

Fig. 2. Analytical IPG-Dalt of very alkaline *Lactobacillus sanfranciscensis* proteins. First dimension: wide range alkaline IPG 6–12. Separation distance: 180 mm. Sample application by cup-loading near the anode. Sample load: 80 μg of protein. Second dimension: vertical SDS-PAGE (15%T). Silver stain (Drews, O., et al., *Proteomics*, **2**, 765–774 (2002)).

and/or running times and analytical sample loads are applied. Streaky 2D patterns may arise with increased sample loads, longer separation distances and narrow-range IPGs (<1–2 pH units wide), e.g. in micropreparative IEF. Several approaches have been proposed to avoid streaking, which is supposed to result from oxidation of protein thiol groups, resulting in inter- and intra-chain disulphide bonds. These remedies include the application of an 'extra' paper strip soaked with dithiothreitol (DTT) near the cathode to compensate for the loss of the reducing agent DTT during IEF [26], alkylation of thiol groups with iodoacetamide prior to IEF [27] or oxidation of thiol groups in proteins with disulfides [28]. All these remedies have their pros and cons, and their success may also depend on the type of sample analysed.

2.2. Zoom gels, non-linear IPGs and extended separation distances for higher resolution and improved detection of low copy number proteins

The number of different protein species in proteomes is likely to be in the range of several thousands for simple prokaryotic organisms and at least 10,000 in complex samples such as eukaryotic cell extracts.

The resolving power of 2D gels primarily depends on the separation length in both dimensions. Using 'standard' 18 cm long IPG gels in the first dimension and 20 cm long second dimension SDS-PAGE gels (Fig. 3), typically no more than 1500–2000 silver stained spots can be resolved. Consequently, with complex samples, 2D PAGE on a single wide-range pH gradient will reveal only a small proportion of the proteome due to its limited resolution and the difficulty to detect low copy number proteins in the presence of highly abundant 'housekeeping' proteins. Current approaches to overcome these problems are sample pre-fractionation procedures (see Section 2.3), and multiple overlapping narrow-range IPGs spanning 1–1.5 pH units ('zoom gels') or, alternatively, but to a lesser extent, non-linear pH gradients. In addition, extended separation distances can be used.

Wide-range, linear pH 3–10 (Fig. 4) or pH 3–12 gradients are often the method of choice for the initial analysis of a new type of sample and provide an overview of the total protein expression. These broad-range IPGs are quite useful for investigating very simple prokaryote proteomes, but considerably more resolving power is needed for the separation of complex eukaryotic proteomes.

This problem can be overcome to some extent with the use of non-linear IPG gels, tailored to the theoretical distribution of p*I* values calculated from the sequenced genomes. Commercial IPG strips (Amersham Biosciences) are available for the pH range 3–10, in which the pH 5–7 region is flattened compared to the pH 7–10 region [29]. For improved resolution of alkaline proteins, a non-linear IPG 4–12 [24] in which the pH gradient is flattened between pH 9 and 12 is recommended.

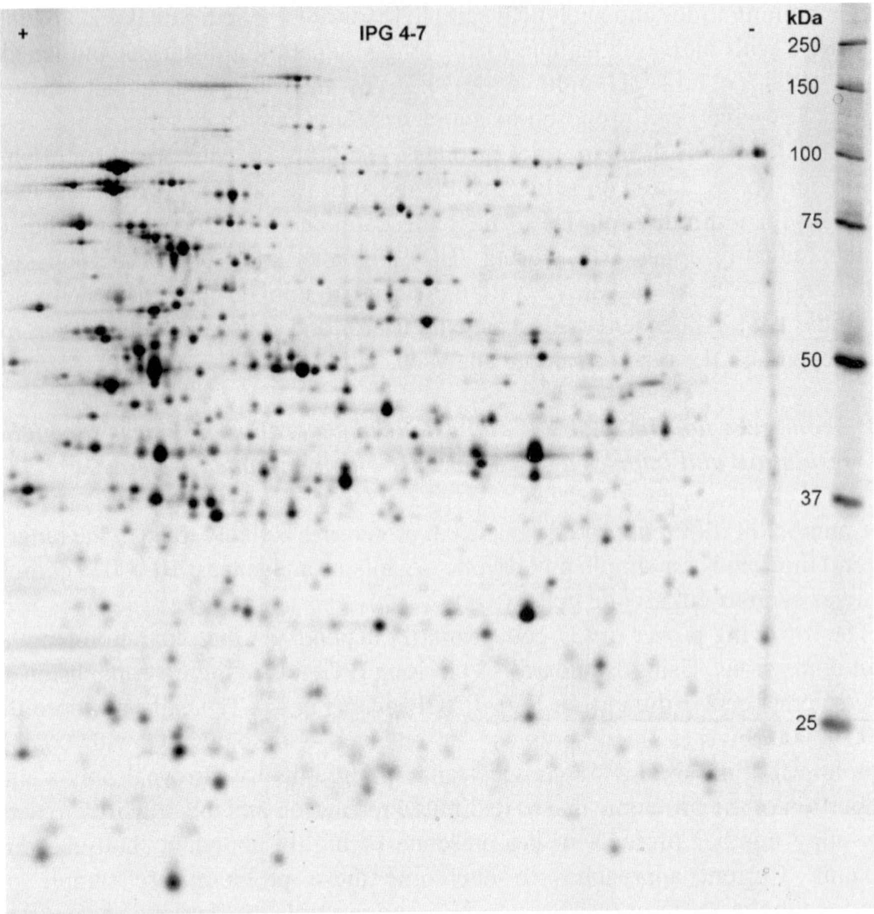

Fig. 3. Analytical IPG-Dalt of *Lactobacillus sanfranciscensis* proteins. First dimension: IPG 4–7. Separation distance: 180 mm. Sample application by cup-loading near the anode. Sample load: 100 μg of protein. Second dimension: vertical SDS-PAGE (12%T). Silver stain (Reprinted with permission from: Drews, O., et al., *Proteomics*, **2**, 765–774 (2002)).

For detailed studies, however, ultrazoom IPG gels are mandatory. The major advantage of these overlapping narrow-range IPGs (e.g. IPG 4–5, 4.5–5.5, 5–6) is the gain in resolution by stretching the protein pattern in the first dimension. The superior ability of very narrow range IPGs to separate different protein isoforms, compared with 3–10 NL and IPG 4–7 gels has been demonstrated [17,30–32]. Furthermore, computer-aided image analysis and protein identification by mass spectrometry are simplified due to the smaller number of co-migrating protein species and the more reliable database search results [30]. Moreover, these ultrazoom gels do not only enhance resolution, but also improve the detection of low-abundance proteins. It has been reported that ultrazoom gels allow

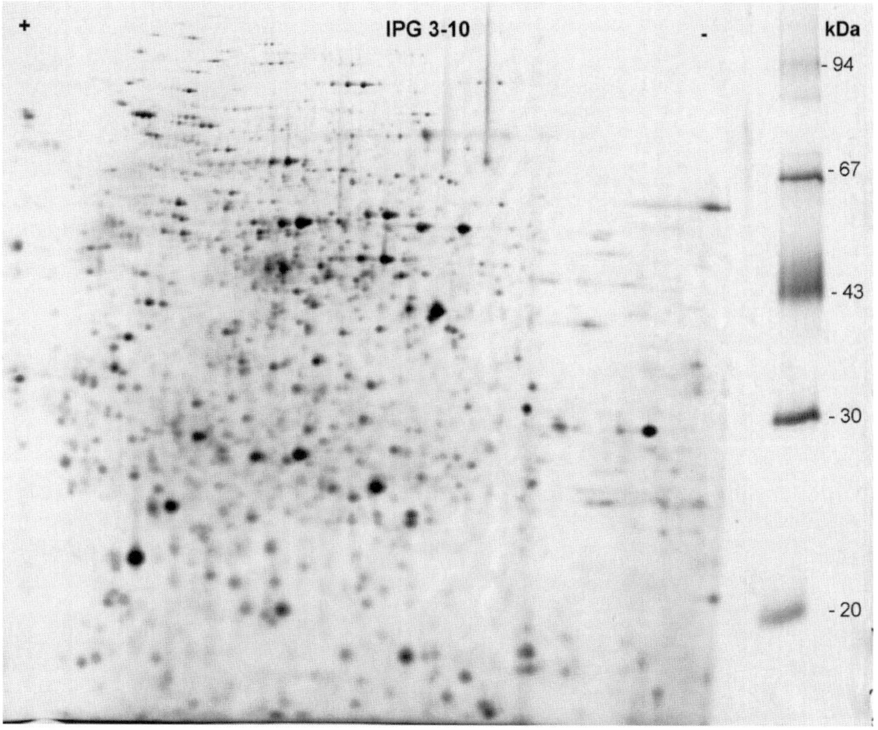

Fig. 4. Analytical IPG-Dalt of *Saccharomyces cerevisiae* proteins. First dimension: wide range IPG 3–10. Separation distance: 180 mm. Sample application by sample in-gel rehydration. Sample load: 80 μg of protein Second dimension: vertical SDS-PAGE (13%T). Silver stain (Wildgruber, R., PhD Thesis, Technical University of Munich, 2002).

the detection of proteins down to 300 copies per cell [31] due to their higher sample loading capacity.

Narrow-range IPG gels in the acidic and neutral pH range between pH 4 and 7 are commercially available and work equally well with both in-gel rehydration and cup-loading (see Section 3.3). These gels are ideal for micropreparative separations where protein loads of several milligrams are applied. In contrast, alkaline narrow-range IPGs, especially in conjunction with micropreparative sample loads, are more difficult to handle and require additional optimisation steps (see Section 2.1).

Very large format gels (>30 cm in both dimensions) have been developed for carrier ampholyte generated 2D gels [33] based on tube gel IEF technique, with the potential to separate more than 5000 proteins *simultaneously*. Unfortunately, this improvement in resolution is achieved at the expense of ease of gel handling and reproducibility [34]. In contrast, handling of 24 cm long IPG gel strips cast on plastic backing does not require special attention, and is, therefore, preferred to generate 2D gels for improved resolution. For example, 24 cm long IPG strips

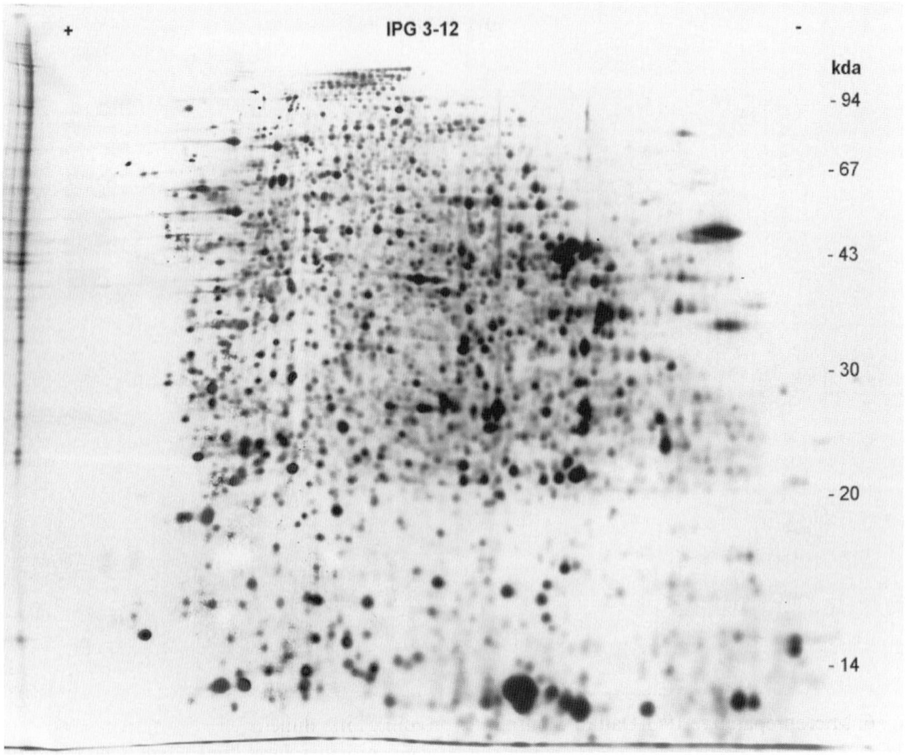

Fig. 5. Analytical IPG-Dalt of mouse liver proteins. First dimension: wide range IPG 3–12. Separation distance: 240 mm. Sample application by cup-loading near the anode. Sample load: 100 μg of protein. Second dimension: vertical SDS-PAGE (13%T). Silver stain (Reprinted with permission from: Görg, A., et al., *Electrophoresis*, **20**, 712–717 (1999)).

have been developed covering both the wide pH ranges 3–12 and 4–12 (Fig. 5), and narrower overlapping pH ranges (e.g. IPG 5–6) (Fig. 6) for the generation of composite, very high resolution 2D gels.

2.3. Protein enrichment and sample pre-fractionation procedures

Due to their wide dynamic range, identification of all proteins expressed in a given cell type is one of the major challenges in proteome analysis. Proteomic studies have revealed that the majority of proteins identified on 2D PAGE gels are highly abundant 'housekeeping' proteins which are present at up to 10^5–10^6 copies per cell, whereas some proteins are present at less than 100 molecules per cell and are usually not detected on 2D gels [35].

There are currently three major approaches to solve this problem: (i) ultrazoom gels (see Section 2.2); (ii) ultrasensitive protein stains (see Section 2.6); and

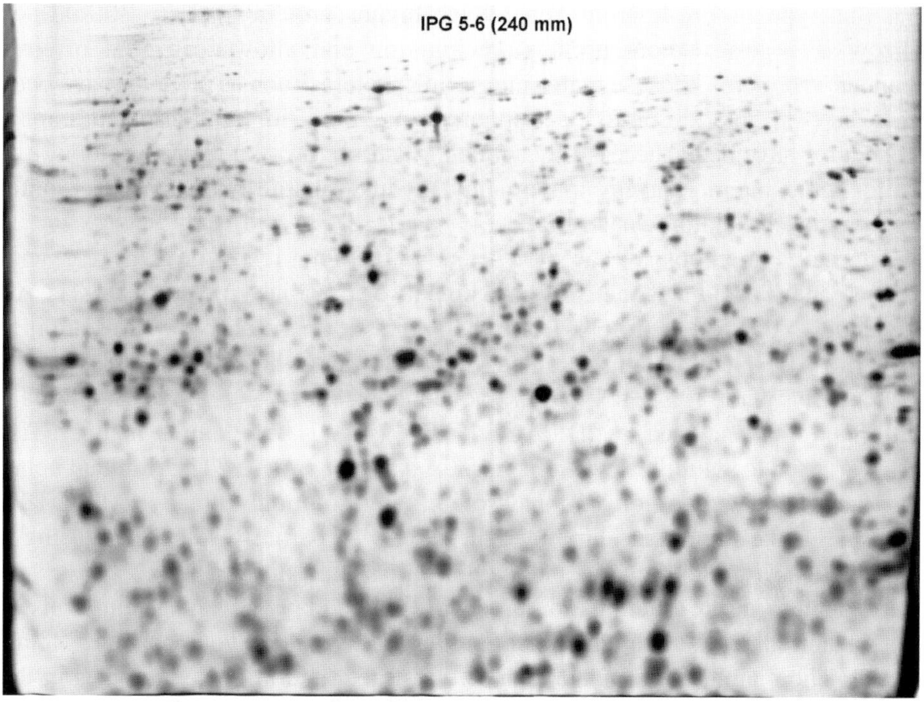

Fig. 6. Micropreparative IPG-Dalt of mouse liver proteins. First dimension: narrow-range IPG 5–6. Separation distance: 240 mm. Sample application by cup-loading near the anode. Sample load: 500 µg of protein. Second dimension: vertical SDS-PAGE (13%T). Silver stain (Reprinted with permission from: Görg, A., et al., *Electrophoresis*, **21**, 1037–1053 (2000)).

(iii) pre-fractionation steps to reduce the complexity of the sample and enrich for low copy number proteins.

The latter may be achieved by methods such as sub-fractionation of cell components, e.g. organelles such as mitochondria, or centrifugation in a sucrose density gradient prior to 2D PAGE analysis [36–39]. Whereas this procedure can be applied for mammalian cells (which do not possess a cell wall) with relative ease, access to these organelles is complicated for most microorganisms because in this case a lysis method is required, which is both efficient in disrupting the cell wall and gentle enough to guarantee that organelles remain intact. For example, sheroblasts are prepared for the isolation of intact organelles from yeast by digesting the cell wall with polysaccharide-cleaving enzymes prior to liberating the cell content by 'gentle' lysis conditions such as hypotonic solutions and/or mechanical treatment [40].

An alternative approach for pre-fractionation is electrophoresis in the liquid phase, e.g. microscale solution IEF [41] or multicompartment electrolysers with isoelectric membranes [42], or IEF in granulated gels [43]. It has been reported that this type of pre-fractionation allows higher protein load (6–30-fold) on narrow IPG gels without protein precipitation and allows detection of low-abundance proteins because major interfering proteins such as albumin have been removed [41]. In addition to electrophoretic methods, different chromatographic procedures have been successfully applied to enrich low-abundance proteins, e.g. hydrophobic interaction chromatography, hydroxyapatite and heparin chromatography or chromatofocusing [44].

Alternative methods are based on selective precipitation of proteins, e.g. with trichloroacetic acid/acetone for enrichment of alkaline proteins [11,21], or sequential extraction of proteins from a cell or tissue on the basis of their solubility properties, e.g. plant proteins obtained from barley and wheat seeds were extracted sequentially with Tris–HCl buffer, aqueous alcohols and urea/CHAPS/DTT buffer and then analysed by 2D PAGE [45,46] (Fig. 7). In a similar manner, Tris base was used to solubilize cytosolic *Escherichia coli* proteins. The resultant pellet was then subjected to conventional solubilizing solutions (urea/CHAPS/DTT), and finally, the membrane protein rich pellet was partially solubilized using a combination of urea, thiourea, and zwitterionic surfactants [47].

In conclusion, pre-fractionation procedures have many advantages in terms of protein enrichment, although in proteomics one would favour a single 'total' extract. The major disadvantage of most pre-fractionation procedures lies in the fact that they are either time consuming, complicated to handle and/or do not allow to process more than a few samples in parallel.

2.4. Low and high molecular mass proteins

Other, albeit, less serious problems are very low M_r (<15 kDa) and high M_r (>150 kDa) proteins, because there is no standard 2D gel system which effectively allows separation of proteins over the entire M_r range between 5 and 500 kDa.

A common approach is to use several gels optimised for the approximate M_r ranges 5–10, 15–200 and >150 kDa. Since conventional Tris–glycine gels do not allow efficient separation of proteins below 15 kDa, Fountoulakis et al. [48] have described a 2D gel system using Tris–tricine gels [49] and two urea concentrations in the second dimension. This gel system allowed efficient and reproducible separation of *Haemophilus influenzae* proteins with molecular masses between 5 and 20 kDa, but may also be suitable for other organisms.

The major problem with high M_r proteins is that a significant proportion of these proteins is quite hydrophobic, and consequently, they do not readily dissolve in 'standard' lysis and/or reswelling solutions used for sample solubilization and IEF. Moreover, even if these proteins can be solubilized, they may not always enter the IEF gel matrix, or the transfer from the first to the second dimension (SDS-PAGE) is impaired.

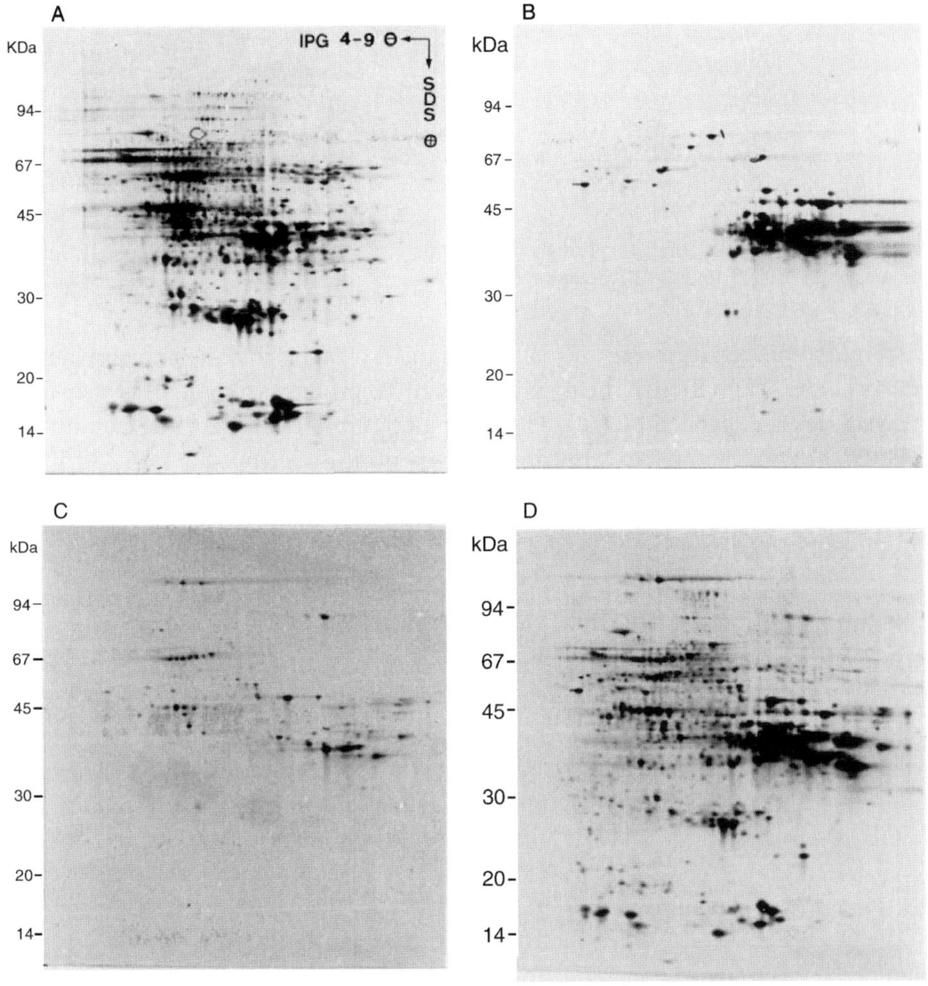

Fig. 7. Sequential extraction procedure of wheat (*Triticum aestivum*) seed proteins. First dimension: IPG 4–9. Separation distance. 110 mm. Second dimension. SDS-PAGE 10%T–15%T. Separation distance: 120 mm. (A)–(C) Protein fractions obtained by sequential extraction of whole-meal flour. (A) Salt-soluble proteins (albumins/globulins). (B) 75% ethanol soluble proteins (gliadins). (C) Urea/CHAPS/DTT-soluble proteins (glutenins). (D) Total proteins extracted with urea/-CHAPS/DTT. Silver stain. Sample loads: (A) 50; (B) 40; (C) 30; (D) 60 μg of protein (Reprinted with permission from: Weiss, W., et al., *Electrophoresis*, **14**, 805–816 (1993)).

Several strategies have been proposed to overcome these obstacles. (i) It has been demonstrated that sample application of high M_r proteins to IPG gels via cup-loading is often superior to 'passive' sample application by sample in-gel rehydration (see Section 3.3). In case the samples are still applied by in-gel rehydration, 'active' reswelling by applying low voltages (30–50 V) during

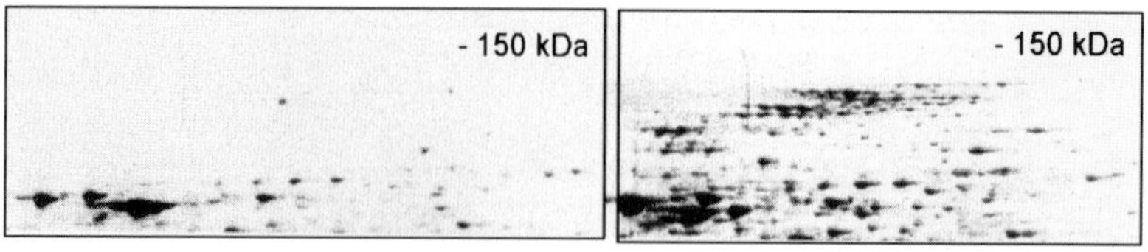

Fig. 8. Improved entry of high M_r proteins by applying low voltage during rehydration (IPGphor). Sample: mouse liver proteins. Silver stain.

the rehydration step improves the entry of high M_r proteins into the polyacrylamide matrix [11,17] (Fig. 8). (ii) The transfer of high M_r proteins from the IPG strip onto the SDS gel is enhanced by sufficiently long equilibration steps (2 × 15 min; see Section 3.3) and comparatively low voltages during the transfer step (50 V for vertical SDS gels and 100 V for horizontal SDS-PAGE systems, respectively) [11]. Interestingly, even the spatial array of IPG strip and SDS gel affects the transfer efficiency, e.g. in horizontal SDS-PAGE setups, best results were obtained when the distance between the cathodic buffer strip and the IPG strip was approximately 3–4 mm (see Section 3.4). (iii) For hydrophobic, very high M_r proteins, Zuo et al. [50] have recommended running this fraction on high-resolution large-pore gradient 1D SDS gels, followed by mass spectrometric identification of proteins.

In conclusion, very low and/or high M_r proteins are not a serious hindrance to proteome analysis. However, the procedure is more elaborate because several different gel systems have to be combined instead of using a single ‘standard’ 2D PAGE system.

2.5. Very hydrophobic proteins

Membrane proteins, which constitute a significant proportion (approximately 30%) of the cell’s proteins, play many key functions in various cellular processes including communication of the cell with its environment, signal transduction or ion transport, and are therefore, important targets for drug development [51].

Recent proteomic studies demonstrated that very hydrophobic proteins and, in particular, membrane proteins are an extremely under-represented group on 2D gels. This under-representation may be attributed to several factors. The most probable explanation is that hydrophobic proteins are inefficiently extracted from the sample and/or are not kept in solution during IEF due to their low solubility in aqueous media used for sample solubilization and IEF. Moreover, many membrane proteins are expressed in low copy numbers and may, therefore, be visualized by highly sensitive protein detection methods only [51].

There are only few reports (reviewed by Santoni et al. [51]) on the successful separation and detection of integral membrane proteins on 2D gels, and in almost all cases, the overall hydrophobicity of these proteins was not too high, because either they were glycosylated or they did not possess more than one transmembrane segment. Alternatively, they had been enriched or purified by pre-fractionation procedures prior to 2D PAGE analysis or had been detected by ultrasensitive stains such as immunoblotting.

Efficient initial solubilization, as well as keeping these proteins in solution during IEF, is a key factor for successful analysis of highly 'insoluble' membrane proteins. Consequently, the development of methods for improved solubilization of very hydrophobic proteins is currently one of the most active subjects of investigation [51–67].

Because of the extremely low solubility of membrane proteins in aqueous media, various chaotropes and detergents are included in sample solubilization buffers. Theoretically, SDS would be the detergent of choice for membrane protein solubilization due to its extremely high solubilizing power. Unfortunately, this negatively charged detergent interferes with subsequent IEF. Therefore it can be applied for initial sample solubilization, but must be displaced from the proteins by dilution with a several-fold excess of non-ionic or zwitterionic detergent prior to IEF.

The most popular buffers for solubilizing proteins in 2D PAGE analysis contain high concentrations (typically 8–9.5 M) of urea, 2–4% of non-ionic (e.g. Triton X-100 or NP-40) or zwitterionic (e.g. CHAPS) detergent, a reducing agent (e.g. DTT) and small amounts of carrier ampholytes. Whereas these buffers work well with hydrophilic proteins, the success for solubilization of very hydrophobic and, in particular, membrane proteins is questionable. Recently, a number of new reagents, including thiourea, were introduced for improved solubilization of hydrophobic proteins [54]. Thiourea, which is a much stronger denaturant than urea, cannot be used as the sole chaotrope due to its low solubility in water. However, since it is more soluble in urea solutions it is applied in the form of urea/thiourea mixtures (usually 5–8 M urea and 2 M thiourea), which exhibit much better solubilizing power for hydrophobic proteins. In addition, more efficient zwitterionic amidosulfobetaine detergents have been synthesized [53] and their superior solubilization power, especially in combination with urea/thiourea mixtures, has also been demonstrated [59]. Despite these recent advances, 2D electrophoretic separation of very hydrophobic integral membrane proteins remains a difficult task, particularly for micropreparative runs, where several milligrams of proteins are applied onto a single IPG strip.

More detailed investigations revealed that the efficiency of zwitterionic detergents for solubilizing hydrophobic proteins depends not only on the nature of the protein itself, but also on the presence and nature of other compounds, in particular lipids. For example, in a recent study Santoni et al. [56] found evidence that the optimal detergent for membrane protein solubilization strongly depends on the lipid content of the sample.

Another approach for improved solubilization of integral membrane proteins is to treat isolated membrane preparations (e.g. obtained by differential centrifugation) with sodium carbonate at alkaline pH [62] or chaotropic salts (e.g. potassium bromide) to remove carry-over cytoplasmatic proteins and only loosely attached peripheral membrane proteins ('membrane washing', 'membrane stripping'). Other procedures to enrich hydrophobic proteins are based on the differential extraction of membrane proteins by organic solvents, e.g. chloroform/methanol mixtures [63,64].

Certain 'loss' of membrane proteins on 2D gels may also be explained by the observation that these proteins, once solubilized, may in fact enter the IPG strips and also focus properly, but do not elute during the transfer step from first to second dimension (see Section 3.4). It has been reported by Pasquali et al. [65] that the bulk of proteins could still be detected in the Coomassie blue stained IPG strip; similar loss of membrane proteins due to hydrophobic interactions with the matrix of the IPG strip was also observed by Adessi et al. [66]. It has been recommended to improve transfer of hydrophobic proteins by including thiourea not only in the sample extraction solution, but also in the equilibration buffer [65].

In conclusion, all these studies provide evidence that there does not seem to exist a single solution for the complex solubility problem of membrane proteins. At least, these findings suggest that the combination of IPG gels with urea/thiourea and zwitterionic detergents is able to solubilize membrane proteins, which cannot be resolved on 2D gels with other systems. Nonetheless, it is equally important to empirically test and optimise the composition of sample solubilization buffer to improve the solubility of specific subsets of membrane proteins [56,65].

2.6. Protein detection and quantitation

The estimated concentrations of individual proteins in a single cell may differ between six or seven orders of magnitude, ranging from several millions of copies/cell for some highly abundant proteins (e.g. glycolytic enzymes) to a few copies/cell for low-abundant proteins. These enormous variations in protein concentrations are a major challenge for almost all currently available protein detection methods applied in proteome analysis.

Chemical stains such as Coomassie blue or silver staining [9,68] have found widespread use for the detection of proteins on 2D gels because of their compatibility with subsequent protein analysis methods. However, in terms of the requirements for proteome analysis, Coomassie blue staining is by far not sensitive enough for the detection of minor components such as low-abundance proteins. Since the detection limit of Coomassie blue is approximately 500 ng protein/spot, typically no more than several hundreds of protein spots are visualized on a 2D gel, even if milligram amounts of protein have been loaded onto the gel.

In contrast, silver staining methods are extremely sensitive (detection limit is as low as 0.1 ng protein/spot), but have a very limited dynamic range only (less than one order of magnitude), and are less reproducible than Coomassie blue staining due to the subjective end-point of the staining procedure. Moreover, silver staining methods are quite laborious and difficult to automate.

Better and more confident results are obtained by protein detection methods relying on fluorescent compounds, or by radioactive labelling of proteins combined with highly sensitive electronic detection methods. For special questions of interest, antibody-based methods are also applied [69], e.g. for the visualization of phosphorylated proteins [70].

Two major approaches to fluorescent detection of proteins on 2D gels are currently practiced. These are: (i) covalent derivatization of proteins with fluorophores prior to IEF; and (ii) post-electrophoretic protein staining by intercalation of fluorophores into the SDS micelles coating the proteins or by direct electrostatic interaction with the proteins [71,72]. Typical examples for pre-electrophoretic stains are monobromobimane [73] and cyanine-based dyes [74] (CyDye; Amersham Biosciences) that react with cysteinyl residues and lysyl residues, respectively. The major problem of many pre-electrophoretic stains is the occurrence of protein size and/or protein charge modifications which may result in altered protein mobility alongside the M_r and/or pI axis. Alternatively, proteins can be stained with a fluorescent dye molecule *after* the completion of electrophoretic separation. The most prominent example is the ruthenium-based dye SYPRO Ruby. Staining is accomplished within a few hours in a single step procedure, which may be easily adapted for use with automated instrumentation. The detection limit is approximately 1–2 ng protein/spot.

Prior to the advent of highly sensitive silver and fluorescent staining methods, detection of proteins labelled with radioisotopes was the only type of sensitive detection for proteins separated on 2D gels. An approach which is often used for proteome analysis of microorganisms and cell culture systems is *in vivo* radiolabelling of samples by the incorporation of radioactive amino acids (such as ^{35}S-methionine, ^{14}C-leucine and/or ^{32}P-phosphotyrosine). Alternatively, it is also possible to radiolabel *in vitro* (e.g. human tissues) by using iodination with ^{131}I or ^{125}I, however, at the risk of formation of artifacts [75].

Radiolabelled proteins separated by 2D PAGE can be detected by autoradiography or fluorography using X-ray films which are exposed to dried gels [76]. Despite their simplicity and the fact that X-ray film images can be quantified by densitometry, these film-based techniques require long exposure times (up to several weeks) if high sensitivity is desired. Moreover, due to non-linear response to radiation, multiple film exposures combined with computer-aided image processing are required to quantitate high as well as low-abundant proteins present in a sample. Nonetheless, even with multiple film detection, only limited dynamic range ($< 10^3$) is achievable.

To overcome the two major limitations inherent in conventional film-based autoradiography, i.e. poor sensitivity and limited, non-linear dynamic range, several electronic methods of detecting radiolabelled proteins in 2D gels have been developed.

In phosphorimaging, the most popular of these methods, X-ray films have been replaced by storage-phosphor screens containing a thin layer of special crystals doped with a europium salt [77]. Radioactive radiation excites electrons in the crystals and a latent image is formed on the plate. Scanning the plate with a He–Ne laser results in the emission of a blue luminescence proportional to the original amount of radiation, which is then quantified with a photomultiplier. The advantages compared to autoradiography using X-ray films are that it is possible to detect very low levels of radioactivity in a considerably shorter time (approximately 10 times faster) and the linear dynamic range is much better (five orders of magnitude). However, the resolution is much lower (150 μm, compared to <50 μm obtainable with X-ray films), but sufficiently high for most purposes. The major disadvantage is the rather high costs for equipment (phosphorimager, imaging screens).

Another detection method for radiolabelled proteins is multiphoton detecion (MPD) [10,78], which is based on detectors for single particles in radioactive decay processes. MPD technology has several advantages compared to conventional autoradiography. These are, in particular, very high linear dynamic range (7–8 orders of magnitude) and high sensitivity. Moreover, dual isotope detection (see later) can be performed, provided the two radiation energies are sufficiently different. Preliminary results presented at recent proteomic conferences demonstrated that the concept works in principle [78]. However, several problems, e.g. low resolution, low throughput (image capture of a single 2D gel takes at least several hours) and high costs of equipment, have to be solved before MPD technology can be routinely applied in proteome analysis.

A bottleneck for high-throughput proteomic studies is image analysis. In conventional 2D methodology, protein samples are separated on *individual* gels, stained and quantified, followed by image comparison with computer-aided image analysis programs (see Section 3.5). Since different images are not always perfectly superimposable due to the multistep 2D PAGE technology, image analysis is frequently very time consuming. To shorten this laborious procedure, Unlü et al. [74] have developed a methodology called DIGE (fluorescent difference gel electrophoresis) where two samples are labelled *in vitro* prior to IEF with different fluorescent cyanine dyes differing in their excitation and emission wavelengths, then mixed and separated on a single 2D gel (see Chapter 3). After consecutive excitation with both wavelengths, the images are overlaid and subtracted, whereby only differences (e.g. up- or down-regulated proteins) between both samples are visualized. Due to co-migration of both samples, methodological variations in spot positions and protein abundance are excluded, and consequently, image analysis is considerably simplified and accelerated.

Another methodology, called 'differential gel exposure' (DifExpo), has been described by Monribot-Espagne and Boucherie [79] and is based on *in vivo* metabolic labelling of samples with ^{14}C and ^{3}H leucine. After 2D PAGE on a single 2D gel, the separated proteins are transferred onto an immobilizing PVDF membrane and then exposed to two different types of phosphorimaging screens which are sensitive to $^{14}C/^{3}H$, and ^{14}C, respectively. Since both samples have been co-separated on the same 2D gel, the images are perfectly superimposable and image analysis is considerably simplified. The major drawback lies in the low specific activity of commercially available ^{14}C- and ^{3}H-labelled amino acids which require much longer exposure times and higher amounts of protein loading compared to ^{35}S labelling experiments [79].

A third approach, termed 'dual channel imaging', has been developed by Bernhardt et al. [80]. By pulse-labelling with ^{35}S-methionine, the protein synthesis pattern (e.g. of stressed cells) can be directly compared with the silver stained protein pattern on the same 2D gel. Because matching of different gels is avoided, this technique is useful for the rapid search for proteins induced or repressed by stress.

In conclusion, all the above-mentioned methods based on fluorescent labelling and/or electronic detection of radiolabelled proteins are quite promising. Fluorescent dye staining procedures have a relatively wide, linear dynamic range ($>10^3$) and are easy to use. Furthermore, most fluorescent staining procedures are compatible with subsequent protein identification methods such as mass spectrometry. The major drawback of most fluorescent staining methods is insufficient absolute sensitivity. Typically, only proteins expressed at greater than 10^3-10^4 copies/cell can be detected on 'standard' 2D gels. In contrast, electronic detection methods based on radiolabelling of proteins are much more sensitive (theoretically, less than a dozen copies of a protein/cell can be detected) and have a much wider dynamic range. Their major disadvantage, however, is the necessity of handling radioactivity.

2.7. Automated procedures

Due to the large number of samples which have to be analysed in high-throughput proteomic studies, there is an increasingly urgent need for automated procedures. Despite improvements such as the possibility to run 10 (or even 20) second dimension SDS gels in parallel (Dalt apparatus) [81], 2D electrophoresis has long been a laborious and difficult-to-automate procedure. Only recently, the situation has changed considerably due to: (i) the availability of ready-made gels (IPG DryStrips, SDS gels) on stable plastic supports; (ii) the introduction of the IPGphor and similar devices for fully automated first dimensional IEF [82];

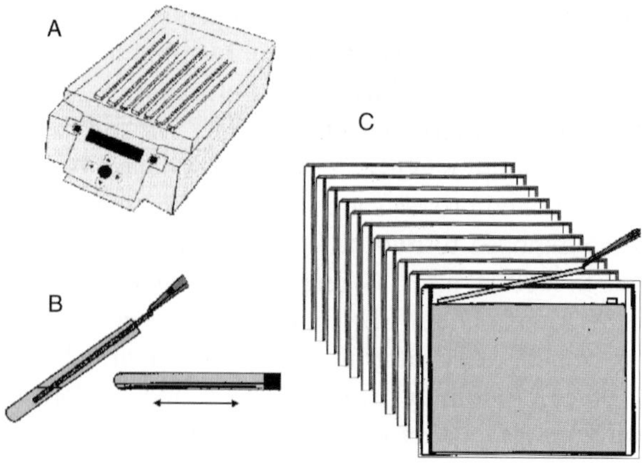

Fig. 9. Simplified procedure of IPG-Dalt (Reprinted with permission from: Görg, A., et al., *Electrophoresis*, **21**, 1037–1053 (2000)). (A) IEF in individual IPG gel strips on the IPGphor. (B) Equilibration of IPG gel strips prior to SDS-PAGE. (C) Loading of the equilibrated IPG gel strips onto multiple vertical SDS gels.

(iii) semi-automated silver staining devices, which allow staining of up to 10 gels in parallel [83,84]; (iv) one-step post-electrophoretic fluorescent protein staining methods (see Section 2.6), which can be easily automated; (v) DIGE (see Section 2.6) due to the extremely simplified image comparison; (vi) improved algorithms and better computer programs for fully automated gel image analysis (see Section 3.5); and (vii) automation of spot excision and protein digestion for mass spectrometric analysis. In particular, the advent of the IPGphor, in combination with ready-made IPG strips, where sample in-gel rehydration and IEF are performed overnight in an unattended operation, was an important step towards automation of the 2D PAGE procedure [17,82] (see Fig. 9). However, despite these improvements, there is currently no fully automated system available that is capable of performing the entire 2D PAGE process.

3. Current technology of 2D electrophoresis with IPGs (IPG-Dalt)

3.1. Protein profile comparisons of microorganisms, cells and tissues using 2D gels

More than a decade ago, a basic protocol of 2D electrophoresis with IPG (IPG-Dalt) was described [8], summarizing the critical parameters inherent to IEF with IPGs and a number of experimental conditions which were not part of the classical 2D electrophoresis repertoire with carrier ampholytes [7,85,86]. In principle, this fundamental protocol is still valid today: IEF is performed in individual IPG gel strips, which are then equilibrated with SDS buffer in the presence of urea, glycerol, DTT and iodoacetamide, and applied onto horizontal or vertical SDS gels in the second dimension. This procedure has become the standard 2D gel method for proteome research, permitting higher resolution [87], improved reproducibility for inter-laboratory comparisons [88,89], higher loading capacity [90] for micropreparative 2D electrophoresis with subsequent spot identification by mass spectrometry or Edman microsequencing [91], as well as the separation of highly alkaline proteins under equilibrium conditions [21–23].

In the following sections, the current technology of IPG-Dalt, including sample preparation, protein visualization and image analysis, will be discussed in more detail. Whereas the protocol for the 2D electrophoretic separation is almost identical for all kinds of samples, the significant difference lies in sample preparation procedures, which must be adapted and optimised for each type of sample.

3.2. Sample preparation

The major problems concerning the visualization of proteins from total cell or tissue extracts lie in the high dynamic range of protein abundance, and the diversity of proteins with respect to molecular weight, isoelectric point and solubility. Although a one-step procedure for protein extraction would be highly desirable with regard to simplicity and reproducibility, there is no single method of sample preparation that can be universally applied to all kinds of samples analysed by 2D PAGE. Although a large number of 'standard' protocols has been published, these protocols have to be adapted and further optimised for the type of sample (e.g. microbial cells or mammalian tissue) to be analysed, as well as for the proteins of interest (e.g. 'soluble' or highly 'insoluble' membrane proteins, respectively). However, some general recommendations can be given. First of all, sample preparation should be as simple as possible to increase reproducibility. Equally important, protein modifications during sample preparation must be minimized since they might result in artifactual spots on 2D gels. In particular, samples containing urea must not be heated in order to avoid charge heterogeneities due to carbamylation of the proteins by isocyanate formed during the decomposition of urea [9,34,75].

3.2.1. General considerations

The fundamental steps in sample preparation are: (i) cell disruption; (ii) inactivation or removal of interfering substances; and (iii) subsequent solubilization of the proteins [9,75].

Briefly, cell disruption can be achieved using single or in combination: osmotic lysis, freeze–thaw cycling, detergent lysis, enzymatic lysis of the cell wall, sonication, grinding with (or without) liquid nitrogen, high pressure (e.g. French press), homogenisation with glass beads and a bead beater, or a rotating blade homogenisation. All these procedures have their pros and cons, and the choice will primarily depend upon the type of sample.

During (or after) cell lysis, interfering compounds (e.g. proteolytic enzymes, salts, lipids, polysaccharides, nucleic acids and/or plant phenols) have to be inactivated and/or removed.

Proteases present within samples have to be inactivated to prevent protein degradation which can result in artifactual spots. To accomplish this goal, protease inhibitors may be added, but this is not always recommended since such reagents may modify proteins, causing charge artifacts [33,75,92]. Other recommendations are boiling the sample in SDS buffer (without urea!) or including high or low pH buffers (e.g. Tris base or TCA). Nonetheless, it should be kept in mind that it may be very difficult to completely inactivate proteases, especially those from certain microorganisms.

High concentrations of *salts* interfere with the electrophoretic separation and may be removed either by (spin)dialysis or by precipitation of proteins (e.g. by TCA or organic solvents). The drawback of these approaches is possible loss of proteins. Alternatives are 2D clean-up kits (Amersham Biosciences) or to dilute the sample below a critical salt concentration and apply a larger volume onto the IPG gel instead, preferably by sample in-gel rehydration (see Section 3.3), thus performing a sort of 'in-gel desalting'.

Lipids may interact with membrane proteins and 'consume' detergents. Delipidation can be accomplished by extraction of the biological material with organic solvents (e.g. ethanol or acetone). However, severe losses in proteins may be experienced, either because certain proteins are soluble in organic solvent or the precipitated proteins do not always resolubilize. In several cases, high-speed centrifugation of lipid-rich material and subsequent removal of the lipid layer has been recommended.

Polysaccharides (especially the charged ones) and nucleic acids can interact with carrier ampholytes and proteins and give rise to streaky 2D patterns. Moreover, these macromolecules may also increase the viscosity of the solutions and clog the pores of the polyacrylamide gels. Unless present at low concentrations, polysaccharides and nucleic acids have to be removed. A common method is precipitation of proteins with acetone or TCA/acetone, but losses in proteins may be experienced due to insufficient resolubilization of proteins. Other recommendations for the removal of nucleic acids are digestion by a mixture of protease-free (!) RNAses and DNAses or by ultracentrifugation and addition of a basic polyamine (e.g. spermine) [93].

Phenols present in plant materials (especially in plant leaves) may interact with proteins and lead to horizontal streaks in 2D gel patterns. It has been recommended to remove polyphenolic compounds either by binding to (insoluble) polyvinylpolypyrrolidone (PVPP) or by protein precipitation with TCA and subsequent extraction with ice-cold acetone.

The ideal sample solubilization procedure for 2D PAGE would result in the disruption of all non-covalently bound protein complexes and aggregates into a solution of individual polypeptides which remain soluble during the 2D electrophoretic separation [9,34,75].

The most popular sample solubilization buffer is based on O'Farrell's [7] lysis buffer. Current protein solubilization procedures include: (i) modified O'Farrell's lysis buffer (9 M urea, 2%–4% CHAPS, 1% DTT, 2% v/v carrier ampholytes); (ii) thiourea/urea lysis buffer (2 M thiourea, 5–7 M urea, 2%–4% v/v CHAPS and/or sulfobetaine detergents, 1% DTT, 2% v/v carrier ampholytes) [52,93]; (iii) boiling with SDS sample buffer, followed by dilution with excess urea or thiourea/urea lysis buffer [75,94–96]; (iv) precipitation of proteins (e.g. by TCA/acetone) and resolubilization in (thiourea/urea) lysis buffer [8,11]; and (v) sequential extraction buffers [45–47].

Whereas satisfying results with modified O'Farrell's lysis buffer are obtained with hydrophilic proteins, this standard IEF sample buffer is not ideal for the solubilization of all protein classes, in particular, membrane or other hydrophobic proteins. Improvement in the analysis of hydrophobic proteins was achieved by thiourea and new zwitterionic detergents. Merits and limits of these new detergents, chaotropes and reducing agents have been thoroughly discussed by Rabilloud [93] and Herbert [98].

Solubilization of proteins in boiling SDS buffer followed by dilution with lysis buffer is a quite useful procedure to increase the solubilization of the majority of proteins and to inhibit protease activities during sample preparation. However, horizontal streaks in the 2D pattern are observed when samples initially solubilized in 1% SDS are not diluted with at least fivefold excess of (thiourea/urea) lysis buffer to displace the SDS from the proteins and replace it with a non-ionic or zwitterionic detergent. Additionally, obtaining sufficient dilution can become a problem when micropreparative protein loads are required.

TCA/acetone precipitation has been found to be very valuable for: (i) inactivation of proteases to minimize protein degradation; (ii) removal of interfering compounds; and especially (iii) for the enrichment of very alkaline proteins such as ribosomal proteins from total cell lysates [21–24]. However, attention has to be paid to protein losses due to incomplete precipitation and/or resolubilization of proteins.

Due to the high dynamic range and diversity of expressed proteins in eukaryotic tissues it is sometimes preferable to carry out a pre-fractionation step to reduce the complexity of the sample and/or to enrich certain proteins (e.g. low copy number proteins or alkaline proteins). Pre-fractionation of proteins can be accomplished by: (i) isolation of cell compartments and/or organelles (e.g. ribosomes, nuclei, mitochondriae or membrane fractions by high speed and/or sucrose gradient centrifugation; (ii) precipitation (e.g. with TCA/acetone); (iii) separation methods such as electrophoresis in the liquid phase or IEF in Sephadex gels, chromatography and/or affinity purification of protein complexes (see Section 2.3); or (iv) sequential extraction procedures with increasingly powerful solubilizing buffers, usually aqueous buffers, organic solvents such as ethanol or chloroform/methanol, and detergent-based extraction solutions. However, cross-contamination between the individual fractions may be a problem, especially with the latter procedure.

Protocols for the solubilization of different kinds of samples are given in Section 3.2.2 and at http://www.expasy.ch/ch2d/protocols/protocols.fm1.html.

3.2.2. Protocols for the extraction and solubilization of cell and tissue samples

In the following section, a brief description of procedures for the extraction and solubilization of cell and tissue samples which have actually been analysed in the laboratory is given. For a more comprehensive review, see refs. [34,75]. In general, frozen cells or tissues (e.g. mouse liver, yeast, plant seeds, etc.) were disrupted by different techniques such as grinding in a liquid nitrogen cooled mortar, sonication, shearing-based methods or homogenisation. Proteins were then solubilized with sonication using: (i) lysis buffer (9 M urea, 1% w/v DTT, 2%–4% w/v CHAPS, 2% v/v carrier ampholytes, pH 3–10 and 10 mM Pefabloc® proteinase inhibitor); (ii) thiourea lysis buffer (2 M thiourea, 7 M urea, 2%–4% w/v CHAPS, 1% w/v DTT and 2% v/v carrier ampholytes, pH 3–10 and 10 mM Pefabloc® proteinase inhibitor); or (iii) hot SDS sample buffer (1% w/v SDS, 100 mM Tris–Cl, pH 7.0) and then diluted with a several-fold excess of lysis buffer or thiourea lysis buffer [94,96].

Tissue samples were collected as soon as possible after the death of the 'donor' and immediately frozen in liquid nitrogen. All samples were disrupted while still frozen. Small tissue pieces were wrapped in aluminium foil, frozen in liquid nitrogen and crushed with a pre-cooled hammer (−20°C), whereas larger tissue pieces were ground under liquid nitrogen using a pestle and mortar and then solubilized in lysis buffer. For enrichment of alkaline proteins, *mouse liver* was ground in a liquid nitrogen cooled mortar, suspended in 20% TCA in acetone (−18°C) containing 0.2% DTT and kept at −18°C overnight in order to ensure complete protein precipitation. Following centrifugation (40,000*g*, 60 min, 15°C), the supernatant was discarded and the pellet resuspended in (−18°C) acetone containing 0.2% DTT. The sample was spun again, and the pellet dried under vacuum and then solubilized in lysis buffer. Following centrifugation, the supernatants were stored in aliquots at −70°C until analysed. Preparation of total *ribosomal proteins* (Tp80S) was according to Madjar [99], and preparation of *histones* from chicken erythrocytes according to Csordas et al. [100].

Microbial cell cultures such as bacteria or yeast need to be standardized and optimised for the growth conditions and the growth phase determined from which the sample is taken, since this has an enormous impact upon the biochemical state of the cells [101]. Because the cells may excrete proteases and other extracellular enzymes into the growth media, they first need to be washed with an isotonic buffer such as phosphate-buffered saline (PBS) or sucrose (at the same temperature as the culture, so as not to cold- or heat-shock the cells) prior to harvesting by centrifugation. Since bacteria and yeast cells are surrounded by a cell wall, protein extraction cannot be simply carried out by an osmotic shock as can be done with mammalian cells. Extensive disruption of cells is required by either vigorously shaking the cells in the presence of glass beads, sonication on ice in the presence of (urea/thiourea) lysis buffer or heating the sample in the presence of SDS [94,96] (Fig. 10). Where necessary, protease-free DNAse and RNAse is added to digest nucleic acids.

Dry *plant seeds* (where proteases are not active) were simply smashed with a hammer and ground with a mortar and pestle (with or without cooling by liquid nitrogen). The plant tissue was then solubilized in lysis buffer, centrifuged and

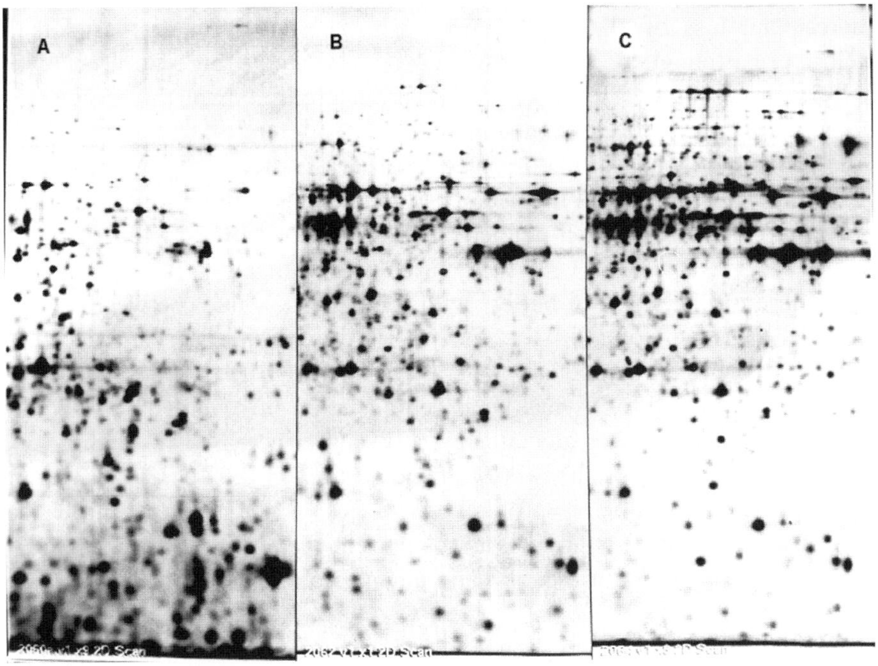

Fig. 10. Comparison of three different solubilization procedures for yeast cell proteins. Yeast cells were sonicated in the presence of (A) urea/CHAPS lysis buffer, (B) SDS, followed by dilution with urea/CHAPS lysis buffer and (C) boiled with SDS, followed by dilution with thiourea/urea/CHAPS lysis buffer. Sample load: 100 μg of protein. Silver stain (Reprinted with permission from: Harder, A., et al., *Electrophoresis*, **20**, 826–829 (1999)).

aliquoted. *Plant leaves* not only contain proteases, but also high concentrations of phenols which can adsorb proteins and cause streaks on the 2D electrophoresis gel. To counter this, cells were first disrupted with a mortar and pestle in the presence of liquid nitrogen. Proteins were then precipitated with 20% TCA in ice-cold acetone [102] identical to the procedure described for mouse liver proteins. After removal of plant phenols by rinsing with ice-cold acetone, proteins were solubilized in lysis buffer, aliquoted and stored at −70°C.

3.3. Two-dimensional electrophoresis with IPGs (IPG-Dalt)

For proteome analysis, it is essential to generate reproducible, high-resolution protein separations. Using the classical 2D PAGE approach of O'Farrell [7], it is, however, often difficult to obtain reproducible results even within a single laboratory, let alone between different laboratories. The problem of limited reproducibility is largely due to the synthetic carrier ampholytes used to generate the pH gradient required for IEF, for reasons such as batch-to-batch variability of carrier ampholytes, pH gradient instability over time, cathodic drift, etc. In practice, carrier ampholyte generated pH gradients rarely extend beyond pH 7.5, with resultant loss of alkaline proteins. For the separation of these alkaline proteins, O'Farrell et al. [20] developed an alternative procedure, known as NEPHGE, however, at the expense of reproducibility, since this procedure is extremely difficult to control and standardize.

The above-mentioned problems have been largely overcome by the development of IPG [103], based on the use of the bifunctional Immobiline® reagents, a series of 10 chemically well-defined acrylamide derivatives with the general structure $CH_2{=}CH{-}CO{-}NH{-}R$, where R contains either a carboxyl or an amino group. These form a series of buffers with different pK values between pK 1 and 13. Since the reactive end is co-polymerised with the acrylamide matrix, extremely stable pH gradients are generated, allowing true steady-state IEF with increased reproducibility, as has been demonstrated in several inter-laboratory comparisons [88,89]. Other advantages of IPGs are increased resolution by the ability to generate (ultra)narrow pH gradients ($\Delta pI = 0.01\ cm^{-1}$) [87], reproducible separation of alkaline proteins [21–24] and increased loading capacity [90]. Consequently, IEF with IPGs is the current method of choice for the first dimension of 2D PAGE for most proteomic applications.

3.3.1. First dimension: IEF with IPGs

The first dimension of IPG-Dalt, IEF, is performed in individual 3 mm wide IPG gel strips cast on GelBond PAGfilm (either ready-made Immobiline DryStrips® or laboratory-made). Samples are applied either by cup-loading or by in-gel rehydration. IPG-IEF can be simplified by use of an integrated system such as the IPGphor [17,82] where rehydration with sample solution and IEF can be performed in a one-step automated procedure.

3.3.1.1. IPG gel casting. IPG slab gels (4%T, 3%C, 0.5 mm thick, 40, 70, 110, 180 or 240 mm long) of the desired pH range are cast, washed, dried and cut into individual 3 mm wide IPG gel strips according to previously published protocols [8,11,104,105]. For casting very alkaline, narrow-range IPG gels (e.g. IPG 9–12 or 10–12), acrylamide should be substituted by DMAA. When DMAA is used as monomer, %T is increased to 6% (instead of 4% acrylamide as usual) [21]. The appropriate amount of Immobiline chemicals, calculated with the computer programs of refs. [106] or [107] or according to published recipes [17,21,22,105, 108,109], is added to the mixture prior to gel polymerisation.

For IPG gel casting, a casting mould consisting of two glass plates, one covered with a GelBond PAGfilm, the other bearing a 0.5 mm thick U-frame is assembled. The latter glass plate has been made hydrophobic by treatment with Repel Silane® in order to prevent the gel from sticking to it. For delayed polymerisation, the cassette is pre-cooled in a refrigerator for 30 min. Meanwhile, two Immobiline® starter solutions (a more acidic one and a more basic one) are prepared as described previously [8,11,105]. They contain not only different amounts of Immobiline® chemicals, but also two different concentrations of glycerol to stabilize the pH gradient during the gel casting procedure by forming a density gradient. For improved polymerisation kinetics, both solutions should be adjusted to pH 7 with sodium hydroxide or acetic acid, respectively. After starting the polymerisation by adding the appropriate amounts of TEMED and ammonium persulfate, the gel solutions are filled into the cassette with the help of a gradient mixer according to the casting procedure of Görg et al. [8] for ultrathin gradient gels. After pouring the gradient, the cassette is kept at room temperature for 10 min to allow adequate levelling of the density gradient before polymerisation is carried out for 1 h at 50°C. After polymerisation, the IPG slab gel is removed from the mould and extensively washed with deionised water (6 × 10 min), then immersed in 2% glycerol (30 min), dried at room temperature in a dust-free cabinet, covered with a plastic film and, if not used immediately, stored at −20°C. The dried gels can be stored frozen for up to 1 year. Prior to use, the IPG gel is cut into individual, 3 mm wide strips with the help of a paper cutter [8,11,110]. Alternatively, a range of ready-made gels (Immobiline DryPlate® or Immobiline DryStrip®) is available commercially. In principle, IPG strips of any desired length can be used, but it should be kept in mind that the larger the separation area of a 2D gel, the higher the number of proteins that can be resolved. IPG strips of 18, 24 or even 54 cm [111] are usually employed for high-resolution separations, while shorter strips (4, 7 or 11 cm) are used for rapid screening applications.

3.3.1.2. Rehydration of IPG strips. Prior to IEF, the dried IPG strips must be rehydrated to their original thickness of 0.5 mm. IPG dry strips are rehydrated either with sample already dissolved in rehydration buffer ('sample in-gel rehydration') or with rehydration buffer without sample, followed by sample application by 'cup-loading'.

For *sample in-gel rehydration* [112,113], cell lysate or tissue sample (1–10 mg protein/ml) is directly solubilized in an appropriate quantity of rehydration buffer. For 180 mm long and 3 mm wide IPG dry strips, 350 μl of this solution is pipetted into the grooves of the reswelling tray or into the IPGphor strip holder (Amersham Biosciences), respectively. For longer or shorter IPG strips, rehydration volume has to be adjusted accordingly. The IPG strips are then laid, gel side-down, into the grooves without trapping air bubbles. The IPG strip, which must still be moveable and not stick to the tray, is then covered with silicone oil (which prevents drying out during reswelling) and rehydrated overnight at approximately 20°C. Higher temperatures (>37°C) hold the risk of protein carbamylation, whereas lower temperatures (<10°C) should be avoided to prevent urea crystallization on the IPG gel.

Sample in-gel rehydration is generally not recommended for samples containing very high molecular weight, very alkaline and/or very hydrophobic proteins, since these are taken up into the gel with difficulty, and is therefore less reliable for quantitative applications. In these cases, *cup-loading* is preferred. For cup-loading, the IPG dry strips are rehydrated overnight in rehydration buffer, either in a reswelling cassette or, more convenient, in the reswelling tray as described above, but without sample.

The rehydration buffer contains 8 M urea (or, alternatively, 2 M thiourea and 6 M urea), 0.5%–4% CHAPS, 0.4% DTT and 0.5% (v/v) Pharmalyte 3–10, and is either prepared fresh before use or stored in aliquots at −70°C. It is important that the urea solution has been deionized with an ion exchange resin prior to adding the other chemicals, because urea in aqueous solution exists in equilibrium with ammonium cyanate, which can react with protein amino groups (e.g. lysine) and introduce charge artifacts, leading to additional spots on the IEF gel. Carrier ampholytes are added for improved protein solubility, but also as a cyanate scavenger.

3.3.1.3. IEF on the Multiphor flat-bed electrophoresis apparatus. After rehydration, but prior to IEF, the rehydrated IPG strips are rinsed with distilled water for a few seconds and then blotted between two sheets of moist filter paper to remove excess rehydration buffer in order to avoid urea crystallization on the gel surface, which can be responsible for prolonged isoelectrofocusing times and 'empty' vertical lanes in the stained 2D pattern.

In the case of *sample application by cup-loading*, up to 40 rehydrated IPG strips may be directly placed, side-by-side and 1–2 mm apart, onto the surface of the kerosene-wetted cooling plate of a horizontal flat-bed electrophoresis apparatus (e.g. Multiphor, Amersham Biosciences). It is of utmost importance that the acidic ends of the IPG gel strips face towards the anode. Electrode paper strips (cut from 1 mm thick filter paper, e.g. MN 440, Machery & Nagel) are then soaked with deionized water, blotted with filter paper to remove excess liquid and placed on top of the aligned IPG gel strips at the cathodic and anodic ends.

When basic IPG gradients are used for the first dimension (e.g. IPG 6–10), horizontal streaking can often be observed at the basic end of 2D protein profiles. This problem may be resolved by applying an extra electrode strip soaked in 20 mM DTT on the surface of the IPG strip alongside the cathodic electrode strip [26]. This has the advantage that the DTT within the gel, which migrates towards the anode during IEF, is replenished by the DTT released from the strip at the cathode. Alternative approaches are to: (i) use the non-charged reducing agent, tributyl phosphine, which does not migrate during IEF [97]; (ii) alkylate the proteins prior to IEF [27]; (iii) substitute DTT in the rehydration buffer by a disulfide such as dimercaptoethanol [28]; or (iv) to apply high voltages for short running times [23].

Samples (20 μl; protein concentration 5–10 mg/ml) are usually applied into silicon rubber frames (size: $2 \times 5\ mm^2$) or special sample cups placed either at the anodic or cathodic end of the IPG strips. As a general advice, it is not recommended to apply proteins at (or proximate to) the pH area which corresponds with their p*I*, because proteins are poorly soluble near their p*I*s and are thus prone to precipitation at the sample application site.

For analytical purposes, typically 50–100 μg of protein are loaded onto a single, 180 mm long IPG gel strip. For micropreparative purposes, up to several milligrams of protein may be applied. If the protein concentration of the sample is unknown, one can do a rough but quick estimation by making a dilution series of both the sample and a known standard (e.g. BSA) and spotting them on to a nitrocellulose membrane followed by staining with Amido Black or Coomassie blue.

A convenient alternative to the procedure described above, particularly for micropreparative IEF, is to use a special strip tray (IPG DryStrip kit, available from Amersham Biosciences) which has a frame that fits the cooling plate of the Multiphor. This tray is equipped with a corrugated plastic plate that contains grooves allowing easy alignment of the IPG strips. In addition, the tray is fitted with bars carrying the electrodes and a bar with sample cups allowing application of samples at any desired point on the gel surface. The advantages are that the cups can handle a larger quantity of sample solution (100 μl can be applied at a time, but it is possible to apply a total of up to 200 μl portion-by-portion, onto a single IPG gel) and the frame allows coverage of the IPG strips with a layer of silicone oil that protects the gel from the effects of the atmosphere (evaporation, etc.) during IEF. In case of very basic pH gradients exceeding pH 10 (e.g. IPG 3–12, 4–12, 6–12, 9–12, 10–12) or narrow-range ('zoom' gels) pH gradients with 1.0 or 1.5 pH units with extended running time (>24 h), the IPG strips must be covered by a layer of silicone oil.

IPG strips *rehydrated with sample solution* are directly placed onto the kerosene-covered cooling plate. When IEF must be performed under a layer of silicone oil, the rehydrated IPG strips are put in the grooves of the strip aligner of the Immobiline DryStrip kit and covered with silicone oil as described for 'cup-loading'.

The initial voltage should be limited to 150 V for 60 min to allow maximal sample entry and then progressively increased until 3500 V is attained. Current and power settings should be limited to 0.05 mA and 0.2 W per IPG strip, respectively. After sample entry, the filter papers beneath the anode and cathode should be replaced with fresh ones. This is because salt contaminants have quickly moved through the gel and have now collected in the electrode papers. For higher salt concentrations, it is recommended to change the filter paper strips more often. In case of IEF with very alkaline, narrow-range IPGs, such as IPG 10–12, this procedure should be performed once an hour.

The time required for the run depends on several factors, including the type of sample, the amount of protein applied, the length of the IPG strips used and the pH gradient being used. The IEF run should be performed at 20°C, as at lower temperatures there is a risk of urea crystallization, and at higher temperatures carbamylation might occur. Precise temperature control is also important since it has been found that otherwise alterations in the relative positions of some proteins on the final 2D patterns may happen [114]. Some typical running conditions are given in Table 1.

After the first dimension (IPG-IEF), IPG strips can be used immediately for the second dimension. Alternatively, strips can be stored between two sheets of plastic film at −70°C for periods of several months.

3.3.1.4. IEF with IPGphor. Recently, an integrated instrument, named the IPGphor (Amersham Biosciences), has been developed to simplify the IPG-IEF dimension of 2D electrophoresis [82]. This instrument features a strip holder that provides rehydration of individual IPG strips with or without sample, optional separate sample cup-loading and subsequent IEF, all without handling the strip after it is placed in the ceramic strip holder. The instrument can accommodate up to 12 individual strip holders and incorporates Peltier cooling with precise temperature control between 19.5°C and 20.5°C and a programmable 8000 V power supply. The IPGphor saves about a day's worth of work by combining sample application and rehydration, as well as by starting the run at pre-programmed times, and by running the IEF at rather high voltages.

In practice, a cell/tissue lysate (0.5–5 mg of protein) is solubilized in 500 μl of a solution containing 8 M urea (or, alternatively, 2 M thiourea and 6 M urea),

Table 1

Running conditions using the Multiphor

Gel length	180 mm
Temperature	20°C
Current max.	0.05 mA per IPG strip
Power max.	0.2 W per IPG strip
Voltage max.	3500 V

I Analytical IEF

Initial IEF

Cup-loading (20–50 μl)	In-gel rehydration (350 μl)
150 V, 1 h	150 V, 1 h
300 V, 1–3 h	300 V, 1–3 h
600 V, 1 h	

IEF to the steady state at 3500 V

1–1.5 pH units	4 pH units	7 pH units
e.g. IPG 5–6, 24 h	IPG 4–8, 10 h	IPG 3–10 L, 6 h
e.g. IPG 4–5.5, 20 h	IPG 6–10, 10 h	IPG 3–10 NL, 6 h

3 pH units	5–6 pH units	8–9 pH units
IPG 4–7, 12 h	IPG 4–9, 8 h	IPG 3–12, 6 h
IPG 6–9, 12 h	IPG 6–12, 8 h	IPG 4–12, 8 h

II Extended separation distances (240 mm)

IEF to the steady state at 3500 V

IPG 3–12, 8 h
IPG 4–12, 12 h
IPG 5–6, 40 h

III Micropreparative IEF

Initial IEF

Cup-loading (100 μl)	In-gel rehydration (350 μl)
50 V, 12–16 h	50 V, 12–16 h
300 V, 1 h	300 V, 1 h

IEF to the steady state at 3500 V
Focusing time of analytical IEF plus approximately 50%

0.5%–4% CHAPS, 0.4% DTT and 0.5% (v/v) Pharmalyte 3–10. The required number of strip holders (up to 12) is put onto the cooling plate's electrode contact area of the IPGphor, and 350 μl of sample-containing rehydration solution (for 180 mm long IPG strips) is evenly pipetted in the strip holder channel, 1–2 cm apart from the electrodes. The IPG strips are lowered, gel side-down, onto the rehydration solution without trapping any air bubbles and overlayered with 1 ml of silicone oil. Rehydration and IEF are carried out automatically according to the programmed settings, preferably overnight. Alternatively, the IPGphor can also be used with a cup-loading procedure, which allows the application of quantities up to 100 μl.

Typical running conditions for IEF using the IPGphor are given in Table 2. As indicated, low voltage (30–50 V) is applied during the rehydration step for improved sample entry of high M_r proteins into the polyacrylamide gel, which otherwise can be a problem with sample in-gel rehydration. Then voltage is increased stepwise up to 8000 V (when IPG strips with separation distances < 11 cm are used, voltage should be limited to 5000 V only). For optimum results, for samples with high salt concentrations or when narrow pH intervals are used, it is beneficial to insert moist filter papers (size: 4 × 4 mm^2) between the electrodes

Table 2

Running conditions using the IPGphor

Gel length	180 mm	
Temperature	20°C	
Current max.	0.05 mA per IPG strip	
Voltage max.	8000 V	
I Analytical IEF		
Reswelling	30 V, 12–16 h	
Initial IEF	200 V, 1 h	
	500 V, 1 h	
	1000 V, 1 h	
IEF to the steady state	Gradient from 1000 to 8000 V within 30 min 8000 V to the steady state, depending on the pH interval used:	
1–1.5 pH units	4 pH units	7 pH units
e.g. IPG 5–6, 8 h	IPG 4–8, 4 h	IPG 3–10 L, 3 h
e.g. IPG 4–5.5, 8 h		IPG 3–10 NL, 3 h
3 pH units	5–6 pH units	8–9 pH units
IPG 4–7, 4 h	IPG 4–9, 4 h	IPG 3–12, 3 h
IPG 9–12, 3 h	IPG 6–12, 4 h	IPG 4–12, 3 h
II Micropreparative IEF		
Reswelling	30 V, 12–16 h	
IEF to the steady state	Focusing time of analytical IEF + additional 50% (approximately)	

and the IPG strip prior to raising the voltage to 8000 V. After termination of IEF, the IPG strips are stored as described above.

3.3.1.5. General guidelines. In the following section, some general guidelines for optimisation of IPG-IEF with respect to sample application, choice of pH gradient and IEF running conditions are given.

Sample application. Samples can be applied to the IPG strip either by cup-loading or by in-gel rehydration. In the former case, samples are typically applied at the pH extremes, i.e. either near the anode or cathode. Sample application near the anode proved to be superior to cathodic application in most cases, with some exceptions. For example, human cardiac proteins were preferably applied at the cathode [88], whereas most kinds of plant proteins, mouse liver, yeast cell proteins, etc., yielded best results when applied at the anode. By using basic pH gradients such as IPGs 6–10, 7–10 or 6–12, anodic application was mandatory for all kinds of samples investigated [11].

The *optimum amount of protein* to be loaded onto a single IPG gel strip for optimum resolution, maximum spot numbers and minimum streaking/background smearing depends on parameters such as pH gradient used (wide or narrow), separation distance and protein complexity of the sample. For *analytical* 2D electrophoresis and silver staining, 50–100 μg protein of a total cell lysate per IPG strip (180 mm long and 3–6 pH units wide) proved to be the optimum for the majority of samples. Samples can be applied by cup-loading or in-gel rehydration. The optimum sample volume for cup-loading is 20–100 μl. Volumes less than 20 μl are not recommended because of the increased risk of protein precipitation at the application point. For *micropreparative* 2D electrophoresis, in-gel rehydration is often preferred, although up to 1 mg of protein (sample volume: 100 μl) was successfully applied by cup-loading, using an IPG 4–7 [11]. By using narrow pH gradients (1 pH unit), up to 10 mg of protein were loaded onto a single IPG gel strip, either by (repeated) sample cup application or by in-gel rehydration. Usually, high protein load, which may lead to overloaded and distorted patterns, is more a problem in the second dimension (SDS-PAGE) than in the first dimension run. Nevertheless, protein concentration in the sample solution should not exceed 10 mg/ml. Otherwise, loss of protein and horizontal or vertical streaking due to protein aggregation and precipitation may occur. In the case of sample in-gel rehydration, IPG dry strips have to be reswollen with a defined sample volume. Optimum sample volume is 350 μl for an 18 cm long and 3 mm wide IPG strip. For longer or shorter strips, the sample volume has to be calculated accordingly.

In general, for proper *sample entry*, voltage has to be limited during the initial stage of IEF, depending on the protein amount, sample volume and conductivity of the sample solution (e.g. salt, carrier ampholytes, etc.). For example, for analytical runs in combination with cup-loading (sample volume: 20 μl), voltage has to be restricted to 150 V during the first 60 min and to 300 V during the next

60 min (see Table 1). When more than 20 μl are applied, IEF at 300 V is prolonged for 30 min each per additional 10 μl. Critical samples with high salt concentrations can be desalted directly in the IPG gel by restricting the voltage to 50–100 V during the first 4–5 h [11] with several changes of the electrode filter paper strips. In case of large sample volumes (micropreparative runs and/or narrow IPGs), voltage should be limited to 100 V overnight for improved sample entry. In case of sample application by in-gel rehydration, protein losses were observed when sample volumes for rehydration of the dry IPG strip significantly exceed the calculated volume of the IPG gel strip because proteins preferably remain in the surplus solution instead of entering the IPG gel matrix. Although in-gel rehydration works with the majority of samples, it has to be confirmed whether alkaline, hydrophobic and/or high M_r proteins have entered the IPG gel properly. By applying low voltage (30 V) during reswelling using the IPGphor, protein entry, especially of high M_r proteins, is improved [11,17]. In-gel rehydration is successfully applied for wide IPGs, e.g. IPG 4–7, 4–9 or 4–10, and narrow IPGs with one pH unit in the pH range between 3.5 and 7.5 [19], whereas for very alkaline IPGs such as 6–12, 9–12 and 10–12 [21–23], cup-loading at the anode is still required.

Choice of pH gradient. The choice of pH gradient primarily depends on factors such as the composition of the sample, or whether the objective is to get an overview of the state of the proteome or to perform a much more detailed analysis.

Wide pH intervals with 3–7 pH units up to pH 10. Wide IPGs up to pH 10, such as IPG 4–7, 4–9 or 3–10 are still run according to the original protocol described in 1988 [8], with minor modifications. These gradients are compatible with cup-loading as well as with in-gel rehydration, and they work fine on the Multiphor and the IPGphor. Wide IPGs up to pH 10 are ideally suited for analytical (sample load: 50–100 μg) and, at least to a certain extent, micropreparative runs (sample load: up to 1 mg).

Narrow (1–1.5 pH units) IPGs. With complex samples such as eukaryotic cell extracts, 2D electrophoresis on a single wide-range pH gradient reveals only a small percentage of the whole proteome because of insufficient spatial resolution and the difficulty detecting low copy number proteins in the presence of abundant proteins. Besides pre-fractionation procedures discussed in Sections 2.3 and 3.2, one remedy is the use of multiple overlapping narrow IPGs in the first dimension ('zoom' gels) [19,29–32] and/or extended separation distances [17,111] to: (i) achieve an optimum resolution in order to avoid multiple proteins in a single spot for unambiguous protein identification; and (ii) facilitate the application of higher protein amounts for the detection of minor components.

'Zoom' gels in the pH range between 4 and 7 (IPG 4–5, IPG 4.5–5.5, IPG 5–6, etc.) work with in-gel rehydration or cup-loading, and may be run on the Multiphor apparatus or on the IPGphor. These gels are typically used for micropreparative purposes with sample loads up to several milligrams. In order to avoid protein precipitation and horizontal streaking, low voltage (approximately 50 V) is applied during the initial stage of IEF. Because of the long focusing time (usually > 24 h), the surface of the IPG strips has to be protected by a layer of silicone oil to prevent them from drying out. For improved results, it is recommended to remove the electrode paper strips after several hours and replace them by fresh ones. This is of particular importance when the sample contains high amounts of salt and to remove proteins with p*I*s outside the chosen pH interval, because they would otherwise 'shorten' the separation distance by 'invading' the gel and hiding protein spots. However, much better results are obtained when pre-fractionated samples are separated on 'zoom' gels [41,43].

Narrow IPGs at the basic extreme up to pH 12. Strongly alkaline proteins such as ribosomal and nuclear proteins with closely related p*I*s between 10.5 and 11.8 were focused to equilibrium by using narrow IPGs 10–12 and 9–12. For highly reproducible 2D patterns, different optimisation steps with respect to pH engineering and gel composition, such as the substitution of dimethylacrylamide for acrylamide and the addition of isopropanol to the IPG rehydration solution, were necessary in order to suppress the reverse electroendosmotic flow which causes highly streaky 2D patterns [21]. Sample cup-loading at the anode and IEF under silicone oil are required. With the advent of the IPGphor, the procedure can be greatly simplified by applying high voltages (8000 V) to shorten run times considerably. In this case, these gradients are run under 'standard' conditions without isopropanol (typical running conditions are given in Table 2) [11,23].

Wide IPGs up to pH 12. Theoretical 2D patterns calculated from sequenced genomes [17,18] indicate that the majority of proteins of a total cell lysate possess isoelectric points between pH 4 and 9, but a considerable number of proteins have p*I*s up to pH 12. In order to obtain an overview of the proteome of a cell or tissue, wide IPGs 3–12 and 4–12 were generated [17,24]. Excellent 2D patterns of cell/tissue extracts and/or TCA/acetone precipitated proteins for the visualization of basic proteins exceeding p*I* 10, which are usually not included in lysis buffer extracts of eukaryotic organisms [11], as well as the Triton-X-100 insoluble cell fraction of *Mycoplasma pneumoniae* [25] were obtained. Moreover, IPG 4–12 which is flattened between pH 9 and 12, proved to be a most useful gradient for the separation of very alkaline proteins such as ribosomal proteins [24]. By using IPGs 3–12 and 4–12, entry and focusing of high M_r proteins is also significantly improved.

Extended separation distances. Very long (> 30 cm) separation distances for maximum resolution of complex protein patterns have been described [33,111]. While size, stability and handling of fragile tube gels [33] is often a problem, handling of longer IPG gel strips cast on plastic backings does not require any special attention. For example, 24 cm long IPG gel strips with pH 3–12 and 4–12 have been successfully used [17]. Moreover, by using 24 cm long IPG strips with narrow pH gradients such as IPG 4–5, 5–6, etc., highly resolved 2D patterns were obtained. Recently, 54 cm long IPG strips were successfully applied [111].

Optimisation of focusing time. In practice, the optimum focusing time required to achieve the best quality and reproducibility is the time needed for the IEF separation pattern to achieve steady state [8,11,110]. If the focusing time is too short, this will result in horizontal streaking. On the other hand, overfocusing should be avoided. Although in contrast to the classical O'Farrell [7] method, overfocusing does not result in migration of proteins towards the cathode (cathodic drift), it does result in excess water exudation at the surface of the IPG gel due to active water transport (reverse electroendosmotic flow). This leads to distorted protein patterns, horizontal streaks at the basic end of the gel and loss of proteins. The optimum focusing time must be established empirically for each combination of protein sample, protein loading and the particular pH range and length of IPG gel strip used. As a guideline, optimum focusing times determined for a number of different wide and narrow pH range IPGs are given in Tables 1 and 2.

Temperature. The effects on spot positions and pattern quality of the temperature used for IEF have been studied in detail [114]. Since IEF in the presence of 8 M urea at low temperatures is not suitable due to the formation of urea crystals, optimum focusing temperature was found to be 20°C rather than 10°C. Elevated temperatures allow more rapid focusing due to lower viscosities and increased mobilities of proteins. Increased temperature also results in improved sample entry, clearer background, less streaking and well-defined spots up to the basic end of the gel. However, high temperatures (>37°C) increase the risk of protein carbamylation. Moreover, temperature not only affects pattern quality, but also has an influence on the spot positions of the 2D polypeptide patterns. Shifts in spot positions were primarily found at the extremes of the pH gradient, whereas in the pH region between 6 and 7, position shifts were less marked [114]. Nevertheless, temperature control is essential in order to allow meaningful comparisons of 2D patterns.

3.3.2. Equilibration of IPG strips

Prior to the second dimension separation, it is essential that the IPG strips are equilibrated to allow the separated proteins to fully interact with SDS. The most common protocol is to incubate the IPG strips for 15 min in 50 mM Tris–HCl buffer, pH 8.8 containing 2% w/v SDS, 1% w/v DTT, 6 M urea and 30% w/v glycerol [8]. Urea and glycerol are used to reduce electroendosmotic effects which otherwise result in reduced protein transfer from the first to the second dimension. This is followed by a further 15 min equilibration in the same solution containing 4% w/v iodoacetamide instead of DTT. The latter step is used to alkylate any free DTT, because DTT migrates through the second dimension SDS-PAGE gel, resulting in an artifact known as point-streaking that can be observed after silver staining [115]. Furthermore, alkylation of protein sulfhydryl groups prevents reoxidation of sulfhydryl groups and is also advantageous for subsequent spot identification by mass spectrometry. Alternative procedures using tributylphosphine (TBP) instead of DTT and iodoacetamide have been successfully applied. Since TBP is uncharged, it has certain advantages over DTT and allows reduction and alkylation of proteins in a one-step procedure. Unfortunately, TBP is only poorly soluble in water, volatile and rather toxic, and therefore, not recommended for general use. However, it is the reagent of choice for reduction of proteins which are characterized by a large number of sulfhydryl groups, such as keratin (e.g. wool filament) proteins [97].

After equilibration, IPG strips are blotted along the edge using filter paper for 1 min to remove excess liquid before application to horizontal second dimension SDS-PAGE gels.

Due to the observation that the focused proteins bind more strongly to the fixed charged groups of the IPG gel matrix than to carrier ampholyte gels, sufficiently long equilibration time (2×10 min, at least) with gentle shaking at room temperature is important for improved protein transfer from the first to the second dimension. Increased temperature (heating) of the IPG strip during the equilibration step did not yield better results. However, protein transfer is considerably improved when urea and glycerol are incorporated in the equilibration buffer [8]. Thiourea is sometimes recommended for more efficient transfer of hydrophobic proteins [65], but may also cause streaks in the 2D pattern (Görg et al., unpublished results).

Loss of proteins during the equilibration step and subsequent transfer from the first to the second dimension is primarily due to: (i) proteins which remain in the IPG strip because of adsorption to the IPG gel matrix and/or insufficient equilibration times; and (ii) wash-off effects. Experiments with radioactive-labelled proteins have shown that up to 20% of the proteins may get lost during equilibration. However, time courses have also revealed that the majority of proteins is lost during the very first minutes of equilibration, whereas protein losses in the second equilibration step are only marginal [116]. Most probably, the proteins lost had been primarily located near the surface of the IPG strip. These studies have also shown that the loss of proteins during equilibration is quite reproducible for a given sample.

3.3.3. Second dimension: SDS-PAGE

SDS-PAGE may be performed on horizontal or vertical systems. Horizontal set-ups are ideally suited for the use of ready-made gels (e.g. ExcelGel SDS™), whereas vertical systems (e.g. the Ettan-Dalt multiple slab gel unit) are preferred for multiple runs in parallel.

The discontinuous buffer system of Laemmli [117] is the most commonly used for the second dimension of 2D PAGE, although for special purposes other buffer systems are employed, e.g. the Tris–Tricine buffer system of Schägger and von Jagow [49] for separation of very low M_r proteins [48] or borate buffers for the separation of highly glycosylated proteins. Gels of either a homogeneous polyacrylamide concentration or a polyacrylamide concentration gradient (to extend the range over which proteins of different molecular mass can be effectively separated) may be used. The IPG gel strips are easy to handle compared with the fragile tube IEF gels used in the classical O'Farrell [7] procedure as they are bound to a flexible plastic support. After equilibration, the IPG strips are applied directly to the surface of the second dimension horizontal or vertical formate SDS-PAGE gels.

3.3.3.1. SDS-PAGE on horizontal systems. Horizontal SDS-PAGE can be carried out on either ready-made (ExcelGel, Amersham Biosciences) or laboratory-made slab gels cast on GelBond PAGfilm by the procedure described by Görg et al. [8] for ultrathin gradient gels. It is essential to use a stacking gel. The gel casting procedure is described in detail in refs. [105,118]. The horizontal second dimension system has the advantage that the gels are attached to a plastic support, thereby preventing alterations in gel size, e.g. during the staining procedure. In addition, spot sharpness can be superior to that obtained using vertical systems due to the decreased gel thickness (typically 0.5 mm compared with 1.0 or 1.5 mm) that allows the application of higher voltages and consequent reduction in protein diffusion due to the reduced running time.

Our preferred gel size is $250 \times 200 \times 0.5$ mm^3. Typically, the laboratory-made SDS gels are composed of a 6%T stacking gel (minimum length: 40 mm) and a 12%T–15%T (or a 10%–17%) linear gradient resolving gel. For certain applications, 13%T or 15%T homogeneous resolving gels are also used. In both stacking and resolving gel, crosslinker concentration is 3%C. The buffer system is 375 mM Tris–HCl, pH 8.8 and 0.1% SDS in both stacking and resolving gel. Paper wicks, soaked in electrode buffer (192 mM glycine, 25 mM Tris, 0.1% w/v SDS) [117] are placed on the gel surface so that they overlap the gel ends by 10 mm while the other end is immersed into the electrode buffer reservoir. (The filter papers must be clean and free from impurities. This can be achieved by laying them in their respective electrode buffers, placing other buffer-moistened filter papers across the cooling plate in contact with the electrode buffers, and running an electric current across them for about 1 h. The electrode papers can then be reused up to 50 times). Paper wicks are substituted by electrode buffer strips made of polyacrylamide when ready-made gels (ExcelGel SDS™, Tris–Tricine buffer system, linear acrylamide gradient from 12%T–14%T) are used.

The equilibrated (see Section 3.3.2) and blotted IPG strips are transferred onto the SDS gel by simply placing one long (240 or 180 mm) or several shorter (e.g. 110 or 70 mm) IPG strips, gel side-down, onto the surface of the stacking gel alongside the cathodic electrode paper wick or buffer strip, respectively. Optimum distance between the IPG strip and the buffer strip (or electrode wick) has been found to be 3–4 mm for improved transfer of high M_r proteins. Protein markers of known molecular weights (e.g. 10–200 kDa) may be run alongside, either loaded with a silicon cup or by a small filter paper square saturated with M_r marker protein solution.

Electrophoresis is performed with a maximum voltage of 100 V for about 60 min until the bromophenol blue dye has migrated off the IPG strip(s) by 3–4 mm. Then the IPG strip is removed from the surface of the SDS gel and the cathodic electrode wick (or buffer strip, respectively) is moved forward by 5 mm so that it now overlaps the former application area of the IPG strip. In case the IPG strip application area is rather wet, excess liquid is removed from the SDS gel surface by blotting the IPG strip application area with a moist filter paper. Electrophoresis is then continued with maximum settings of 600 V (30 mA, 30 W) when using laboratory-made gels, and 800 V (40 mA, 30 W maximum) in the case of ExcelGel™ SDS. The temperature of the cooling plate is set to 20°C. Total running time is approximately 4–5 h.

3.3.3.2. Multiple SDS-PAGE on vertical systems. Large-scale proteome analysis usually requires the simultaneous electrophoresis of batches of second dimension SDS-PAGE gels in order to maximize the reproducibility of 2D PAGE protein profiles. This requirement is most easily met using multiple vertical second dimension SDS-PAGE systems, such as provided by the Ettan-Dalt (Amersham Biosciences) (see Fig. 9) or the Investigator system (Genomic Solutions). It is not necessary to use stacking gels with the vertical format as the protein zones within the IPG strips are already concentrated and the non-restrictive (low polyacrylamide concentration) IEF gel can be considered to act as a stacking gel [75].

Multiple SDS slab gel casting is performed as described earlier [81,105]. Alternatively, ready-made SDS gels on plastic backing may be used (Amersham Biosciences). The preferred gel thickness is 1 mm, acrylamide concentration typically 13%T or 15%T (homogenous) and crosslinker concentration 2.7%C. No stacking gel is required. For loading the equilibrated IPG strip, the gel cassettes are placed in an upright position and filled with 2–3 ml of hot (75°C) agarose solution containing 0.5% agarose in electrode buffer (25 mM Tris, 192 mM glycine, 0.1% w/v SDS and 0.03% w/v bromophenol blue). Then the IPG strip is inserted between the glass plates with a spatula and brought in close contact with the upper edge of the SDS gel. A filter paper square, soaked with M_r marker protein solution (5–7 μl), can also be inserted. After the agarose has set, 10 gels are run simultaneously overnight in the Ettan-Dalt tank (settings: 200 V, 150 mA, 20°C). In vertical gels, the IPG strip does not need to be removed.

3.4. Protein visualization

Detection methods of proteins on 2D gels can be divided into five major types. These are detection by: (i) anionic dyes (e.g. Coomassie blue); (ii) negative staining with metal cations (e.g. zinc stain); (iii) silver staining; (iv) fluorescence; and (v) radioactive isotopes, using autoradiography, fluorography or phosphorimaging [68].

In this section, those methods which are preferentially applied in the laboratory will be discussed in more detail. Briefly, after SDS-PAGE, the resolved polypeptides are usually fixed in ethanol/acetic acid/water (4/1/5) for at least several hours (but usually overnight). Analytical gels are typically stained with silver nitrate, whereas micropreparative gels are usually stained with Coomassie brilliant blue (CBB) or imidazole–zinc. Post-electrophoretic fluorescent staining is preferably done with SYPRO Ruby™. Proteins which have been radiolabelled with ^{35}S-methionine are visualized by phosphorimaging. Silver and Coomassie blue stained gels are digitised using flat-bed scanners or a laser scanning densitometer, whereas SYPRO Ruby™ and ^{35}S-methionine labelled proteins are visualized using the Typhoon 9400 Variable Mode Imager (Amersham Biosciences) or the Molecular Imager FX (BioRad).

3.4.1. Coomassie brilliant blue staining

CBB is an anionic triphenylmethane dye which binds to the proteins non-covalently. This post-electrophoretic stain requires an acidic medium for electrostatic interaction between the dye molecules and the amino groups of the proteins. After a short fixation (30–60 min) of proteins, e.g. in 20% TCA, the gel is first saturated with the dye solution (0.1% CBB R-250 in 40% ethanol and 10% acetic acid) for 2–3 h (depending on the gel thickness) and then destained with 40% ethanol and 10% acetic acid. It is not recommended, however, to destain the gel completely, but rather to immerse it in distilled water (or 1% acetic acid) for 24–48 h for increased sensitivity. Detection limit of this stain is slightly better than 1 μg protein/spot. It has been reported that sensitivity can be increased by a factor of five by staining with a dimethylated form of the dye, CBB G-250, in a colloidal dispersion (0.1% CBB G-250 dispersed in 2% phosphoric acid, 10% ammonium sulfate and 20% methanol) according to Neuhoff et al. [119]. Staining takes between 24 and 48 h, but destaining is usually not necessary. The detection limit is as low as 0.2 μg protein/spot. Both CBB staining procedures are compatible with most protein analysis methods (i.e. mass spectrometry and Edman microsequencing).

3.4.2. 'Negative' protein staining with imidazole–zinc

Protein-bound salts (metal cations) are less reactive than the free salt in the gel. Thus, the precipitation of an insoluble salt is slower on the sites occupied by proteins than in the protein-free background. This difference in precipitation speed is exploited in 'negative' staining methods where the proteins stay transparent while the background becomes white due to salt precipitation. Zinc or imidazole–zinc stains [120] are currently the most popular negative staining methods. A typical protocol is as follows: after 2D electrophoresis, the gel is transferred in 1% sodium carbonate for 5 min; then the solution is replaced by 0.2 M imidazole containing 0.1% SDS; the gel is incubated for 15 min, rinsed for 10 s in water; and then transferred to 0.2 M zinc acetate and agitated for 60 s. The gel background becomes white at this stage and protein spots stay transparent. Finally, the gel is thoroughly rinsed with water several times [121].

This method is rapid, simple and rather sensitive (detection limit: 20 ng/spot). If necessary, the gel can be destained with chelators such as EDTA. The major disadvantage of zinc staining is its very poor linear range, which means that this procedure cannot be used for detecting quantitative variations on 2D gels. It is, however, quite useful for micropreparative 2D PAGE and subsequent protein identification by mass spectrometry.

3.4.3. Silver staining

Silver staining techniques are up to several hundred times more sensitive than CBB. The detection limit is about 0.1 ng protein per spot. Although there are dozens of silver staining protocols, silver stains can be classified into two major types. The simplest is the silver nitrate stain where the gel is soaked with silver nitrate and the colour is developed by reduction with formaldehyde at alkaline pH [122]. The second, and slightly more sensitive, silver stain is the diamine stain in which the silver is complexed with ammonia [123]. Most procedures include one or several step(s) to increase the sensitivity, usually prior to silver impregnation of the gel. Generally speaking, protocols using fixation with glutardialdehyde are more sensitive, especially for the detection of alkaline proteins, and also slightly more reproducible. Since silver staining is a surface phenomenon, the inner part of a silver stained protein has not necessarily been damaged by the staining process, allowing subsequent protein identification with methods where the protein is enzymatically cleaved into peptides (e.g. peptide mass fingerprinting). However, it should be remembered that the use of glutardialdehyde as sensitising agent precludes any identification of proteins from silver stained gels due to protein cross-linking! Several silver staining protocols compatible with mass spectrometry have been published [124,125].

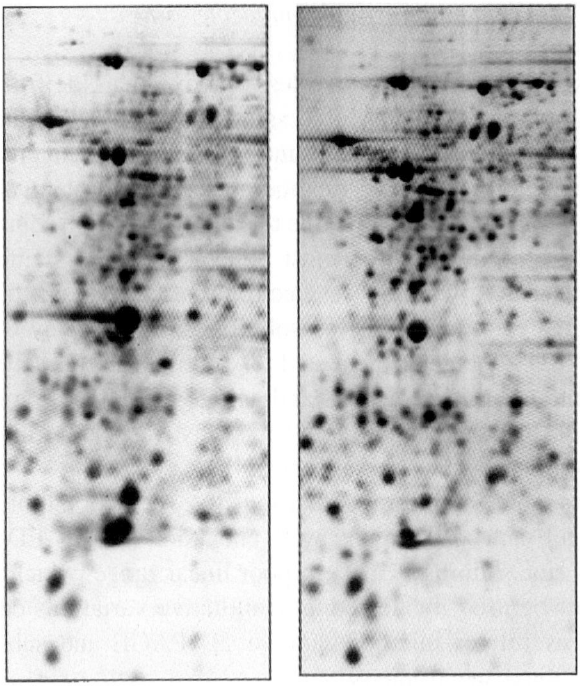

SYPROB Ruby stain Silver stain

Fig. 11. Comparison of protein detection with SYPRO Ruby (left) and silver staining (right). Sample: mouse liver proteins. Sample load: 100 μg of protein (Reprinted with permission from: Görg, A., et al., *Electrophoresis*, **21**, 1037–1053 (2000)).

A comparison of a silver- and SYPRO Ruby stained (see Section 3.4.4) 2D gel is shown in Fig. 11.

3.4.4. Visualization of proteins in 2D gels with the fluorescent dye SYPRO Ruby

Post-electrophoretic fluorescent dyes usually detect the proteins by binding non-covalently, either directly to the proteins or to the detergent (SDS) coating the proteins [71]. The proteins are stained by simply soaking the gel in aqueous solution containing a fluorophore. These methods are sensitive, have wide dynamic range, give good linearity of detection and are easy to use. In addition, these dyes do not interfere with post-electrophoretic analysis such as mass spectrometry. Sensitivity is slightly worse than silver staining. Prominent examples are the SYPRO™ dye series (available from Molecular Probes). Currently, SYPRO™ Ruby is the most sensitive dye of this series.

Briefly, after removal of the gel from the electrophoretic cassette, the gel is placed in a polypropylene box (do not use a glass vessel!) and fixed in acetic acid/ethanol/water (5/30/65) for 30 min. Then up to four gels are simultaneously stained for a minimum of 3 h, but preferably overnight, in 500 ml of SYPRO Ruby staining solution with gentle agitation. Destaining is not absolutely necessarily, but the background (or speckling) can be decreased by washing the gel with water or in methanol/acetic acid/ water (10:7:83) for 30 min. The gel is then placed on a Pyrex glass plate and the image is captured using an appropriate fluorescence imager. Detection limit is approximately 1–2 ng per spot.

3.4.5. Protein visualization by autoradiography, fluorography and phosphorimaging

Proteins radiolabelled with ^{32}P, ^{125}I, ^{14}C or ^{35}S can be detected by direct *autoradiography* [75] in which the dried gels are placed in contact with an autoradiographic film and exposed between 1 day and 1 month. More efficient detection of weak beta-emitting isotopes such as ^{3}H, ^{14}C and ^{35}S can be accomplished with *fluorography* where the fixed wet gel is impregnated with a scintillant, e.g. 2,5 diphenyloxazole (PPO), before drying. The scintillant converts the low energy radiations into visible light, which is then detected with an X-ray film. Moreover, for ^{3}H, ^{14}C or ^{35}S label sensitivity can be increased 10-fold by exposure at $-70°C$. Due to disadvantages such as long exposure times and/or limited, non-linear dynamic range of film response, these X-ray film based methods have been widely replaced by film-less electronic detection methods, the most popular of which is phosphorimaging (see Section 2.6).

A brief description of the procedure for detection of ^{35}S-methionine labelled proteins by phosphorimaging is as follows: after SDS-PAGE, the 2D gel is directly dried onto a sheet of filter paper with the help of a vacuum gel dryer; the dried gel is then exposed to a phosphorimaging plate (Kodak) for 24–48 h; after exposure, the phosphorimaging screen is inserted into the phosphorimager and scanned; and when scanning is complete, the image is saved and can be exported to an image analysis system.

3.5. Computer-aided image analysis

The major steps of computer-aided image analysis include: (i) data acquisition; (ii) spot detection and quantitation; (iii) pattern matching; and (iv) database construction. A more comprehensive description of computer-aided image analysis is given by Garrels [126] and Dunn [127].

Briefly, once the gel has been stained, the image is digitised with flat-bed scanners, CCD cameras, laser scanners, fluoroimagers or phosphorimagers, depending on the protein visualization method used. Prior to image analysis, the digitised image is subjected to several 'clean-up' steps to reduce background smear and remove horizontal and/or vertical streaks. This procedure is usually quite fast and does not require much user interaction. Then the individual spots on the 2D pattern are detected and quantified. This step is also performed automatically. Unfortunately, most image analysis programs do not identify all spots correctly, particularly when the overall quality of the electrophoretic separation is low (e.g. when artifacts or crowed areas and overlapping spots due to improper sample preparation or insufficient resolution are present on the gel). Therefore, manual spot editing with reference to the original stained 2D gel (or image) is still necessary. Depending upon the number of spots, the quality of the 2D electrophoretic separation and the algorithms applied for spot detection, this process may be quite laborious and time-consuming. After spot editing, each spot on one 2D gel must be matched to its counterpart on the other gels, usually by means of a reference ('master') gel. For this, in most computer-aided 2D image analysis programs, several 'landmark' spots (which should be evenly distributed over the entire gel area) are manually identified on each gel by the operator. Starting from these landmark spots, the program proceeds to match the other spots automatically. Again, mismatches must be carefully checked and edited manually [127].

Typically, at least two 2D gel patterns are matched (e.g. a stress induced versus a non-stressed control) and then compared to each other with respect to qualitative and/or quantitative differences between the 2D patterns (e.g. up- or down-regulated proteins). In most cases, however, many gels from different experiments have to be compared, usually by establishing a hierarchical 2D pattern database (see refs. [126,127]). Once the 2D gel database has been established, information stored in it can be exploited by addressing questions such as 'How do changes in environmental conditions (e.g. heat-, cold- or osmotic shock) affect the expression of individual proteins?', 'Can particular proteins be identified that are associated with a certain disease or disease state (e.g. disease markers)?', or 'What is the function of a particular protein?' [127].

Currently, several 2D image analysis software packages are commercially available. Programs have been continuously improved and enhanced over the years in terms of faster matching algorithms with lesser manual intervention and with focus on automation and better integration of data from various sources. New 2D software packages have also emerged which offer completely new approaches to image analysis and novel algorithms for more reliable spot detection and matching. Several programs include options such as control of a spot cutting robot, automated import of protein identification results from mass spectrometry,

Fig. 12. 2D Database of alkaline *Saccharomyces cerevisiae* proteins. First dimension: IPG 6–12. Second dimension: SDS-PAGE, 15%T. Sample load: 100 μg of protein. Silver stain (Reprinted with permission from: Wildgruber, R., et al., *Proteomics*, **2**, 727–732 (2002)) (http://www.weihenstephan.de/blm/deg/2ddb/).

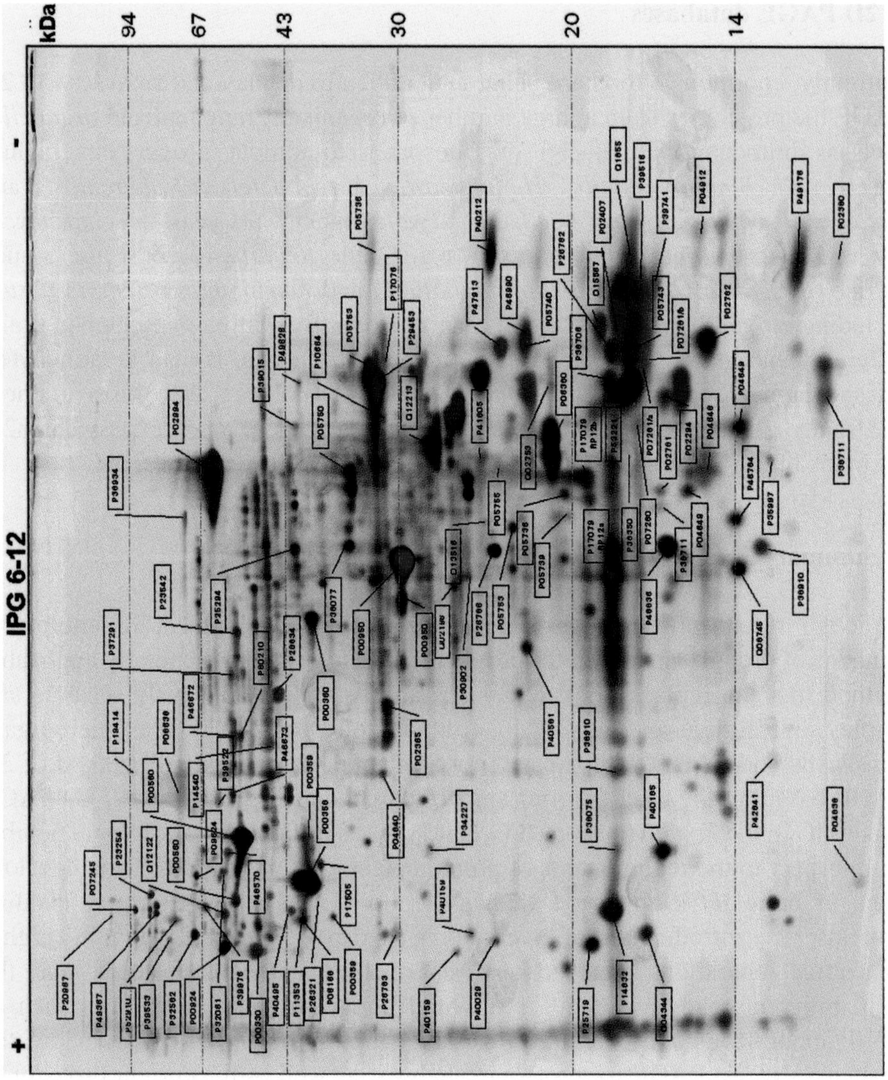
IPG 6-12
kDa
94
67
43
30
20
14
+
-

superior annotation flexibility (e.g. protein identity, mass spectrum, intensity/quantity, links to the Internet) and/or multichannel image merging of up to three different images to independent colour channels for fast image comparison.

4. 2D PAGE databases

Currently, enormous efforts are being undertaken to display and analyse with 2D PAGE the proteomes from a large number of organisms, ranging from *organelles* such as mitochondriae, nuclei or ribosomes to simple *procaroytes* including *E. coli*, *Bacillus subtilis*, *H. influenzae*, *Mycobacterium tuberculosis* and *Helicobacter pylori*, to *single-celled eukaryotes* such as the yeast *Saccharomyces cerevisiae*, to *multicellular organisms*, e.g. *Caenorhabditis elegans*, *plants* such as rice (*Oryza sativa*) or *Arabidopsis thaliana* and *mammalian cells and tissues* including rat and human heart, mouse and human liver, mouse and human brain, different cancer cell lines, HeLa cells, human fibroplasts, human keratinocytes, rat and human serum, etc. [23,33,38,48,96,128–137] (Fig. 12). Most of these and many other studies in progress are summarized at www.expasy.ch/ch2d/2d-index.html ('WORLD-2DPAGE Index to 2D PAGE databases').

5. Summary

Although recent progress has been made in the development of alternative methods of protein separation for proteomics, there is still no generally applicable method that can replace 2D PAGE in its ability to simultaneously separate and display several thousands of proteins from complex samples such as microorganisms, cells and tissues. In comparison with the classical O'Farrell method of 2D PAGE based on the use of carrier ampholyte generated pH gradients, 2D PAGE using IPG in the first dimension (IPG-Dalt) [8] has proven to be extremely flexible with respect to the requirements of proteome analysis. In particular, the development of basic IPGs up to pH 12 has facilitated the analysis of very alkaline proteins; the introduction of overlapping narrow-range IPGs permits higher resolution separations and the analysis of less abundant proteins; and the availability of ready-made IPG strips and integrated running devices such as the IPGphor have contributed towards the goal of automation. In this review, the challenges of 2D PAGE for proteome analysis, as well as the current protocol of IPG-Dalt, in particular the merits and limits of different methods for sample solubilization and sample application (by cup-loading or in-gel rehydration) with respect to the pH interval used for IPG-IEF are critically discussed, and guidelines for running conditions of analytical and micropreparative IPG-Dalt, using wide IPGs up to pH 12 for overview patterns or narrow IPGs for zoom-in gels and extended separation distances for optimum resolution and detection of minor components are given. Limitations remain in the field for the analysis of hydrophobic and/or membrane proteins, as well as in the lack of sensitive and reliable methods for protein quantitation, although recent developments in fluorescent dye technology and ultrasensitive electronic detection methods of radiolabelled proteins are quite promising. In conclusion, although by no means perfect, IPG-Dalt remains the core technology of choice for separating complex protein mixtures in proteomic projects at least for the near future.

Acknowledgements

The proteomic research in AG's laboratory is supported by grants from the European Community Biotech Programme, the Deutsche Forschungsgemeinschaft (DFG) and the Bundesministerium für Bildung und Forschung (BMBF). We would like to thank Günther Boguth, Oliver Drews, Angelika Köpf, Gerold Reil and Robert Wildgruber for contributing to this review.

References

1. Gygi, S.P., Rist, B., Gerber, S.A., Turecek, F., Gelb, M.H. and Aebersold, R., Quantitative analysis of complex protein mixtures using isotope-coded affinity tags. *Nat. Biotechnol.*, **17**, 994–999 (1999).
2. Link, A.J., Eng, J., Schieltz, D.M., Carmack, E., Mize, G.J., Morris, D.R., Garvik, B.M. and Yates, J.R. III, Direct analysis of protein complexes using mass spectrometry. *Nat. Biotechnol.*, **17**, 676–682 (1999).
3. Washburn, M.P., Wolters, D. and Yates, J.R. III, Large-scale analysis of the yeast proteome by multidimensional protein identification technology. *Nat. Biotechnol.*, **19**, 242–247 (2001).
4. Ducret, A., van Oostveen, I., Eng, J.K., Yates, J.R. III and Aebersold, R., High throughput protein characterization by automated reverse-phase chromatography/electrospray tandem mass spectrometry. *Protein Sci.*, **7**, 706–719 (1998).
5. Figeys, D., Gygyi, S.P., McKinnon, G. and Aebersold, R., An integrated microfluidics-tandem mass spectrometry system for automated protein analysis. *Anal. Chem.*, **70**, 3728–3734 (1999).
6. Haynes, P.A. and Yates, J.R. III, Proteome profiling—pitfalls and progress. *Yeast*, **17**, 81–87 (2000).
7. O'Farrell, P.H., High resolution two-dimensional electrophoresis of proteins. *J. Biol. Chem.*, **250**, 4007–4021 (1975).
8. Görg, A., Postel, W. and Günther, S., The current state of two-dimensional electrophoresis with immobilized pH gradients. *Electrophoresis*, **9**, 531–546 (1988).
9. Dunn, M.J., Two-Dimensional Polyacrylamide Gel Electrophoresis, In: Chrambach, A., Dunn, M.J. and Radola, B.J. (Eds.), Advances in Electrophoresis, Vol. 1. VCH, Weinheim, 1987, pp. 1–109.
10. Godovac-Zimmermann, J. and Brown, L.R., Perspectives for mass spectrometry and functional proteomics. *Mass Spectrom. Rev.*, **10**, 1–57 (2001).

11. Görg, A., Obermaier, C., Boguth, G., Harder, A., Scheibe, B., Wildgruber, R. and Weiss, W., The current state of two-dimensional electrophoresis with immobilized pH gradients. *Electrophoresis*, **21**, 1037–1053 (2000).
12. Celis, J.E. and Gromov, P., 2D protein electrophoresis: can it be perfected? *Curr. Opin. Biotechnol.*, **10**, 16–21 (1999).
13. Rabilloud, T., Two-dimensional gel electrophoresis in proteomics: old, old fashioned, but it still climbs up the mountains. *Proteomics*, **2**, 3–10 (2002).
14. Fey, S.J. and Larsen, P.M., 2D or not 2D. Two-dimensional gel electrophoresis. *Curr. Opin. Chem. Biol.*, **5**, 26–33 (2001).
15. Ong, S.E. and Pandey, A., An evaluation of the use of two-dimensional gel electrophoresis in proteomics. *Biomol. Eng.*, **18**, 195–205 (2001).
16. Lilley, K.S., Razzaq, A. and Dupree, P., Two-dimensional gel electrophoresis: recent advances in sample preparation, detection and quantitation. *Curr. Opin. Chem. Biol.*, **6**, 46–50 (2002).
17. Görg, A., Obermaier, C., Boguth, G. and Weiss, W., Recent developments in two-dimensional gel electrophoresis with immobilized pH gradients: wide pH gradients up to pH 12, longer separation distances and simplified procedures. *Electrophoresis*, **20**, 712–717 (1999).
18. Link, A.J., Robison, K. and Church, G.M., Comparing the predicted and observed properties of proteins encoded in the genome of *Escherichia coli* K-12. *Electrophoresis*, **18**, 1259–1313 (1997).
19. Wildgruber, R., Harder, A., Obermaier, C., Boguth, G., Weiss, W., Fey, S.J., Larsen, P.M. and Görg, A., Towards higher resolution: two-dimensional electrophoresis of *Saccharomyces cerevisiae* proteins using overlapping narrow immobilized pH gradients. *Electrophoresis*, **21**, 2610–2616 (2000).
20. O'Farrell, P.Z., Goodman, H.M. and O'Farrell, P.H., High resolution two-dimensional electrophoresis of basic as well as acidic proteins. *Cell*, **12**, 1133–1141 (1977).
21. Görg, A., Obermaier, C., Boguth, G., Csordas, A., Diaz, J.J. and Madjar, J.J., Very alkaline immobilized pH gradients for two-dimensional electrophoresis of ribosomal and nuclear proteins. *Electrophoresis*, **18**, 328–337 (1997).
22. Görg, A., IPG-Dalt of very alkaline proteins. *Methods Mol. Biol.*, **112**, 197–209 (1999).
23. Wildgruber, R., Reil, G., Drews, O., Parlar, H. and Görg, A., Web-based two-dimensional database of *Saccharomyces cerevisiae* proteins using immobilized pH gradients from pH 6 to pH 12 and matrix-assisted laser desorption/ionization-time of flight mass spectrometry. *Proteomics*, **2**, 727–732 (2002).
24. Görg, A., Boguth, G., Obermaier, C. and Weiss, W., Two-dimensional electrophoresis of proteins in an immobilized pH 4–12 gradient. *Electrophoresis*, **19**, 1516–1519 (1998).
25. Regula, J.T., Boguth, G., Görg, A., Hegermann, J., Mayer, F., Frank, R. and Herrmann, R., Defining the mycoplasma 'cytoskeleton': the protein composition of the Triton X-100 insoluble fraction of the bacterium *Mycoplasma pneumoniae* determined by 2D gel electrophoresis and mass spectrometry. *Microbiology*, **147**, 1045–1057 (2001).
26. Görg, A., Boguth, G., Obermaier, C., Posch, A. and Weiss, W., Two-dimensional polyacrylamide gel electrophoresis with immobilized pH gradients in the first dimension (IPG-Dalt): the state of the art and the controversy of vertical versus horizontal systems. *Electrophoresis*, **16**, 1079–1108 (1995).
27. Altland, K., Becher, P., Rossmann, U. and Bjellqvist, B., Isoelectric focusing of basic proteins: the problem of oxidation of cysteines. *Electrophoresis*, **9**, 474–485 (1988).
28. Olsson, I., Larsson, K., Palmgren, R. and Bjellqvist, B., Organic disulfides as a means to generate streak-free two-dimensional maps with narrow range basic immobilized pH gradient strips as first dimension. *Proteomics*, **2**, 1630–1632 (2002).

29. Bjellqvist, B., Sanchez, J.C., Pasquali, C., Ravier, F., Paquet, N., Frutiger, S., Hughes, G.J. and Hochstrasser, D.F., A nonlinear wide-range immobilised pH gradient for two-dimensional electrophoresis and its definition in a relevant pH scale. *Electrophoresis*, **14**, 1357–1365 (1993).
30. Westbrook, J.A., Yan, J.X., Wait, R., Welson, S.Y. and Dunn, M.J., Zooming-in on the proteome: very narrow-range immobilised pH gradients reveal more protein species and isoforms. *Electrophoresis*, **22**, 2865–2871 (2001).
31. Hoving, S., Voshol, H. and van Oostrum, J., Towards high performance two-dimensional gel electrophoresis using ultrazoom gels. *Electrophoresis*, **21**, 2617–2621 (2000).
32. Sabounchi-Schutt, F., Astrom, J., Eklund, A., Grunewald, J. and Bjellqvist, B., Detection and identification of human bronchoalveolar lavage proteins using narrow-range immobilized pH gradient DryStrip and the paper bridge sample application method. *Electrophoresis*, **22**, 1851–1860 (2001).
33. Gauss, C., Kalkum, M., Lowe, M., Lehrach, H. and Klose, J., Analysis of the mouse proteome. (I) Brain proteins: separation by two-dimensional electrophoresis and identification by mass spectrometry and genetic variation. *Electrophoresis*, **29**, 575–600 (1999).
34. Dunn, M.J. and Görg, A., Two-dimensional polyacrylamide gel electrophoresis for proteome analysis, In: Pennington, S.R. and Dunn, M.J. (Eds.), *Proteomics —from Protein Sequence to Function*. BIOS, Oxford, 2001, pp. 43–63.
35. Corthals, G.L., Wasinger, V.C., Hochstrasser, D.F. and Sanchez, J.C., The dynamic range of protein expression: a challenge for proteomic research. *Electrophoresis*, **21**, 1104–1115 (2000).
36. Cordwell, S.J., Nouwens, A.S., Verrills, N.M., Basseal, D.J. and Walsh, B.J., Subproteomics based upon protein cellular location and relative solubilities in conjunction with composite two-dimensional electrophoresis gels. *Electrophoresis*, **21**, 1094–1103 (2000).
37. Hanson, B.J., Schulenberg, B., Patton, W.F. and Capaldi, R.A., A novel subfractionation approach for mitochondrial proteins: a three-dimensional mitochondrial proteome map. *Electrophoresis*, **22**, 950–959 (2001).
38. Fountoulakis, M., Berndt, P., Langen, H. and Suter, L., The rat liver mitochondrial proteins. *Electrophoresis*, **23**, 311–328 (2002).
39. Nouwens, A.S., Cordwell, S.J., Larsen, M.R., Molloy, M.P., Gillings, M., Willcox, M.D. and Walsh, B.J., Complementing genomics with proteomics: the membrane subproteome of *Pseudomonas aeruginosa* PAO1. *Electrophoresis*, **21**, 3797–3809 (2000).
40. Zinser, E. and Daum, G., Isolation and biochemical characterization of organelles from the yeast *Saccharomyces cerevisiae*. *Yeast*, **11**, 493–536 (1995).
41. Zuo, X. and Speicher, D.W., Comprehensive analysis of complex proteomes using microscale solution isoelectrofocusing prior to narrow pH range two-dimensional electrophoresis. *Proteomics*, **2**, 58–68 (2002).
42. Herbert, B. and Righetti, P.G., A turning point in proteome analysis: sample prefractionation via multicompartment electrolyzers with isoelectric membranes. *Electrophoresis*, **21**, 3639–3648 (2000).
43. Görg, A., Boguth, G., Köpf, A., Reil, G., Parlar, H. and Weiss, W., Sample prefractionation with Sephadex isoelectric focusing prior to narrow pH range two-dimensional gels. *Proteomics*, **2**, 1652–1657 (2002).
44. Fountoulakis, M., Takacs, M.F. and Takacs, B., Enrichment of low-copy-number gene products by hydrophobic interaction chromatography. *J. Chromatogr. A*, **833**, 157–168 (1999).
45. Weiss, W., Postel, W. and Görg, A., Application of sequential extraction procedures and glycoprotein blotting for the characterization of the 2D polypeptide patterns of barley seed proteins. *Electrophoresis*, **13**, 770–773 (1992).

46. Weiss, W., Vogelmeier, C. and Görg, A., Electrophoretic characterization of wheat grain allergens from different cultivars involved in bakers' asthma. *Electrophoresis*, **14**, 805–816 (1993).
47. Molloy, M.P., Herbert, B.R., Williams, K.L. and Gooley, A.A., Extraction of *Escherichia coli* proteins with organic solvents prior to two-dimensional electrophoresis. *Electrophoresis*, **20**, 701–704 (1999).
48. Fountoulakis, M., Juranville, J.F., Roder, D., Evers, S., Berndt, P. and Langen, H., Reference map of the low molecular mass proteins of *Haemophilus influenzae*. *Electrophoresis*, **19**, 1819–1827 (1998).
49. Schägger, H. and von Jagow, G., Tricine-sodium dodecyl sulfate-polyacrylamide gel electrophoresis for the separation of proteins in the range from 1 to 100 kDa. *Anal. Biochem.*, **166**, 368–379 (1987).
50. Zuo, X., Echan, L., Hembach, P., Tang, H.Y., Speicher, K.D., Santoli, D. and Speicher, D.W., Towards global analysis of mammalian proteomes using sample prefractionation prior to narrow pH range two-dimensional gels and using one-dimensional gels for insoluble and large proteins. *Electrophoresis*, **22**, 1603–1615 (2001).
51. Santoni, V., Molloy, M. and Rabilloud, T., Membrane proteins and proteomics: un amour impossible? *Electrophoresis*, **21**, 1054–1070 (2000).
52. Rabilloud, T., Adessi, C., Giraudel, A. and Lunardi, J., Improvement of the solubilization of proteins in two-dimensional electrophoresis with immobilized pH gradients. *Electrophoresis*, **18**, 307–316 (1997).
53. Chevallet, M., Santoni, V., Poinas, A., Rouquie, D., Fuchs, A., Kieffer, S., Rossignol, M., Lunardi, J., Garin, J. and Rabilloud, T., New zwitterionic detergents improve the analysis of membrane proteins by two-dimensional electrophoresis. *Electrophoresis*, **19**, 1901–1909 (1998).
54. Rabilloud, T., Use of thiourea to increase the solubility of membrane proteins in two-dimensional electrophoresis. *Electrophoresis*, **19**, 758–760 (1998).
55. Rabilloud, T., Blisnick, T., Heller, M., Luche, S., Aebersold, R., Lunardi, J. and Braun-Breton, C., Analysis of membrane proteins by two-dimensional electrophoresis extracted form normal or *Plasmodium falciparum*-infected erythrocyte ghosts. *Electrophoresis*, **20**, 3603–3610 (1999).
56. Santoni, V., Rabilloud, T., Doumas, P., Rouqui, D., Mansion, M., Kieffer, S., Garin, J. and Rossignol, M., Towards the recovery of hydrophobic proteins on two-dimensional electrophoresis gels. *Electrophoresis*, **20**, 705–711 (1999).
57. Santoni, V., Kieffer, S., Desclaux, D., Masson, F. and Rabilloud, T., Membrane proteomics: use of additive main effects with multiplicative interaction model to classify plasma membrane proteins according to their solubility and electrophoretic properties. *Electrophoresis*, **21**, 3329–3344 (2000).
58. Santoni, V., Doumas, P., Rouquie, D., Mansion, M., Rabilloud, T. and Rossignol, M., Large scale characterization of plant plasma membrane proteins. *Biochimie*, **81**, 655–661 (1999).
59. Molloy, M.P., Herbert, B.R., Slade, M.B., Rabilloud, T., Nouwens, A.S., Williams, K.L. and Gooley, A.A., Proteomic analysis of the *Escherichia coli* outer membrane. *Eur. J. Biochem.*, **267**, 2871–2881 (2000).
60. Molloy, M.P., Two-dimensional electrophoresis of membrane proteins using immobilized pH gradients. *Anal. Biochem.*, **280**, 1–10 (2000).
61. Frisco, G. and Wikstrom, L., Analysis of proteins from membrane-enriched cerebellar preparations by two-dimensional gel electrophoresis and mass spectrometry. *Electrophoresis*, **20**, 917–927 (1999).

62. Fujiki, Y., Hubbard, A.L., Fowler, S. and Lazarow, P.B., Isolation of intracellular membranes by means of sodium carbonate treatment: application to endoplasmic reticulum. *J. Cell Biol.*, **93**, 97–102 (1982).
63. Molloy, M.P., Herbert, B.R., Williams, K.L. and Gooley, A.A., Extraction of *E. coli* proteins with organic solvents prior to two-dimensional electrophoresis. *Electrophoresis*, **20**, 701–704 (1999).
64. Ferro, M., Seigneurin-Berny, D., Rolland, N., Chapel, A., Salvi, D., Garin, J. and Joyard, J., Organic solvent extraction as a versatile procedure to identify hydrophobic chloroplast membrane proteins. *Electrophoresis*, **21**, 3517–3526 (2000).
65. Pasquali, C., Fialka, I. and Huber, L.A., Preparative two-dimensional gel electrophoresis of membrane proteins. *Electrophoresis*, **18**, 2573–2581 (1997).
66. Adessi, C., Miege, C., Albrieux, C. and Rabilloud, T., Two-dimensional electrophoresis of membrane proteins: a current challenge for immobilized pH gradients. *Electrophoresis*, **18**, 127–135 (1997).
67. Wilkins, M.R., Gasteiger, E., Sanchez, J.-C., Bairoch, A. and Hochstrasser, D.F., Two-dimensional gel electrophoresis for proteome projects: the effect of protein hydrophobicity and copy number. *Electrophoresis*, **19**, 1501–1505 (1998).
68. Rabilloud, T., Detecting proteins separated by 2-D gel electrophoresis. *Anal. Chem.*, **72**, 48A–55A (2000).
69. Bini, L., Liberatori, S., Magi, B., Marzocchi, B., Raggiaschi, R. and Pallini, V., Protein blotting and immunoblotting, In: Rabilloud, T. (Ed.), *Proteome Research: Two-Dimensional Electrophoresis and Identification Methods*. Springer, Berlin, 2000, pp. 127–141.
70. Kaufmann, H., Bailey, J.E. and Fussenegger, M., Use of antibodies for detection of phosphorylated proteins separated by two-dimensional gel electrophoresis. *Proteomics*, **1**, 194–199 (2001).
71. Patton, W.F., A thousand points of light: the application of fluorescence detection technologies to two-dimensional gel electrophoresis and proteomics. *Electrophoresis*, **21**, 1123–1144 (2000).
72. Herick, K., Jackson, P., Wersch, G. and Burkovski, A., Detection of fluorescence dye-labeled proteins in 2-D gels using an Arthur 1442 Multiwavelength Fluoroimager. *Biotechniques*, **1**, 146–149 (2001).
73. Urwin, V.E. and Jackson, P., Two-dimensional polyacrylamide gel electrophoresis of proteins labelled with the fluorophore monobromobimane prior to first-dimensional isoelectric focusing; imaging of the fluorescent protein spot patterns using a cooled charge-coupled device. *Anal. Biochem.*, **209**, 57–63 (1993).
74. Unlü, M., Morgan, M.E. and Minden, J.S., Difference gel electrophoresis: a single gel method for detecting changes in protein extracts. *Electrophoresis*, **18**, 2071–2077 (1997).
75. Dunn, M.J., Gel Electrophoresis: Proteins, BIOS, Oxford, 1993.
76. Link, A.J., Autoradiography of 2-D gels. *Methods Mol. Biol.*, **112**, 285–290 (1999).
77. Patterson, S.D. and Latter, G.I., Evaluation of storage phosphor imaging for quantitative analysis of 2-D gels using the Quest II system. *Biotechniques*, **15**, 1076–1083 (1993).
78. Drukier, A.K., MPD enables ultra-sensitive methods for proteomics, In: Hochstrasser, D.F., Sanchez, J.-C., Bini, L. and Pallini, V. (Eds.), *From Genome to Proteome*, 4th Siena Meeting, 2000, pp. 123–124.
79. Monribot-Espagne, C. and Boucherie, H., Differential gel exposure, a new methodology for the two-dimensional comparison of protein samples. *Proteomics*, **2**, 229–240 (2002).
80. Bernhardt, J., Büttner, K., Scharf, C. and Hecker, M., Dual channel imaging of two-dimensional electropherograms in *Bacillus subtilis*. *Electrophoresis*, **20**, 2225–2240 (1999).

81. Anderson, N.G. and Anderson, N.L., Analytical techniques for cell fractions. XXII. Two-dimensional analysis of serum and tissue proteins: multiple gradient-slab gel electrophoresis. *Anal. Biochem.*, **85**, 341–354 (1978).
82. Islam, R., Ko, C. and Landers, T., A new approach to rapid immobilised pH gradient IEF for 2-D electrophoresis. *Sci. Tools*, **3**, 14–15 (1998).
83. Granier, F. and de Vienne, D., Silver staining of proteins: standardized procedure for two-dimensional gels bound to polyester sheets. *Anal. Biochem.*, **155**, 45–50 (1986).
84. Sinha, P., Poland, J., Schnölzer, M. and Rabilloud, T., A new silver staining apparatus and procedure for matrix-assisted laser desorption/ionization-time of flight analysis of proteins after two-dimensional electrophoresis. *Proteomics*, **1**, 835–840 (2001).
85. Klose, J., Protein mapping by combined isoelectric focusing and electrophoresis of mouse tissues. A novel approach to testing for induced point mutations in mammals. *Humangenetik*, **26**, 231–243 (1975).
86. Scheele, G.A., Two-dimensional gel analysis of soluble proteins. Charaterization of guinea pig exocrine pancreatic proteins. *J. Biol. Chem.*, **250**, 5375–5385 (1975).
87. Görg, A., Postel, W., Weser, J., Patutschnick, W. and Cleve, H., Improved resolution of pi (alpha I-antitrypsin) phenotypes by a large scale immobilized pH gradient. *Am. J. Hum. Genet.*, **37**, 922–930 (1985).
88. Corbett, J.M., Dunn, M.J., Posch, A. and Görg, A., Positional reproducibility of protein spots in two-dimensional polyacrylamide gel electrophoresis using immobilised pH gradient isoelectric focusing in the first dimension: an interlaboratory comparison. *Electrophoresis*, **15**, 1205–1211 (1994).
89. Blomberg, A., Blomberg, L., Norbeck, J., Fey, S.J., Larsen, P.M., Roepstorff, P., Degand, H., Boutry, M., Posch, A. and Görg, A., Interlaboratory reproducibility of yeast protein patterns analyzed by immobilized pH gradient two-dimensional gel electrophoresis. *Electrophoresis*, **16**, 1935–1945 (1995).
90. Bjellqvist, B., Sanchez, J.C., Pasquali, C., Ravier, F., Paquet, N., Frutiger, S., Hughes, G.J. and Hochstrasser, D., Micropreparative two-dimensional electrophoresis allowing the separation of samples containing milligram amounts of proteins. *Electrophoresis*, **12**, 1375–1378 (1993).
91. Hanash, S.M., Strahler, J.R., Neel, J.V., Hailat, N., Melhem, R., Keim, D., Zhu, X.X., Wagner, D., Gage, D.A. and Watson, J.T., Highly resolving two-dimensional gels for protein sequencing. *Proc. Natl Acad. Sci. USA*, **88**, 5709–5713 (1991).
92. Olivieri, E., Herbert, B. and Righetti, P.G., The effect of protease inhibitors on the two-dimensional electrophoresis pattern of red blood cell membranes. *Electrophoresis*, **22**, 560–565 (2001).
93. Rabilloud, T., Solubilization of proteins in 2-D electrophoresis. An outline. *Methods Mol. Biol.*, **112**, 9–19 (1999).
94. Harder, A., Wildgruber, R., Nawrocki, A., Fey, S.J., Larsen, P.M. and Görg, A., Comparison of yeast cell protein solubilization procedures for two-dimensional electrophoresis. *Electrophoresis*, **20**, 826–829 (1999).
95. Harder, A., Wildgruber, R., Behr, J., Braig, C., Mann, M. and Görg, A., IPG-Dalt of glucan and chitin linked yeast cell wall proteins. In: *Proteomic Forum 2001. International Meeting on Proteome Analysis*. PSP München, 2001, p. 186.
96. Boucherie, H., Dujardin, G., Kermorgant, M., Monribot, C., Slonimski, P. and Perrot, M., Two-dimensional protein map of *Saccharomyces cerevisiae*: construction of a gene–protein index. *Yeast*, **11**, 601–613 (1995).
97. Herbert, B.R., Molloy, M.P., Gooley, A.A., Walsh, B.J., Bryson, W.G. and Williams, K.L., Improved protein solubility in two-dimensional electrophoresis using tributyl phosphine as reducing agent. *Electrophoresis*, **19**, 845–851 (1998).

98. Herbert, B., Advances in protein solubilisation for two-dimensional electrophoresis. *Electrophoresis*, **20**, 660–663 (1999).
99. Madjar, J.J., Preparation of Ribosomes and Ribosomal Proteins from Cultured Cells, In: Celis, J.E. (Ed.), Cell Biology. A Laboratory Handbook, Vol. 1. Academic Press, New York, 1994, pp. 657–661.
100. Csordas, A., Pedrini, M. and Grunicke, H., Suitability of staining techniques for the detection and quantitation of nonhistone high mobility group proteins. *Electrophoresis*, **11**, 118–123 (1990).
101. Futcher, B., Cell cycle synchronization. *Methods Cell. Sci.*, **21**, 79–86 (1999).
102. Granier, F., Extraction of plant proteins for two-dimensional electrophoresis. *Electrophoresis*, **9**, 712–718 (1988).
103. Bjellqvist, B., Ek, K., Righetti, P.G., Gianazza, E., Görg, A., Postel, W. and Westermeier, R., Isoelectric focusing in immobilized pH gradients: principle, methodology and some applications. *J. Biochem. Biophys. Methods*, **6**, 317–339 (1982).
104. Görg, A., Fawcett, J.S. and Chrambach, A., The current state of electrofocusing in immobilized pH gradients. *Adv. Electrophoresis*, **1**, 1–44 (1988).
105. Görg, A. and Weiss, W., 2D electrophoresis with immobilized pH gradients, In: Rabilloud, T. (Ed.), *Proteome Research: Two-Dimensional Electrophoresis and Identification Methods*. Springer, Berlin, 2000, pp. 57–106.
106. Altland, K., IPGMAKER: a program for IBM-compatible personal computers to create and test recipes for immobilized pH gradients. *Electrophoresis*, **11**, 140–147 (1990).
107. Righetti, P.G. and Tonani, C., Immobilized pH gradients (IPG) simulator — an additional step in pH gradient engineering: II. Nonlinear pH gradients. *Electrophoresis*, **12**, 1021–1027 (1991).
108. Righetti, P.G., Immobilized pH Gradients: Theory and Methodology, Elsevier, Amsterdam, 1990.
109. Righetti, P.G. and Bossi, A., Isoelectric focusing in immobilized pH gradients: an update. *J. Chromatogr. B. Biomed. Sci. Appl.*, **699**, 77–89 (1997).
110. Görg, A., Two-dimensional electrophoresis. *Nature*, **349**, 545–546 (1991).
111. Poland, J., Cahill, M.A. and Sinha, P., Isoelectric focusing in long immobilized pH gradient gels to improve protein separation in proteomic analysis. *Electrophoresis*, **24**, 1271–1275 (2003).
112. Rabilloud, T., Valette, C. and Lawrence, J.J., Sample application by in-gel rehydration improves the resolution of two-dimensional electrophoresis with immobilized pH gradients in the first dimension. *Electrophoresis*, **15**, 1552–1558 (1994).
113. Sanchez, J.C., Rouge, V., Pisteur, M., Ravier, F., Tonella, L., Moosmayer, M., Wilkins, M.R. and Hochstrasser, D.F., Improved and simplified in-gel sample application using reswelling of dry immobilized pH gradients. *Electrophoresis*, **18**, 324–327 (1997).
114. Görg, A., Postel, W., Friedrich, C., Kuick, R., Strahler, J.R. and Hanash, S.M., Temperature-dependent spot positional variability in two-dimensional polypetide patterns. *Electrophoresis*, **12**, 653–658 (1991).
115. Görg, A., Postel, W., Weser, J., Günther, S., Strahler, S.R., Hanash, S.M. and Somerlot, L., Elimination of point streaking on silver stained two-dimensional gels by addition of iodoacetamide to the equilibration buffer. *Electrophoresis*, **8**, 122–124 (1987).
116. Zuo, X. and Speicher, D.W., Quantitative evaluation of protein recoveries in two-dimensional electrophoresis with immobilized pH gradients. *Electrophoresis*, **21**, 3035–3047 (2000).
117. Laemmli, U.K., Cleavage of structural proteins during the assembly of the head of bacteriophage T4. *Nature*, **221**, 680–685 (1970).

118. Görg, A. and Weiss, W., Horizontal SDS-PAGE for IPG-Dalt. *Methods Mol. Biol.*, **112**, 235–244 (1999).
119. Neuhoff, V., Arold, N., Taube, D. and Ehrhardt, W., Improved staining of proteins in polyacrylamide gels including isoelectric focusing gels with clear background at nanogram sensitivity using Coomassie Brilliant Blue G-250 and R-250. *Electrophoresis*, **9**, 255–262 (1988).
120. Castellanos-Serra, L., Proenza, W., Huerta, V., Moritz, R.L. and Simpson, R.L., Proteome analysis of polyacrylamide gel-separated proteins visualized by reversible negative staining using imidazole–zinc salts. *Electrophoresis*, **20**, 732–737 (1999).
121. Rabilloud, T. and Charmont, S., Detection of proteins on two-dimensional electrophoresis gels, In: Rabilloud, T. (Ed.), *Proteome Research: Two-Dimensional Electrophoresis and Identification Methods*. Springer, Berlin, 2000, pp. 107–126.
122. Merril, C.R., Goldman, D., Sedman, S.A. and Ebert, M.H., Ultrasensitive stain for proteins in polyacrylamide gels shows regional variation in cerebrospinal fluid proteins. *Science*, **211**, 1437–1438 (1981).
123. Oakley, B.R., Kirsch, D.R. and Morris, N.R., A simplified ultrasensitive silver stain for detecting proteins in polyacrylamide gels. *Anal. Biochem.*, **105**, 361–363 (1980).
124. Shevchenko, A., Wilm, M., Vorm, O. and Mann, M., Mass spectrometric sequencing of proteins on silver-stained polyacrylamide gels. *Anal. Chem.*, **68**, 850–858 (1996).
125. Mortz, E., Krogh, T.N., Vorum, H. and Görg, A., Improved silver staining protocols for high sensitivity protein identification using matrix-assisted laser desorption/ionization-time of flight analysis. *Proteomics*, **1**, 359–363 (2001).
126. Garrels, J.I., The QUEST system for quantitative analysis of two-dimensional gels. *J. Biol. Chem.*, **264**, 5269–5282 (1989).
127. Dunn, M.J., The analysis of two-dimensional polyacrylamide gels for the construction of protein databases, In: Bryce, C.F.A. (Ed.), *Microcomputers in Biochemistry: A Practical Approach*. IRL Press, Oxford, 1992, pp. 215–242.
128. Scharfe, C., Zaccaria, P., Hoertnagel, K., Jaksch, M., Klopstock, T., Dembowski, M., Lill, R., Prokisch, H., Gerbitz, K.D., Neupert, W., Mewes, H.W. and Meitinger, T., MITOP, the mitochondrial proteome database: 2000 update. *Nucl. Acids Res.*, **28**, 155–158 (2000).
129. VanBogelen, R.A., Abshire, K.Z., Moldover, B., Olson, E.R. and Neidhardt, F.C., *Escherichia coli* proteome analysis using the gene–protein database. *Electrophoresis*, (18), 1243–1251 (1997).
130. Antelmann, H., Bernhardt, J., Schmid, R., Mach, H., Volker, U. and Hecker, M., First steps from a two-dimensional protein index towards a response-regulation map for *Bacillus subtilis*. *Electrophoresis*, **18**, 1451–1463 (1997).
131. Langen, H., Takacs, B., Evers, S., Berndt, P., Lahm, H.W., Wipf, B., Gray, C. and Fountoulakis, M., Two-dimensional map of the proteome of *Haemophilus influenzae*. *Electrophoresis*, **21**, 411–429 (2000).
132. Mollenkopf, H.J., Jungblut, P.R., Raupach, B., Mattow, J., Lamer, S., Zimny-Arndt, U., Schaible, U.E. and Kaufmann, S.H., A dynamic two-dimensional polyacrylamide gel electrophoresis database: the mycobacterial proteome via Internet. *Electrophoresis*, **20**, 2172–2180 (1999).
133. Bumann, D., Meyer, T.F. and Jungblut, P.R., Proteome analysis of the common human pathogen *Helicobacter pylori*. *Proteomics*, **1**, 473–479 (2001).
134. Costanzo, M.C., Hogan, J.D., Cusick, M.E., Davis, B.P., Fancher, A.M., Hodges, P.E., Kondu, P., Lengieza, C., Lew-Smith, J.E., Lingner, C., Roberg-Perez, K.J., Tillberg, M., Brooks, J.E. and Garrels, J.I., The yeast proteome database (YPD) and *Caenorhabditis*

elegans proteome database (WormPD): comprehensive resources for the organization and comparison of model organism protein information. *Nucl. Acids Res.*, **28**, 73–76 (2000).

135. Perrot, M., Sagliocco, F., Mini, T., Monribot, C., Schneider, U., Shevchenko, A., Mann, M., Jeno, P. and Boucherie, H., Two-dimensional gel protein database of *Saccharomyces cerevisiae* (update 1999). *Electrophoresis*, **20**, 228022–228098 (1999).
136. Dunn, M.J., Corbett, J.M. and Wheeler, C.H., HSC-2DPAGE and the two-dimensional gel electrophoresis database of dog heart proteins. *Electrophoresis*, **28**, 2795–2802 (1997).
137. Fountoulakis, M., Juranville, J.F., Berndt, P., Langen, H. and Suter, L., Two-dimensional database of mouse liver proteins. An update. *Electrophoresis*, **22**, 1747–1763 (2001).
138. Drews, O., Weiss, W., Reil, G., Parlar, H., Wait, R. and Görg, A., High pressure effects step-wise altered protein expression in *Lactobacillus sanfranciscensis*. *Proteomics*, **2**, 765–774 (2002).

Proteome Analysis. Interpreting the Genome.
D.W. Speicher (editor)

Chapter 3

Protein profiling using two-dimensional gel electrophoresis with multiple fluorescent tags

WILLIAM A. HANLON* and PATRICK R. GRIFFIN[1]

Department of Molecular Profiling Proteomics, Merck Research Laboratories, Rahway, NJ 07065, USA

*Corresponding author. Address: 126 E. Lincoln Avenue, P.O. Box 2000, RY33-200, Rahway, NJ 07065, USA. Tel.: +1-732-594-4931; fax: +1-732-594-1030. E-mail: william_hanlon@merck.com (W.A. Hanlon).

[1] Current address: ExSAR Corporation, Monmouth Junction, NJ 08852, USA.

1. Introduction

Reproducibility of protein separation between gels using two-dimensional electrophoresis (2DE, see Chapter 2) is hindered by many factors such as inhomogeneities in acrylamide polymerization along with temperature inequities and power fluctuations. These variables and others prevent 2DE gel images from being directly superimposable and require warping capability to overlay and compare them. Therefore, detection of differences between samples by 2DE requires a high level of skill and sophisticated spot-detection and matching software along with appropriately powered experimental design to compensate the technical and biological variation in order to increase the statistical significance of the data.

Fluorescent 2D difference gel electrophoresis (DIGE), first described by Unlu et al. [1] has made it possible to identify and compare quantitative differences between multiple experimental samples resolved in the same gel. The basis of this technology is the use of mass- and charge-matched fluorescent dyes (*N*-hydroxy succinimidyl ester derivatives of Cy3 and Cy5), which have different excitation and emission spectra. These dyes undergo a nucleophilic substitution reaction with the epsilon amine group of lysine residues to form an amide. The dyes are positively charged to match the charge on lysine that is lost due to the labeling reaction. Samples are covalently labeled separately with different cyanine dyes, mixed together and resolved on a single 2DE gel. Proteins that are present in both samples will migrate to the same location in the gel eliminating the reproducibility problem. The samples contained within the same gel are scanned separately at their appropriate excitation and emission wavelengths. Because the images originated from the same gel they can be overlaid and compared directly without warping. This technology affords a reduction in the number of gels that need to be run to address variability and increases the confidence associated with differences that are identified and quantitated between samples. This technology was recently validated by Tonge et al. [2] with proteomic studies on mouse liver, Zhou et al. [3] with the identification of esophageal scans cell cancer-specific protein markers, and Gharbi et al. [4] analyzed the effects of the hepatotoxin, *N*-acetyl-*p*-aminophenol, on mouse liver protein expression.

Profiling studies using 2DE technology routinely involve the analysis of more than two samples for comparison. Often, time-dependent or concentration-dependent effects are of interest for evaluation. Such experiments incorporate a third cyanine dye (Cy2) with excitation and emission wavelengths independent of Cy3 and Cy5 to allow comparison of three samples in one gel. With proper experimental design, as discussed later in the chapter, a third dye can provide more robust statistical analysis of protein expression differences across many gels. This is accomplished by generating a 'pooled standard' that consists of an equal protein amount from every sample in the experiment. The pooled standard is run on every gel to serve as a common anchor for gel-to-gel matching and normalization.

2. Mechanics of DIGE technology

2.1. Sample preparation

When using DIGE technology, samples can be prepared by the same methods used for conventional 2DE approaches (see Chapter 2) providing that buffer components are compatible with the cyanine dye labeling procedure. For the studies described later, tissue samples were prepared using either Polytron disruption on ice in tissue lysis buffer (8 M urea, 4% 3-[(3-cholamidopropyl)-dimethylammonio]-1-propane sulfonate (CHAPS), 5 mM magnesium acetate and 10 mM Tris–Cl (pH 8.0)) or by pulverizing frozen tissues into powder using a liquid nitrogen cooled mortar and pestle. The powdered tissue was stored at −80°C until needed. Tissue lysates were prepared from the frozen powder by extracting the pulverized tissue into tissue lysis buffer. Following tissue solubilization, protein concentration was determined and an equal volume of labeling buffer (5 M urea, 3 M thiourea, 4% CHAPS, 5 mM magnesium acetate and 10 mM Tris–Cl (pH 8.5)) was added to all the samples. The samples were kept on ice for an additional 15 min and insoluble material was removed by centrifugation. Tissue culture samples can be prepared by extracting a pellet of washed cells directly in tissue lysis buffer. Finally, the pH of the samples was adjusted as needed to be between 8.0 and 8.5 for optimal labeling.

2.2. Cyanine dye labeling

The tissue lysates were labeled with the cyanine dyes Cy2, Cy3 and Cy5 (Amersham Biosciences, Inc.) through nucleophilic substitution reactions with the epsilon amine group of lysine residues to form an amide. In the labeling reaction, the dye:protein ratio needs to be kept low ensuring that proteins are labeled with a single dye molecule [2]. The stock cyanine dyes were diluted 1:5 with fresh anhydrous *N,N*-dimethylformamide (DMF) and incubated with tissue lysates at a final ratio of 200 pmol dye to 50 μg protein for 30 min on ice. The reaction was quenched by adding 10 nmol of L-lysine per 200 pmol of dye, vortexed and kept on ice for an additional 10 min. After labeling, an equal volume of reducing buffer (6.5 M urea, 2 M thiourea, 4% CHAPS, 20 mg/ml DTT and 1% pharmalytes (pH 3–10)) was added to each sample and the samples returned to ice for 10 min. To ensure optimal labeling efficiency, it was important that the protein concentration of the samples was at least 1 mg/ml (5–10 mg/ml is optimal) and that a pH between 8 and 8.5 was maintained [2]. With some tissues, excessive amounts of contaminants such as nucleic acids, salt and especially lipids can contribute to poor labeling efficiencies. Under these conditions, it was necessary to improve the quality of the samples by acetone precipitation prior to labeling. Labeled samples were either frozen until needed or combined with other cyanine dye labeled samples and the volume adjusted for the appropriate size IPG strip with 1D buffer (6.5 M urea, 2 M thiourea, 4% CHAPS, 2 mg/ml DTT and 1% pharmalytes (pH 3–10)).

2.3. Cyanine image visualization

Analytical gels (20 cm × 25 cm) were cast between low fluorescent glass (Hoefer) to maximize sensitivity during scanning. Following 2DE electrophoresis, the unopened gel cassettes were scanned using a 2920-2D Master Imager (Amersham Biosciences). The gels were scanned for each cyanine dye image separately using filters specific for each dye's excitation and emission wavelength. The 2920-2D Master Imager uses a 150 W xenon lamp (ozone free). The light was focused through the bottom edge of the gel using a fiber-optic fishtail assembly. The emitted light was measured by a Peltier cooled 16-bit CCD camera in multiple 5 cm × 7 cm size frames that were stitched together by the imager software. Images were exported from the Imager software files as 16-bit tagged image file format (TIFF) files.

Mini precast gels in plastic cassettes have also been used with this technology. Following electrophoresis, gels need to be removed from the cassette and placed between low fluorescent glass plates. The gels can then be imaged in a similar manner as larger size analytical gels. This process was more prone to artifactual features as a result of dust, fluid bubbles and air bubbles that occur during the reassembly process. With a little practice, very good results using this approach could be obtained.

2.4. Gel image analysis

Gel TIFF images can be analyzed by a variety of different 2D image analysis programs. The evaluation of the DIGE technology described here was done using the DeCyder (Amersham Biosciences) 2D analysis software, which was specifically designed to leverage the existence of an internal pooled standard run on each gel. The benefit of having a pooled standard is addressed later. DeCyder 2D analysis software consists of two different modules. Gel TIFF images were first analyzed using the DeCyder-DIA (difference in-gel analysis) module. This module compared any two images generated from the same gel in a pairwise manner. When using an internal standard, a pooled sample was compared to each of the individual sample images from the same gel. If two dyes (Cy3 and Cy5) were used in one gel, then one DIA analysis was performed comparing the two images directly (Cy3:Cy5). If three dyes (Cy2, Cy3 and Cy5) were used, then two separate DIA analyses were performed (Cy2:Cy3 and Cy2:Cy5), where Cy2 was used to label the pooled standard.

The DIA module electronically overlays and combines the two images to be compared using common *x*/*y* coordinates resulting from running the respective samples on the same gel. Spots were then detected as a combined image generating a single spot map with an associated spot number as a unique identifier. These spot maps were then applied to both images separately and the resulting volumes were calculated from the pixel information contained within each spot outline of the map. The identical spot map on both images provided an immediate, robust and flawless matching of the images without warping. If a protein did not exist in one of the two samples and therefore did not appear in one of the two images, a volume measurement was still calculated from the background pixels contained within the spot outline on the image missing the protein (Fig. 1). This approach helped to reduce unique protein expression changes that might have been missed in a defined experimental condition when an image from that condition was not selected as the reference gel. This benefit was further realized in multiple gel studies by the fact that each DIA analysis contained a pooled standard image run on every gel. A volumetric ratio was calculated for every spot on the paired gel images. The ratio provided a means for normalizing differences in image intensities, sample loading or other technical variations. This was accomplished automatically by the DIA module through comparison of the actual distribution of the log spot ratios with the log values of a normal distribution (Fig. 2). The difference between the centers of the two distributions was used to correct and normalize the two samples. If only two samples were compared, then the log spot ratios that occurred above a defined threshold (i.e. fold change or 1, 2 or 3 standard deviations (SD)) were considered either up or down-regulated.

For experiments with multiple gels, the DeCyder-DIA results were imported into the DeCyder-BVA (biological variation analysis) module. This module matched spots of the DIA gel image pairs to a reference (master) gel. The DIA image pair that contained the most spots was automatically assigned by the DeCyder-BVA as the reference gel. When the need for multiple BVA files was necessary to analyze large data sets, a common master DIA image pair was manually selected so that spot numbers were consistent between the BVA files. Normalization of gel-to-gel differences was accomplished by equating the sample to pooled standard ratios from the DIA module results to 1.00. The ratio-metric differences of the pooled standard to each individual sample gel image were then used to determine differences between samples and statistical differences between experimental conditions.

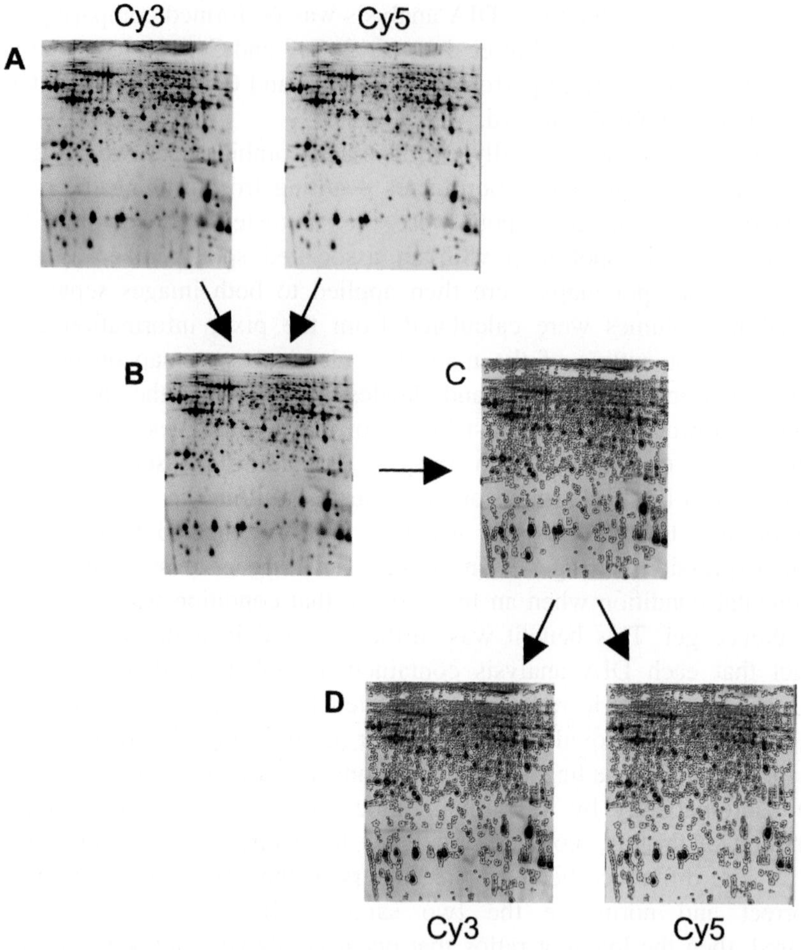

Fig. 1. The process used by DeCyder-DIA to create a common spot map for two images generated from a single gel. Two separate images generated from a single gel (A) are virtually overlaid (B) and spots detected. A single spot map (C) is then applied separately to the original two images (D).

3. Characterization of DIGE technology

3.1. Cyanine dye labeling bias

When employing DIGE technology for profiling experiments it is very important to be sure that artifacts are not introduced as a result of selective labeling of one or more proteins with one dye over another. To test this, tissue samples from at least three different animals were prepared as described earlier and the tissue lysate labeled with each of the three cyanine dyes separately. Following completion of the labeling reaction, the Cy2, Cy3 and Cy5 labeled aliquots from each animal

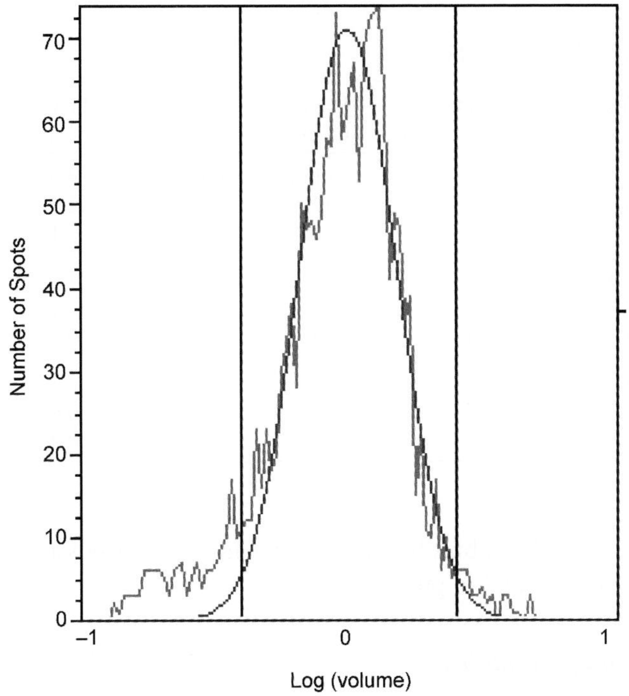

Fig. 2. A graph of the DeCyder-DIA ratio data. Ratios (individual sample volume/pooled standard volume) are generated for each spot detected in a spot map. The log of these ratios is graphed producing an observed distribution of the data set. A normal distribution centered around 0 is also shown. The difference between the median of the observed and normal distribution is used to normalize the entire data set. The vertical black lines represent user-defined threshold boundaries. Spots with ratios that fall to the right (up-regulation) or left (down-regulation) of the thresholds are considered to be different.

were combined separately and run on three different 2DE gels. Using the DeCyder-DIA module the protein patterns of the Cy2 vs. Cy3 labeled samples along with the Cy2 vs. Cy5 and Cy3 vs. Cy5 labeled images were separately compared for each tissue/gel. Volumetric ratios greater than 3 SD from the mean were determined and compared (Table 1). Cy2 images have a higher background than Cy3 and Cy5 images because the excitation and emission wavelengths for Cy2 are closer to the natural fluorescence of acrylamide. This leads to more artifact features that show relatively large differences between images. These spots were quickly evaluated and eliminated by looking at their location on the gel image. The Cy3 vs. Cy5 images usually had a very clean comparison with less than 10 spots identified on average that varied in intensity by more than 3 SD from the mean (Table 1). Comparison of these spots across the three different gels (animals) indicated that none of these variations were reproducible from gel to gel. This indicated that whatever variability did occur was not reproducible. Although

Table 1

Analysis of cyanine dye labeling bias

Gel no.	Cy dye pair comparison	Spots increasing	Percent of spots	Spots decreasing	Percent of spots	3 SD ratio change
1	Cy2:Cy3	13	1.0	65	4.8	1.345
	Cy2:Cy5	6	0.5	37	2.9	1.418
	Cy3:Cy5	4	0.2	1	0.1	1.367
2	Cy2:Cy3	71	13.7	144	27.7	1.193
	Cy2:Cy5	64	10.7	210	35.1	1.231
	Cy3:Cy5	1	0.1	2	0.2	1.374
3	Cy2:Cy3	3	0.5	14	2.1	1.450
	Cy2:Cy5	3	0.5	4	0.6	1.719
	Cy3:Cy5	6	0.5	2	0.2	1.638

a reproducible dye-related bias has not been identified, this possibility is tested on every new tissue type that is profiled.

3.2. Assessment of variability

The ability to detect small changes between protein expression on 2DE with high significance is dependent upon technical and biological variability. To assess the extent of the combined variability in this system, heart ventricles from 10 rats were profiled in duplicate. Hearts were obtained from 10 healthy rats and snap frozen in liquid nitrogen. Lysates were prepared as described earlier and samples were separated simultaneously by 2DE using the same batch of IPG strips and the same casting of gels to minimize technical variability. A pooled standard was created (20 gels worth) using equal protein aliquots of all 10 samples and labeled with Cy3. Each individual sample was labeled with Cy5 in duplicate. The Cy5 labeled samples were combined with an equal amount of Cy3 labeled pooled standard and separated on 2DE gels. Spot detection and matching was performed on the 20 gels using DeCyder software. Analysis of the two groups of 10 gels was performed using Student's T-test. In order to be considered a significant difference, spots had to appear in at least half of the gels, with an average change in abundance of greater than 20% and a p-value < 0.001. A total of 1213 spots appeared on at least half the gels in the study; however, none of the spots met both of the other two criteria. When the data from six of the 10 replicates were randomly analyzed and the same constraints were applied as before, the same number of spots appeared on at least half the gels in the study. Three spots had p-values less than 0.001 but only one spot had greater than 20% change (29%). This spot was of very low abundance (average volume $< 30{,}000$). In two separate analyses using different gels, the data from three of the 10 replicates were randomly analyzed and the same constraints were applied as before. The same number of spots appeared on at least half the gels in the study in both evaluations. In the first evaluation, only one spot had a *p*-value less than 0.001, however, it had only a 15% change in abundance. In the second evaluation, three spots had *p*-values less than 0.001 of which two spots had a greater than 20% change in abundance (22% and 31%). These data suggested that variability of the DIGE platform was low. Although it was not possible to separate the technical and biological variability in the described study, the data demonstrated that it would have been possible to detect changes induced by experimental conditions at as little as 30% change in abundance with high significance. The biological variability in this study was dependent upon the species and the tissue analyzed. However, similar studies have been performed on other species and tissues and similar results were observed.

3.3. Limit of detection of DIGE

Cyanine dyes have a literature-cited limit of detection of 2–10 ng, ammoniacal silver staining 1–5 ng and Sypro Ruby 1–5 ng [5–7]. Traditionally, limits of detection have been determined by performing serial dilutions of purified proteins resolved on conventional SDS-PAGE. In order to determine a working limit of detection, cyanine dye labeling, 2D-DIGE separation and DeCyder detection of two different purified proteins in the background of a complex mixture were used. A mouse liver lysate was prepared as described earlier and split into three equivalent aliquots. The first aliquot was labeled with Cy3 at 200 pmol of dye per 50 μg of total protein. The second aliquot was labeled with Cy5 at 200 pmol of dye per 50 μg of total protein. The final aliquot was spiked with 3000 ng of bovine albumin (66 kDa, p*I* 5.1) and carbonic anhydrase (31 kDa, p*I* 6.0; Calbiochem) per 75 μg of liver cell lysate and labeled with Cy5 at 200 pmol of dye per 50 μg total protein. The spiked aliquot of liver lysate was serially diluted (1:3) with the non-spiked liver lysate labeled with Cy5 to generate seven fractions of decreasing amounts of albumin and carbonic anhydrase with a constant amount of liver lysate. An eighth fraction contained only Cy5 labeled liver lysate and no spiked proteins. Each of the eight fractions was combined with an equal amount of the Cy3 labeled liver lysate and separated using 2D-DIGE on eight different gels. The DeCyder-DIA module was used to match and compare the Cy5 (spiked) and Cy3 liver images from the same gels. The DeCyder-BVA was used to match the spots on each gel. The Cy3 labeled liver image was used in place of the pooled standard.

For bovine albumin, five spots of similar molecular weight but with different p*I*s were identified (Fig. 3A and C). For carbonic anhydrase, three spots with similar molecular weight and different p*I*s were identified (Fig. 3B). The collective volumes were calculated for the groups of spots comprising each protein on each of the gels where these were detected. The average percentage of the individual

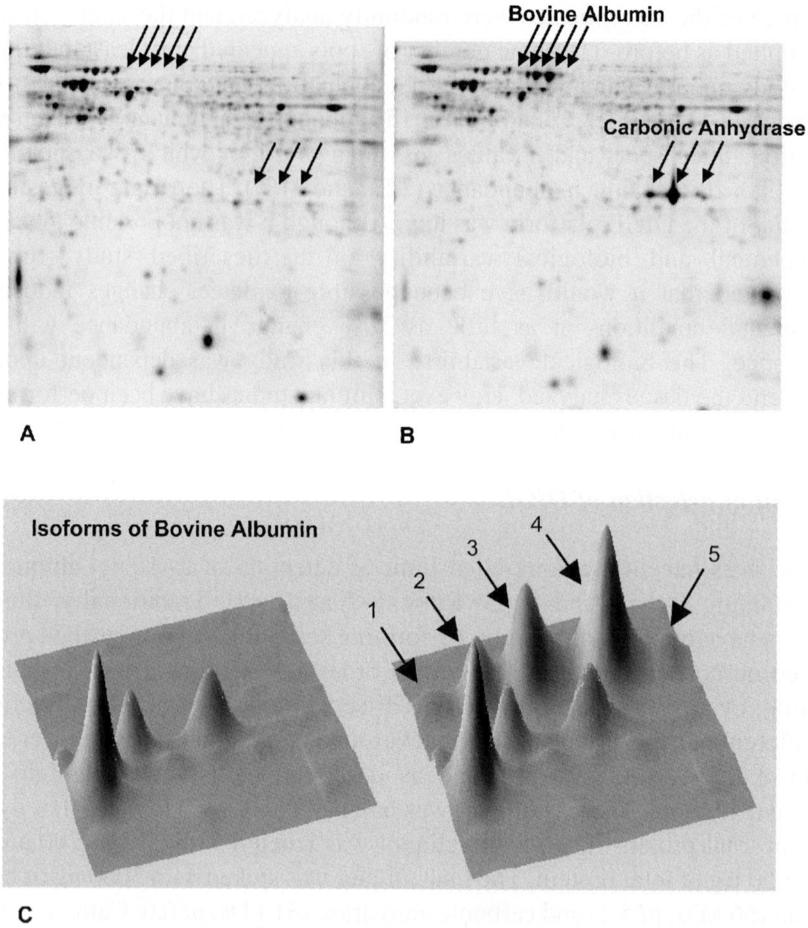

Fig. 3. Large format (20 cm × 20 cm) cyanine dye labeled gel images with and without spiked proteins. Mouse liver cell lysate was analyzed with and without a serial dilution of both bovine albumin and carbonic anhydrase spiked into it. Five isoforms of bovine albumin (A) and three isoforms of carbonic anhydrase (B) were observed. A 3D view of a magnified region of the bovine albumin isoforms is illustrated (C).

spots comprising each spiked protein was determined on all the gels (Table 2). The limit of detection was determined for each of the spots by observing the gel with the lowest dilution of spiked protein in which the spot could still be detected (Table 2). The average limit of detection for bovine albumin was 13 ng/spot or 0.2 pmol. The average limit of detection for carbonic anhydrase was 4.7 ng/spot or 0.15 pmol. These results of spiked protein recovery from a complex mixture agreed well with the results that have been generated using conventional SDS-PAGE.

Table 2

Determination of the limit of detection of spiked proteins, serially diluted into a complex cell lysate using DeCyder-DIA spot detection and DeCyder-BVA spot matching

Gel no.		1	2	3	4	5	6	7	8	Average %T	Average concentration (ng)
Total spike concentration (ng)		3000	1000	333.3	111.1	37.0	12.3	4.1	0		
Bovine albumin isoforms											
1	Vol	1.1×10^6	5.0×10^5	NT	NT	NT	NT	NT	NT	3.4	13
	%T	3.5	4.2								
2	Vol	3.7×10^6	1.9×10^6	4.3×10^5	6.3×10^4	NT	NT	NT	NT	13.3	13
	%T	12.3	15.6	14.3	12.6						
3	Vol	9.5×10^6	4.4×10^6	1.3×10^6	1.4×10^5	2.8×10^4	NT	NT	NT	33.3	11
	%T	31.3	36.4	41.5	27.8	36.7					
4	Vol	1.4×10^7	4.3×10^6	1.1×10^6	2.7×10^5	4.5×10^4	NT	NT	NT	42.7	17
	%T	46	35.8	35.8	54.4	58.2					
5	Vol	2.1×10^6	9.6×10^5	2.5×10^5	2.6×10^4	NT	NT	NT	NT	7.2	8
	%T	6.9	8.0	8.3	5.3						
Average											12.4 ng, 0.2 pmol
Carbonic anhydrous isoforms											
1	Vol	6.4×10^6	4.5×10^6	6.0×10^5	1.5×10^5	4.4×10^4	NT	NT	NT	11.4	3.2
	%T	10.8	11.5	9.5	12.3	13.0					
2	Vol	3.1×10^7	3.3×10^7	5.7×10^6	1.0×10^6	3.0×10^5	5.0×10^4	NT	NT	87.5	2.6
	%T	85.9	86.5	90.5	87.7	87.0					
3	Vol	2.0×10^6	7.9×10^5	NT	NT	NT	NT	NT	NT	2.8	8.4
	%T	3.3	2.0								
Average											4.7 ng, 0.15 pmol

Volume numbers were generated using DeCyder-DIA spot detection, normalization and quantitation. For the most abundant isoforms at the highest concentration of spiked proteins, spots were split at times into multiple spots due to shouldering effects caused by imperfect separation in both dimensions. This caused difficulties with absolute quantitation of each spiked concentration. Vol, volume calculated by DeCyder; %T, percent of total isoform volume; NT, not detected.

3.4. Importance of a pooled standard

Even using the most reproducible systems, gel-to-gel variability is still a major source of error that reduces confidence in differences observed when conducting comparative 2DE. One unique benefit of the DIGE technology is the ability to run multiple samples in one gel thereby reducing the number of gels required. In addition, having the availability of three different dyes allows one to run a common sample (pooled standard) in all gels to aid in spot matching and normalization while still reducing the number of gels required by half. The pooled standard is created using an equal amount of protein from every sample to be evaluated in an experiment. Therefore, the pooled standard theoretically contains a collection of all proteins that are present in every sample from an experiment. Proteins of low abundance that are unique to a small subset of samples within large experiments may fall below the limits of detection in the pooled standard due to a dilution effect. The magnitude of the dilution effect is dependent upon the number of samples in the experiment that are to be combined, the number of conditions in which the protein is present and the abundance of the protein in each sample.

In an attempt to evaluate the benefit of having a pooled standard, the volume data for 1023 spots were rank-ordered twice by their ANOVA generated p-values. The first ranking was performed using individual sample volumes normalized with the pooled standard volumes on a spot by spot basis. The second ranking was performed without considering the pooled standard. In the second analysis, gels were normalized to one another using the raw background subtracted volumes. Identification of the 50 most significant spots either using the pooled standard or without using the pooled standard showed that only 16 spots were found in common. When the top 200 most significant spots were evaluated using both methods, only 90 spots were in common. In addition, the top 10 most significant spots determined when using the pooled standard were also found within the top 200 most significant spots when the pooled standard was not used. This suggested that the most dramatic changes observed were detected using both methods. However, they were not found as the most dramatic in both rankings. In addition, there was a lack of consistency in the data sets.

The 50 most significant spots that were determined using the pooled standard were evaluated and this showed that 17 of those spots were not found in the top 200 spots identified without using a pooled standard. Four of the 17 were gel artifacts (unresolved spots at the edge of the gel or parts of streaks) and of no interest. The remaining 13 were nicely resolved spots that should have been considered. Therefore, these were false negatives in the evaluation done without the pooled standard. When the top 50 most significant spots that were identified without using a pooled standard were evaluated, 19 of those spots were not found in the top 200 spots determined using a pooled standard. All 19 spots were determined to be artifacts (unresolved, part of streaks, etc.) in the gel. Therefore, they were false positives in the evaluation without the pooled standard.

In this experiment, the pooled standard helped to reduce the rate of false negatives and false positives detected. The pooled standard provides a protein-by-protein, specific, internal standard for every spot on the gels, thereby improving the accuracy of quantitation from gel to gel. A recent publication by Alban et al. [8] also demonstrates, in a similar manner, the benefit of incorporating a pooled standard using DIGE technology. This result confirms our observations and provides support in the advantage of having multiple samples run on the same 2DE gel.

3.5. Comparison to other 2D gel methods

Pairwise or multi-label comparative approaches using 2DE have been attempted for some time. This has mostly occurred using radionucleotides [9]. For example, ProteoSys (Mainz, Germany) claims success using 125I and 131I to label two different samples mixed together and which are then resolved on 2DE with a limit of sensitivity as low as 1 amol.

The efficiency of labeling protein samples with cyanine dyes is an important consideration when using DIGE technology. A number of reagents compatible with other 2DE gel methods have been shown to reduce the labeling efficiency of cyanine dyes. (1) Labeling efficiency is reduced with some detergents such as Triton X-100. Other 2D gel methods are compatible with non-ionic or zwitterionic detergents containing non-linear alkyl tails [10,11]. Only a limited number of detergents have been proven to be compatible with cyanine dye labeling, therefore, solubility problems of some proteins could be an issue due to the limited number of compatible detergents. Each detergent of interest needs to be titrated for its effect on cyanine dye labeling. (2) Although the reasons are not clear, we found that the reducing agent, tris-(2-carboxyethyl)phosphine (TCEP), in place of dithiothreitol (DTT) is not compatible with cyanine dye labeled proteins during the IEF separation. However, TCEP performs well as a reducing agent after cyanine dye labeling and IEF focusing when used in the equilibration buffer prior to the second dimension separation. (3) The use of the irreversible serine protease inhibitor 4-(2-aminoethyl)-benzenesulfonyl fluoride (Pefabloc SC, AEBSF) by itself or in an inhibitor cocktail requires the use of a 'protector' reagent (Pefabloc SC Plus — Boehringer-Mannheim) to prevent the inhibitor from binding to the cyanine dyes.

Post-electrophoretic staining procedures (Coomassie, silver, fluorescence, etc.) are required for all 2DE gel methods except the DIGE technology. This attribute is the basis for several advantages of DIGE over other 2DE technologies. Efficiency and reproducibility of protein labeling with cyanine dyes can be easily monitored using multiple methods prior to running 2DE. Because DIGE gels are cast between low fluorescent glass, gels are scanned without removing the gels from between the plates. This eliminates the possibility of damaging and contaminating a gel through handling. Gels can be scanned for different lengths of time to regulate sensitivity of detection for abundant and lower abundant proteins.

Because proteins are labeled by covalent modification with cyanine dyes, electrotransfer of proteins from 2DE gels to PVDF allows one to scan the PVDF membrane for the cyanine protein pattern observed in the gel. This allows one to overlay a Western blot image directly on the cyanine dye image from the same membrane. The benefit of this is that it is easy to identify which protein spots from the total protein background are immunoreactive.

The total number of spots detected using cyanine dye labeling, as expected, is dependent upon the sample complexity, purity, quantity, tissue type, software used for detection and the detection parameters selected. In our experience with the DIGE technology, we routinely identify between 1000 and 1500 spots per gel image. Sypro Ruby, although reported to be more sensitive, in our hands has been less sensitive (800–1200 spots) and silver staining, although slightly more sensitive (1300–1800 spots), is much more variable, takes longer, not compatible with MS analysis and has the greatest variability (depending upon the method used). We suspect that the lower sensitivity with Sypro Ruby imaging may be due to the instrument and filters being used (2920-2D Master Imager, Amersham Biosciences).

Overall, cyanine labeling is not the most sensitive staining method. However, for experiments comparing only a couple of samples, this technology has the advantage of having samples run on the same gel, detected in a single step with identical spot maps and volumetric comparisons made on the fly. For larger studies with 10–100 samples, a pooled standard run on all the gels provides both image landmarking for matching as well as normalization of protein abundance, necessary due to all the gel-to-gel variables inherent with 2DE technologies. We believe these benefits, as well as others, make DIGE more attractive over conventional 2DE staining techniques and the loss in sensitivity acceptable.

4. MS identification of cyanine labeled proteins

4.1. Excision of cyanine labeled proteins

Protein spots that were determined to be reproducibly and statistically different between two or more experimental conditions were excised from subsequently prepared preparative gels containing only a single sample with between 300 and 500 μg total protein. Because cyanine dyes label less than 5% of each protein, and the molecules that were labeled shifted in mass (0.5 kDa) from the bulk of the protein spot, the ability to accurately excise spots from gels progressively decreased for lower molecular weight proteins. To circumvent this problem, samples of interest were run on preparative gels and stained post-electrophoretically with Sypro Ruby. Sypro Ruby images were visible on the Propic Spot Cutter (Genomic Solutions, Inc.) and spots can be selected for excision. For robotic spot cutters without a fluorescent camera, strategically placed white light visible/-fluorescent markers along with the use of triangulation software can be used for blind selection of protein spots for excision.

4.2. Protein identification by LC–MS/MS

Protein-containing spots of interest were excised from the gel and digested in situ with trypsin. The resulting proteolytic fragments were extracted from the gel slices and subjected to liquid chromatography–tandem mass spectrometry (MS/MS) analysis. The MS/MS spectra of tryptic fragments were subjected in an automated batch mode to SEQUEST, a direct database correlation algorithm. SEQUEST was running on a large cluster of Unix servers and the job queue was controlled using Load Share Facility. Identified proteins were verified by manual inspection of the sequence ions present in each spectrum as described previously [10].

5. Summary

This chapter describes and characterizes a new enabling technology from Amersham Biosciences called Difference In Gel Electrophoresis (DIGE). This technology leverages the use of fluorescent cyanine dyes that are used to label proteins prior to electrophoresis. This helps reduce the overall background during scanning, and provides more reproducible labeling. The existence of three different cyanine dyes with separate excitation and emission wavelengths also enables analysis of up to three samples in a single gel. Multiple samples per gel reduce the number of gels necessary for each study thereby reducing the need to match as many gels to the master gel. The use of an internal pooled standard run on every gel aids in reducing false positives and false negatives. This is evident by the ability to observe significant changes in protein expression as low as 30%. Cyanine dyes cause a mass shift, without effecting p*I* upon labeling, and less than 5% of the total protein complement in the sample is typically labeled. Therefore, excision of proteins of interest requires subsequent preparative gels stained post-electrophoretically with Sypro Ruby. Robotic spot cutters can be utilized for excision, and identification is made via traditional mass spectrometry approaches after in-gel trypsin digestion.

Although cyanine dye labeling can be less sensitive than some post-electrophoretic staining techniques, the ability to have an internal standard on every gel provides reproducibility and confidence in differential protein expression analysis for large numbers of samples. The internal standard also provides accurate detection of small differences in protein amounts between samples.

The DIGE technology coupled with prefractionation of samples, as described elsewhere in this book, allows lower abundant proteins to be visualized and compared with greater reliability.

Acknowledgements

We are grateful to Dr Margaret McCann and the members of her laboratory for their pharmacology support, Anita Hsieh, Linda Kochanski and Jin Qian for their technical 2DE expertise, Dr Ellen Rohde for her mass spectroscopy expertise and Dr Bert Gunter for his statistical assessment of our data. We would like to thank Dr Paul Guest for his critical review of this chapter and we would also like to thank Amersham Biosciences for their technical support with the DIGE technology and DeCyder software.

References

1. Unlu, M., Morgan, M.E. and Minden, J.S., Difference gel electrophoresis: a single gel method for detecting changes in protein extracts. *Electrophoresis*, **18**, 2071–2077 (1997).
2. Tonge, R., Shaw, J., Middleton, B., Rowlinson, R., Rayner, S., Young, J., Pognan, F., Hawkins, E., Currie, I. and Davison, M., Validation and development of fluorescence two-dimensional differential gel electrophoresis proteomics technology. *Proteomics*, **1**, 377–396 (2001).
3. Zhou, G., Li, H., DeCamp, D., Chen, S., Shu, H., Gong, Y., Flaig, M., Gillespie, J.W., Hu, N., Taylor, P.R., Emmert-Buck, M.R., Liotta, L.A., Petricoin, E.F. III and Zhao, Y., 2-D differential in-gel electrophoresis for the identification of esophageal scans cell cancer-specific protein markers. *Mol. Cell. Proteom.*, **1**, 117–123 (2002).
4. Gharbi, S., Gaffney, P., Yang, A., Zvelebil, M.J., Cramer, R., Waterfield, M.D. and Timms, J.F., Evaluation of two-dimensional differential gel electrophoresis for protein expression analysis of a model breast cancer cell system. *Mol. Cell. Proteom.*, **1**, 91–98 (2002).
5. Berggren, K., Chernokalskaya, E., Steinberg, T., Kemper, C., Lopez, M., Diwu, Z., Haugland, R. and Patton, W., Background-free, high sensitivity staining of proteins in one- and two-dimensional sodium dodecyl sulfate–polyacrylamide gels using a luminescent ruthenium complex. *Electrophoresis*, **21**, 2509–2521 (2000).
6. Patton, W.F., A thousand points of light: the application of fluorescence detection technologies to two-dimensional gel electrophoresis and proteomics. *Electrophoresis*, **21**, 1123–1144 (2000).

7. Yan, J.X., Harry, R.A., Spibey, C. and Dunn, M.J., Postelectrophoretic staining of proteins separated by two-dimensional gel electrophoresis using SYPRO dyes. *Electrophoresis*, **21**, 3657–3665 (2000).
8. Alban, A., David, S.O., Bjorkesten, L., Andersson, C., Sloge, E., Lewis, S. and Currie, I., A novel experimental design for comparative two-dimensional gel analysis: two-dimensional difference gel electrophoresis incorporating a pooled internal standard. *Proteomics*, **3**, 36–44 (2003).
9. Vuong, G.L., Weiss, S.M., Kammer, W., Priemer, M., Vingron, M., Nordheim, A. and Cahill, M.A., Improved sensitivity proteomics by postharvest alkylation and radioactive labeling of proteins. *Electrophoresis*, **21**, 2594–2605 (2000).
10. Dunn, M.J. and Burghes, A.H.M., High resolution two-dimensional poly-acrylamide electrophoresis. I. Methodological procedures. *Electrophoresis*, **4**, 97–116 (1983).
11. Rabilloud, T., Gianazza, E., Catto, N. and Righetti, P.G., Amidosulfobetaines, a family of detergents with improved solubilization properties: application for isoelectric focussing under denaturing conditions. *Anal. Biochem.*, **185**, 94–102 (1990).
12. Roy, R.S., Yang, P., Kodali, S., Xiong, Y., Kim, R., Griffin, P.R., Onishi, R., Kohler, J., Silver, L.L. and Chapman, K., Direct interaction of a vancomycin derivative with bacterial enzymes involved in cell wall biosynthesis. *Chem. Biol.*, **139**, 1–12 (2001).

Chapter 4

Electrophoretic prefractionation for comprehensive analysis of proteomes

XUN ZUO, KIBEOM LEE and DAVID W. SPEICHER*

The Wistar Institute, 3601 Spruce Street, Philadelphia, PA 19104, USA

Abbreviations: 1D, one dimension; 1DE, one-dimensional polyacrylamide gel electrophoresis; 2D, two dimension; 2DE, two-dimensional polyacrylamide gel electrophoresis; FFE, free flow electrophoresis; IEF, isoelectric focusing; IPG, immobilized pH gradient; p*I*, isoelectric point; PAGE, polyacrylamide gel electrophoresis; μsol-IEF, microscale solution isoelectrofocusing.

* Corresponding author. Tel.: +1-215-898-3972. E-mail: speicher@wistar.upenn.edu (D.W. Speicher).

1. Introduction

Two-dimensional polyacrylamide gel electrophoresis (2DE) has dominated protein profile analysis for more than 25 years, and it is still the method of choice in many laboratories for quantitatively comparing changes of proteins for proteome analysis experiments. Unfortunately, the existing 2DE method has inadequate resolution and insufficient dynamic range when used for separating complex proteomes. A typical 'full-size' 2D gel ($\sim 18 \times 20\ cm^2$) can resolve only up to ~1000–1500 protein spots from a cell extract using high-sensitivity stains, while about 10,000 genes are typically expressed at one time in a single mammalian cell [1]. Furthermore, the total number of protein components in higher eukaryotic cells greatly exceeds the number of expressed genes due to mRNA alternative splicing and post-translational modifications [2]. Hence, it is highly likely that at least 20,000–50,000 or more unique protein components comprise typical proteomes from individual mammalian cell types, and the total unique protein species represented in a single tissue probably exceed 100,000. In addition, due to very divergent protein expression levels in cells or tissues, protein levels in eukaryotic proteomes cover wide ranges. In human cells, for example, the most abundant protein is often actin, which is present at about 10^8 molecules per cell. On the other hand, some cellular receptors, transcription factors, and other low abundant proteins are present at only about 100–1000 molecules per cell, resulting in a dynamic range of about 10^6 [3]. The dynamic range is even wider in some physical fluids such as plasma, where albumin is present at >30 mg/ml and trace level proteins are present at less than pg/ml levels for a dynamic range of about 10 orders-of-magnitude [4].

A conceptually attractive approach for increasing 2DE protein profiling capacities is direct analysis of proteome samples using multiple narrow pH range 2D gels (e.g. a series of ~1.0–1.2 pH unit 'zoom' gels) [3,5,6]. It is advantageous to use a series of slightly overlapping narrow pH 2D separations to be combined into a composite image, since separation distance per pH unit (spatial resolution) is greatly increased on the zoom gels compared with a single broad pH range gel. Theoretically, this zoom gel strategy should result in a dramatic increase of total number of spots detected, because this strategy can increase total IEF separation distance by 5-fold or more compared with a single broad pH range gel. In practice, however, using a series of narrow pH range gels without sample prefractionation only results in a moderate increase in the total proteins detected compared to using a single gel with same pH range. This is because narrow pH range gels work reasonably well at very low protein loads, but severe artifacts including horizontal streaking rapidly become limiting as protein loads are increased. When high protein loads of unfractionated samples are analyzed on narrow pH range IPG gels, some proteins with isoelectric points (p*I*s) outside the pH range of the gels cause extensive precipitation and aggregation, which causes co-precipitation of some proteins with p*I*s within the fractionation range [7,8].

The great complexity and high dynamic range of most eukaryotic proteomes together with limitations of existing 2DE as well as alternative MS/MS methods suggest that sample prefractionation is essential for more comprehensive coverage and reliable detection of low abundant proteins in complex proteomes. Ideally, such prefractionation methods should be capable of resolving, detecting, and quantitatively comparing the majority of unique protein components present in mammalian cells or tissues, including discrimination of protein isoforms and different post-translational modifications [9,10].

Over the past several years, multiple research groups have attempted to expand the resolving power of 2D gels using various prefractionation methods to increase the number of proteins separated and to detect less abundant proteins. Prefractionation methods that are at least partially orthogonal to the separation modes of 2DE include sequential extractions with increasingly stronger solubilization solutions [11], subcellular fractionation [12], selective removal of the most abundant protein components [13], and fractionation of eukaryotic cell extracts using different chromatographic techniques [14–16]. Unfortunately, most of these methods are relatively low-resolution techniques that result in substantial and often variable cross-contamination of many proteins between two or more fractions, which complicates quantitative comparisons of proteins profiles. In addition, liquid chromatography methods usually result in large sample volumes, which need to be concentrated prior to 2D gel analysis. Concentration of multiple samples can be time consuming and sample recovery is often low and variable.

The limited resolution of chromatographic and alternative prefractionation methods suggest that preparative IEF methods could be more practical for proteome prefractionation. High-resolution IEF separations that are closely analogous to the actual first dimension analytical IPG separation of 2D PAGE might be especially advantageous when combined with subsequent high sample loading on slightly overlapping narrow pH range IPG strips [17]. The ideal prefractionation method should be able to reproducibly resolve complex protein mixtures such as extracts of eukaryotic cells or tissues into a small number of well-resolved fractions that are compatible with subsequent 2DE analysis. The number of fractions should be small because 2D PAGE and alternative LC–MS/MS analyses are relatively time consuming. In addition, a high-resolution prefractionation method is necessary to minimize cross-contamination of proteins in adjacent fractions.

Several preparative solution IEF methods have been reported for fractionation of complex proteomes prior to 2D PAGE or other downstream analysis methods. Generally, these approaches use one of two separation principles. They are: (i) ‘free solution IEF’ where soluble carrier ampholytes are used to create and maintain pH gradients across a single focusing chamber and

(ii) 'multi-compartment solution IEF' where pH selective partitions with protein-scale pores divide a series of tandem chambers containing the sample.

This chapter reviews preparative IEF methods and separation devices that can be used for prefractionation of complex proteomes prior to comprehensive proteome analysis. In addition, simple strategies for analysis of insoluble and large proteins, which are usually excluded from 2D gels, are described. Although the primary focus for subsequent fraction analysis is on gel-based methods in this chapter, alternative MS-based approaches such as multi-dimensional LC–MS/MS and ICAT methods can be used for downstream analysis of fractionated proteins.

2. Electrophoretic prefractionation methods

2.1. *Rotofor*

The Rotofor is a commonly used device for preparative IEF that separates proteins based on their p*I* in solution. The Rotofor method was originally described by Bier and co-workers [18,19] as a procedure for the purification of a small number of proteins. Commercial versions, the Standard Rotofor Cell (60 ml chamber) and the Mini Rotofor Cell (18 ml chamber), are produced by Bio-Rad Laboratories (Hercules, CA, USA). The focusing chamber is a rotating tube that is divided into 20 compartments by open grids that are screens made of woven polyesters. Liquid can easily move through these screens, and therefore, in practice there is a single liquid chamber. The purpose of the screens is to minimize convection currents. Soluble ampholytes are used to produce a pH gradient to separate proteins in solution. Protein mixtures are initially dispersed uniformly throughout the chamber and specific proteins migrate to positions that are at pH values equal to their p*I*s. The Rotofor uses rotation around its horizontal axis during IEF to inhibit gravitationally induced convection, maintain even cooling, maintain relative stabilization of focused protein zones, and prevent the clogging of screens by precipitated proteins. In addition, the device has a cooling system that allows IEF be completed within a short time (~4 h) using constant power (e.g. 12 W) at 4°C. Operation of the Rotofor fractionation system is relatively simple.

The Rotofor approach has been widely applied for initial protein purification under either native or denaturing conditions. This device has been used for the first dimension separation of soluble protein mixtures, followed by the second dimension separation using preparative 1D PAGE, constituting a preparative 2D PAGE method for purifying protein(s) of interest [20,21]. The Rotofor approach has also been successfully used as the initial step in 2D liquid-phase separation of proteins, followed by second dimension separations using non-porous reversed-phase HPLC [22].

Some investigators have used the Rotofor method to reduce sample complexity for proteome studies [20,21,23,24]. However, the Rotofor device has several limitations when used for analytical proteome fractionation. First, it uses carrier ampholytes to generate a pH gradient, which limits fractionation to mid-pH ranges, making it difficult to obtain fractions at extremely low or high pH. Second, it requires large sample volumes (18 or 60 ml), which can be problematic for many biological applications where the quantity of sample is limited. As a result of the large volume, fractions often must be concentrated prior to downstream analyses, which are difficult due to the presence of high concentrations of ampholytes, denaturing reagents, and detergents. Third, adjacent fractions can be easily mixed during the separation procedure or during sample removal because the large opening in the screens allow liquid flow.

2.2. Free flow electrophoresis

Free flow electrophoresis (FFE) is an electrophoretic technique that has been used for separation of various kinds of cells for more than three decades [25–27]. The major component of FFE devices is a rectangular chamber. The sizes of this separation chamber may vary considerably, with lengths ranging from 20 to 100 cm, widths ranging from 4 to 20 cm, and depths ranging from 2 to 20 mm. However, there was very limited utilization of FFE for protein fractionation and separation of organelles until the Octopus FFE device was manufactured by Dr Weber (Kirchheim, Germany). This device was a compact unit containing a chamber with a size of 500 mm × 100 mm × 0.4 mm and was operated in a vertical position. Recently, Weber and co-workers have refined the FFE process for separation of proteins and subcellular particles by introducing new buffer systems and segmented buffer films [28–30]. Subsequently, Tecan Group Ltd (Munich, Germany) acquired this technology and a redesigned instrument, the ProTeam™ FFE, was marketed in 2002. This is currently the only commercially available FFE device.

FFE can be operated in several separation modes including 'free solution IEF' where carrier ampholytes are used to create and maintain pH gradients. Preparative solution IEF is the most commonly used separation mode for prefractionation of proteomes. Soluble proteins migrate to positions between electrodes where the pH is equal to their p*I*s, while sample flows continuously in a thin film (0.2 mm) perpendicular to the electrodes. Either native or denaturing conditions can be used for protein separations. The FFE device is similar to the Rotofor in separation principle, i.e. IEF using pH gradients established by soluble ampholytes in a single liquid compartment. However, the FFE device prevents mixing of separated components by using laminar flow rather than the porous grids and axial rotation. In addition, the FFE device can be used for either preparative or analytical separation with wide ranges in sample loads and fraction number can range from a few to 96 to meet different research goals.

A major application of FFE is prefractionation of proteomes prior to 2D PAGE to reduce sample complexity and enrich minor proteins [28]. Recently, the FFE device has also been used as part of a 2D non-gel-based proteome analysis strategy, by combining FFE with 1D PAGE to produce protein arrays followed by protein identification by RP-HPLC/mass spectrometry (LC/MS) [31]. In addition, FFE can be used to separate other charged particles, such as whole cells, organelles, and membrane particles [27].

2.3. IsoPrime and related multicompartment electrolyzers

Preparative IEF using separation membranes that restrict liquid flow and contain immobilized buffers was originally described by Righetti and co-workers [32]. These original separation membranes were formed using thin immobiline/acrylamide gels cast on glass–fiber discs, which provided mechanical support. Proteins were segregated by p*I* ranges into different chambers bracketed by the pH-selective separation membranes or partitions, thus achieving fractionation of proteins. Prototype devices referred to as 'Multi-Chambered Electrofocusing Units', were initially used for final stage purification of individual proteins under native conditions starting with partially purified proteins [33,34]. A commercial device based on this method of '*Iso*electrofocusing' and '*Pr*eparative *I*soelectric *M*embrane *E*lectrophoresis' (IsoPrime) is available from Amersham Biosciences.

The IsoPrime is a large and complicated instrument that uses peristaltic pumps to circulate samples through each separation chamber. Very high-resolution separations can be achieved because membrane partitions can be selectively made with different precise pHs and proteins with p*I*s differing by as little as 0.01 pH units can be separated [33]. However, several factors complicate the facile use of this instrument for proteome prefractionation where sample sizes are limited, including (i) each fraction volume is of the order of 30 ml; (ii) unless very high protein loads, e.g. >100 mg, are used, the final large dilute fractions require extensive concentration prior to downstream protein profile analysis; (iii) the complex plumbing system for re-circulating samples through the separation chambers makes operation using denaturing conditions with high levels of urea and detergent problematic; and (iv) separation times >24 h are typically required to prevent excessive heating and rupturing of the partition membranes at high voltages.

Recently, Herbert and Righetti described a modified IsoPrime-type apparatus, that was referred to as a multicompartment electrolyzer (MCE), and was used to prefractionate *Escherichia coli* lysates and human serum samples prior to 2D gel analysis [35]. This new device was relatively simpler and smaller than the Isoprime unit because the re-circulating tubing and peristaltic pumps were eliminated and replaced with intra-chamber stirring. This device is therefore much easier to use with denaturing conditions.

2.4. Microscale solution isoelectrofocusing combined with narrow pH range 2D PAGE

2.4.1. The μsol-IEF/zoom IEF prefractionation method and device

As described earlier, existing preparative IEF devices have one or more potential drawbacks when applied to sample prefractionation for comprehensive proteome analysis. The most common limitation associated with these devices is their relatively large separation chambers. This requires either a large amount of initial sample or results in high volume, dilute fractions that must be concentrated prior to most downstream analysis methods. An ideal, efficient, high-resolution prefractionation would seamlessly interface with subsequent narrow pH range 2D PAGE and alternative downstream analysis methods for comprehensive protein profile comparisons. A prefractionation method, microscale solution isoelectrofocusing (μsol-IEF), was independently developed at about the same time as the MCE device [17]. The μsol-IEF device utilized the basic separation principle originally described by Righetti et al. [32] for IsoPrime, but unlike the IsoPrime it was: (i) miniaturized with 500–700 μl sealed separation chambers; (ii) recirculation or active mixing was eliminated; and (iii) cooling was not needed. In addition, acrylamide/immobiline partitions with larger pores and stronger supports were used. In early experiments, glass–fiber filters (~1.5 mm thickness) were used to support partition membranes [17], but subsequently hydrophilic polyethylene (1.5 mm thickness) discs were found to be superior [8,36]. Finally, thinner polyethylene discs (~0.67 mm) were used to minimize partition gel volumes. The μsol-IEF device can be loaded with at least 2 or 3 mg of a complex proteome sample and fractionation is typically completed within a few hours under denaturing conditions. Separated fractions can be loaded directly onto narrow range 2D gels or high-resolution 1D gels without sample concentration. Typically 5–7 separation chambers have been used to prefractionate eukaryotic cell extracts as schematically illustrated in Fig. 1. In Fig. 1A, the region from pH 4.5–6.5 is divided into four very narrow pH ranges (each 0.5 pH unit) since the major population of proteins in mammalian cellular proteomes typically have p*I*s in this pH range. The μsol-IEF technology developed in the author's laboratory led to the development of a more convenient commercial device, the 'Zoom IEF Fractionator' (Invitrogen Life Technologies, Carlsbad, CA, USA) shown in Fig. 2.

The μsol-IEF prefractionation has several key advantages compared with alternative methods for enhancing comprehensive proteome analysis. This method is simple, inexpensive, uses small sample volumes and can cover the full pH range. Discrete well-resolved fractions with minimal cross-contamination and minimal

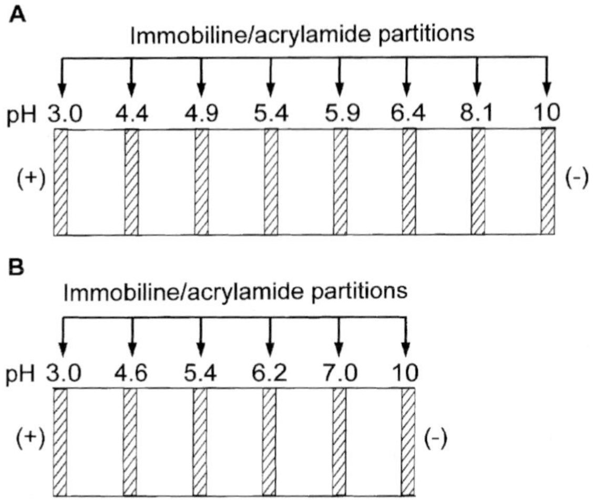

Fig. 1. Two schemes for μsol-IEF prefractionation of mammalian proteomes. (A) A fractionation scheme using seven separation chambers and custom made immobiline/acrylamide partitions. (B) A fractionation scheme using five separation chambers and commercially available immobiline/acrylamide partitions (Invitrogen Life Technologies). The small volume chambers (each ~700 μl) are separated by partition membranes, which are prepared by casting immobiline/acrylamide gels on porous polyethylene disc supports (0.67 mm thickness) to produce fraction boundaries that are buffered at specific pHs as indicated. All partition membranes are large pore size polyacrylamide gels (3%–4%), to allow proteins of all sizes including large proteins (up to ~500 kDa) to migrate through the partitions while restricting bulk liquid flow.

sample loss are produced, which is essential for quantitative comparisons of protein profiles. The number of chambers for the devices can be easily varied to allow fractionation of a complex proteome into as few as two or as many as seven or more pH ranges where the pH boundaries of each fraction are tailored to specific experimental requirements. Even very large proteins can be effectively separated by the device due to the large pore size of the gels (~3%–4%) used to form the pH partition membranes. Hence, 100–500 + kDa proteins can be separated, recovered in good yield, and quantitatively compared using large pore, high-resolution 1D SDS gels [8]. The prefractionation method has been used to separate *E. coli* cell extracts [17], mouse serum proteins [7,36] and human cancer cell extracts [7,8] to enhance comprehensive proteome analyses.

2.4.2. Analysis of μsol-IEF fractionated proteins on narrow pH range 2D gels

One practical approach for more comprehensive proteome analysis is to combine μsol-IEF prefractionation with subsequent separation on narrow pH range 2D gels. The acrylamide/immobiline gels for μsol-IEF device partitions can be made

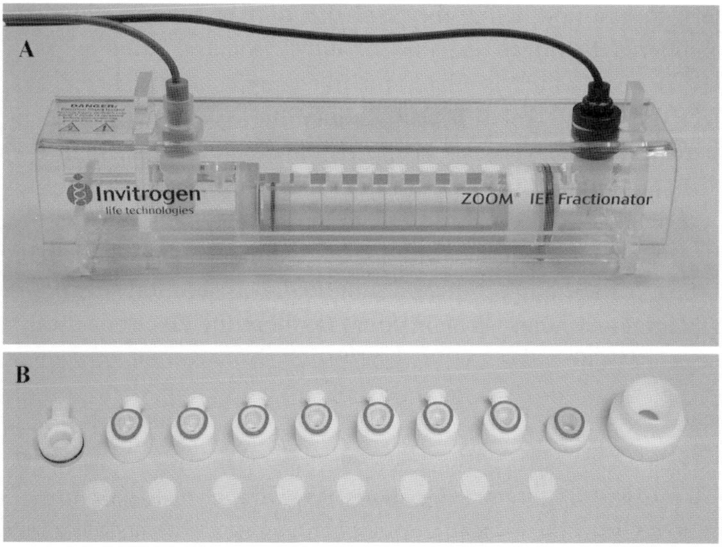

Fig. 2. A photograph of the commercial μsol-IEF device, the 'Zoom IEF Fractionator', which is produced by Invitrogen Life Technologies. (A) A completely assembled device with seven chambers. (B) Components of the device shown in A: top row (from left to right) — Anode End Sealer, seven oval shaped separation chambers (oval shape) with caps and O-rings, Cathode End Sealer, and Cathode End Screw Cap; bottom row — eight porous polyethylene discs to support acrylamide/immobiline gel partitions.

with the same pH precision as is inherent in IPG technology [37]. This allows reproducible fractionation of a complex proteome into precisely defined pH ranges that will be reliably separated on IPG gels with appropriate pH ranges.

Segregating a complex proteome into separate pools by the initial μsol-IEF separation in solution effectively minimizes 'non-ideal' behavior on subsequent narrow range IPG gels (e.g. precipitation, aggregation, protein–protein interactions). The prefractionation also dramatically improves protein solubility during this initial separation because proteins are focused in far larger volumes of the same solubilization buffer compared with the focused protein volume in IPG gel strips. The simpler samples resulting from μsol-IEF prefractionation allow higher loads of proteins within the separation range to be applied to the IPG strips and these simpler samples focus better than unfractionated samples on narrow range IPG strips. Typically, 10- to 50-fold more protein can be applied to the narrow pH range 2D gels when a complex proteome such as serum is initially prefractionated using μsol-IEF as compared to the optimal loads without prefractionation [36]. As a result of this higher sample load capacity, less abundant spots can be detected and the dynamic detection range is increased. The reduced incorrect isoelectric focusing of some proteins at high protein loads by μsol-IEF prefractionation also allows more reliable quantitative comparisons. In addition, using the μsol-IEF prefractionation conserves protein samples compared to direct analyses of unfractionated samples on a series of narrow pH 2D gels, which is important for samples with limited availability.

The effectiveness of μsol-IEF prefractionation for enhanced protein detection on narrow pH range 2D gels has been demonstrated using human breast cancer cells [8]. When a small amount (~20 μg) of an unfractionated cell extract was separated on pH 5.0–6.0 2D gels, reasonably good resolution was obtained, but few spots were detected (Fig. 3). More spots were detected with a 10-fold higher load (~200 μg) of the unfractionated sample, but resolution was poor and many proteins appeared as smears. Protein streaking near the electrodes was substantial

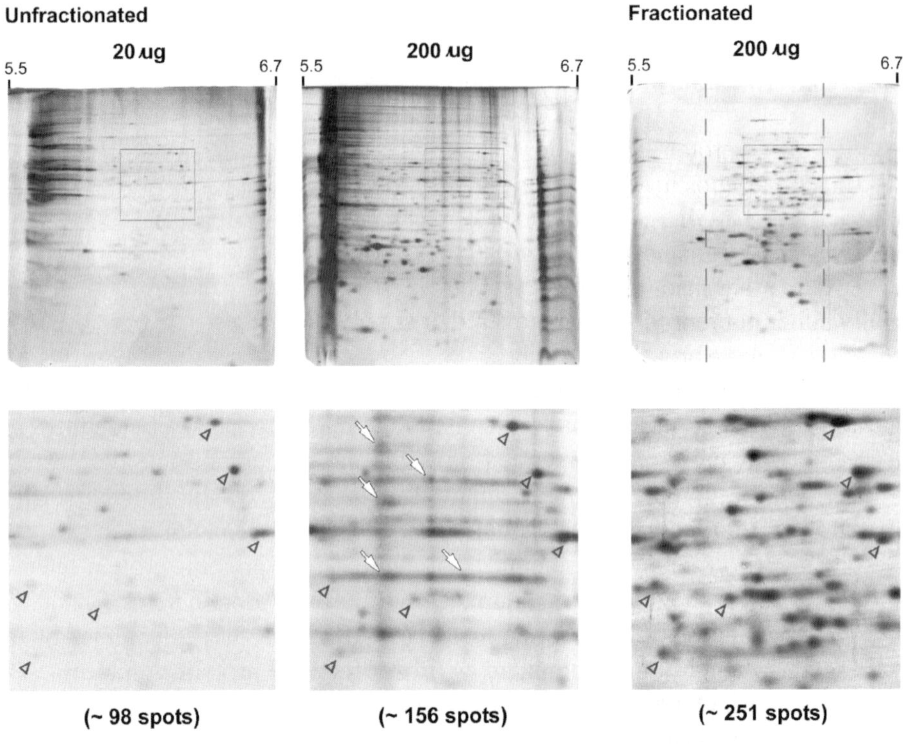

Fig. 3. Comparison of unfractionated and fractionated human breast cancer cell extracts separated on full-sized silver stained narrow pH range 2D gels. An unfractionated extract (20 and 200 μg) is compared with the fraction (pH 6.0–6.5) equivalent to 200 μg of cell extract resulting from a μsol-IEF prefractionation. Samples were focused on pH 5.5–6.7 IPG strips (18 cm), followed by separation on 10% Tris Tricine SDS gels (18 × 19 cm^2). Upper panels: complete 2D gel images with a highlighted 0.3 pH wide region. Lower panels: enlargements of the highlighted regions. Open triangles indicate landmark proteins to facilitate visual comparisons, and arrows highlight some poorly focused proteins in the higher load unfractionated sample that co-migrated with either vertical or horizontal streaks and were detected at incorrect positions [Reprinted with permission from: Zuo, X., et al., *J. Chromatogr. B*, **782**, 253–265 (2002)].

at the 20 μg load and became very severe at the 200 μg load because many proteins with p*I*s outside the pH separation range migrated toward the electrodes and precipitated there. In contrast, resolution was much better on the 2D gel of the fractionated sample and a greater number of total spots in the pertinent pH range were detected. Differences between the unfractionated and fractionated samples were particularly evident in the enlarged areas of the gels shown in the lower panels of Fig. 3. Loading 200 μg of the unfractionated sample resulted in the loss of ~38% of the spots detected in the fractionated sample. This loss was due to co-precipitation near the electrodes of these proteins with proteins that had p*I*s outside the pH range of the IPG strip. These results show that μsol-IEF prefractionation enables use of greatly increased protein loads on narrow pH range 2D gels while maintaining good resolution.

3. Strategies for analysis of large soluble proteins and insoluble proteins

Comprehensive protein profile analysis requires proteomic techniques capable of detecting all proteins in a sample regardless of size, solubility, or abundance level. However, 2D PAGE is incapable of reliably detecting some types of proteins including very large and insoluble proteins. Proteins that are larger than 100 kDa and those that are insoluble in lysis buffer used to prepare proteome samples are usually either not reproducibly recovered or simply excluded from 2D gels. These large and insoluble proteins constitute about 20%–25% of the total protein mass in most cells or tissues. High-throughput, high-resolution 1D SDS gels can be readily used to detect these proteins in parallel with analysis of μsol-IEF fractions on slightly overlapping narrow range 2D gels.

3.1. Detection of insoluble proteins

While different extraction buffers vary in the number and types of proteins extracted, no single IEF-compatible extraction solution can reliably extract all proteins from complex samples such as mammalian cells and tissues. In addition, use of very harsh extraction solutions, e.g. an extraction method that uses a small amount of SDS that can be tolerated in the IEF separation, may result in excessive precipitation artifacts when the SDS is removed during focusing. Protein profile analyses will obviously be incomplete if the pellet fraction is discarded and the insoluble proteins are ignored.

To address this problem, a mild Tris–CHAPS extraction method was used to solubilize human breast cancer cell proteins and to produce distinct pellet and supernatant fractions [7]. It was observed that many major proteins in the supernatant were not in the pellet and a number of major proteins in the pellet were missing or present as minor bands in the supernatant. While some co-distribution of proteins in both the supernatant and pellet is inevitable, the reciprocal nature of multiple protein bands suggests an effective separation. In addition, protein patterns from different experiments are similar, indicating that reproducible separation is obtained. To further evaluate the effectiveness of the 1D separation approach, several bands from the pellet were identified using LC–MS/MS and most samples were single proteins, indicating that the pellet was a fairly simple mixture that could be used for quantitative comparisons. This study indicates that 1DE is an effective method to separate insoluble proteins to complement analysis of soluble proteins that are analyzed on slightly overlapping narrow pH range 2D gels, and on large pore 1D gels as described later.

3.2. Detection of large soluble proteins

Large proteins (> 100 kDa) present insurmountable problems for 2D gels because they typically focus poorly and variably due to poor solubility near their p*I*s. In addition, their greater size restricts their diffusion into IPG gels and they may be eluted from IPG gels with poor efficiency. As a result, investigators usually simply

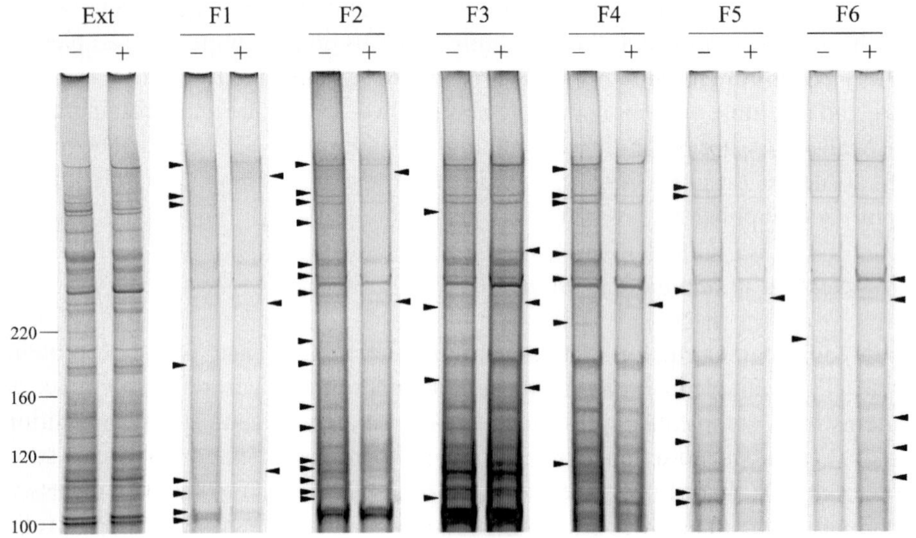

Fig. 4. Comparison of large proteins in closely related human breast cancer cells with low and high metastatic potential using μsol-IEF and large pore 1D gels. The > 100 kDa regions from a silver-stained 3%–8% Tris–Acetate NuPAGE mini gel shows the comparison of MCF-7/AZ cells (low metastatic potential (−)) with MCF-7/6 cells (high metastatic potential (+)). Ext, cell extracts without prefractionation (4 μg); F1–F6, μsol-IEF fractions 1–6, respectively (see Fig. 1A), equivalent to ~15 μg of cell extract. Protein bands were quantitatively analyzed and compared using Discovery Series Quantity One (version 4.2.0, Bio-Rad) software. Triangles highlight proteins with > 2-fold increase in density compared with the alternative cell line.

'write-off' large proteins and select SDS gels for the second dimension that optimize separation of proteins between 10 and 100 kDa.

An effective method of analyzing large proteins (> 100 kDa) is using high-resolution large-pore 1D SDS gels. However, unlike insoluble proteins, unfractionated soluble protein bands on 1D gels typically contain a number of proteins, which greatly complicates quantitative protein comparisons based upon staining intensities [7]. A solution to the complexity problem is to use μsol-IEF prefractionation prior to 1D gel analysis. When μsol-IEF was used to prefractionate the soluble proteins from human breast cancer cells into seven narrow pH pools (see Fig. 1A) and the large proteins (100–500 kDa) were analyzed on large pore size SDS gels, the complexity of individual bands was minimized [8]. A comparison of μsol-IEF fractionated large proteins (> 100 kDa) from closely related human breast cancer cell lines with low and high metastatic potential is shown in Fig. 4. Few differences were observed between the two cell lines when the unfractionated cell extracts were compared due to sample complexity with multiple proteins present in individual bands. In contrast, many quantitative differences were readily detected in the fractionated samples because the μsol-IEF effectively separated the large proteins into discrete well-resolved pH pools that substantially reduced sample complexity.

4. Downstream proteome analysis after sample fractionation

4.1. Narrow pH range 2D PAGE

As described earlier, prefractionation of complex proteomes followed by separation of individual fractions on an appropriate series of narrow pH range IPG strips can improve detection, resolution, and dynamic range of proteins that can be resolved on 2D gels. Utilization of a series of narrower pH range gels is better than a broader pH gel due to the much larger IEF separation distance.

Prefractionation of proteomes using μsol-IEF interfaces well with 2D PAGE because the same solubilization buffer is used for the μsol-IEF and the IPG gel separations, and at most, only sample dilution is required between these two steps. For example, a 700 μl fraction volume from μsol-IEF prefractionation of 2.0 mg of a cell lysate will allow duplicate narrow pH range gels to be run at high protein loads proportional to 1.0 mg of initial sample when 18-cm IPG strips are used (2 × 350 μl per strip). If μsol-IEF pH ranges are chosen so that the complexity and hence the total number of spots present in each resulting narrow pH range 2D gel are similar, 1500 or more protein spots may be detectable on 2D gels of most pH ranges. When eukaryotic proteomes are divided into seven or more fractions, it should be feasible to detect and quantitatively compare at least 10,000 or more proteins.

To maximize spot separation and resolution of the fractionated proteins on narrow pH range 2D gels, separation distances over the pH range of each fraction should be maximized. Ideally, the IPG gels should have pH ranges slightly greater (typically ±0.1 pH units) than the fraction pH range of each. The slightly wider pH range in the IPG strips will prevent loss of proteins near the sides of the 2D gels, while maximizing separation distance as much as possible. Custom IPG strips can be made using immobilines to produce the exact desired pH range. But it is time-consuming to prepare IPG strips for all the fractions and laboratory-to-laboratory consistency is likely to be reduced for custom-made IPG gels compared to commercial products. Since various pH ranges and sizes of IPG strips are currently commercially available, a preferred strategy is to use existing narrow range IPG strips to analyze μsol-IEF fractions where possible. When μsol-IEF fractions are substantially narrower than available commercial IPG strips, strips can be trimmed to fit the next smaller size 2D gel. This strategy maximizes separation distances as well as throughput because running second dimension gels is the major bottleneck in 2D PAGE and smaller second dimension gels require less time and reagents than larger gels. This approach is illustrated in Fig. 5 for 0.5 pH unit μsol-IEF fractions. Hence, 24 cm one pH unit IPG strips can be trimmed after IEF to remove the 'unused' separation areas. These 0.67 pH unit wide trimmed strips can then be run on 18 cm wide second dimension gels without loss of resolution compared with more costly and time consuming 24 cm wide gels (Fig. 5A). Similarly, if 18 cm one pH unit strips are used, they can be trimmed to fit Bio-Rad Criterion gels which are particularly easy to run in large numbers (Fig. 5B). These gels require much less sample for a given staining method compared with larger 2D gels; however, resolution is generally somewhat lower than that with larger gels and the range for optional sample amounts is much lower for these $13 \times 9\ cm^2$ gels (Fig. 5).

A strategy for comprehensive 2D PAGE analysis of all μsol-IEF fractions using commercial IPG strips and high-throughput narrow pH range 2D gels for maximum separation of fractionated proteins is shown in Fig. 6. Briefly, μsol-IEF

Fig. 5. Effect of IPG gel length and second dimension gel size on resolution and sensitivity when analyzing 0.5 pH unit μsol-IEF fractions. A 0.5 pH wide fraction (pH 4.7–5.2) was obtained from prefractionation of a human breast cancer MCF-7/6 cell extract and separated on commercial 24- or 18-cm IPG strips (pH 4.5–5.5 L). After IEF, different sized second dimension gels were evaluated. IPG strips were either used in directly matching second dimension gels or the excess IPG gel regions (pHs outside the fraction pH range) were trimmed so that the IPG gels (~0.7 pH units) could be separated in the next size smaller second dimension gels. All second dimension separations were performed on 10% Tris Tricine SDS gels and proteins were visualized using silver staining. Protein loads were adjusted for the different gel volumes so that similar staining intensities were obtained. The fractionated sample equivalent to the following amounts of original cell extract was used (left to right): 150, 100, 100, and 25 μg.

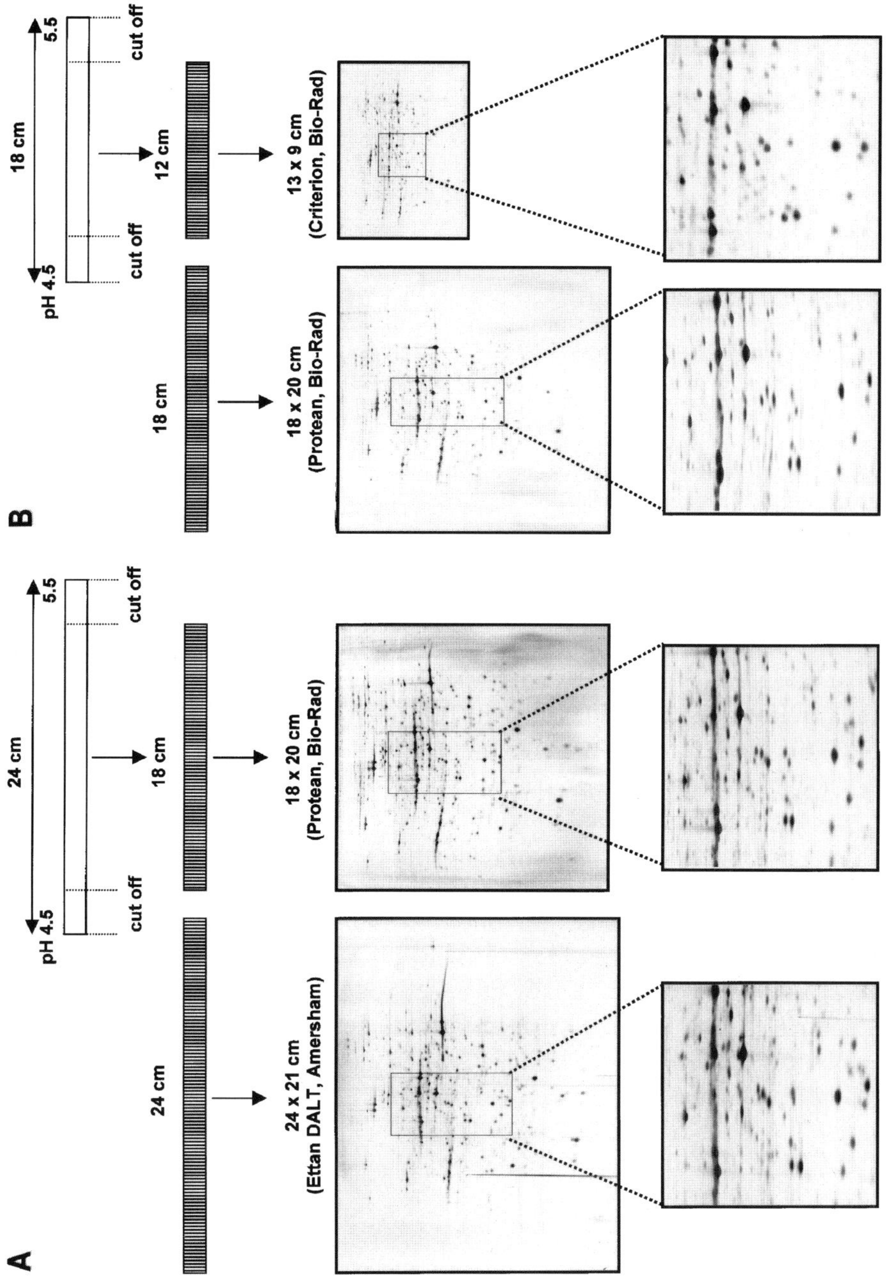
A
24 cm
pH 4.5
5.5
cut off
cut off
24 cm
18 cm
24 x 21 cm
(Ettan DALT, Amersham)
18 x 20 cm
(Protean, Bio-Rad)
B
18 cm
pH 4.5
5.5
cut off
cut off
18 cm
12 cm
18 x 20 cm
(Protean, Bio-Rad)
13 x 9 cm
(Criterion, Bio-Rad)

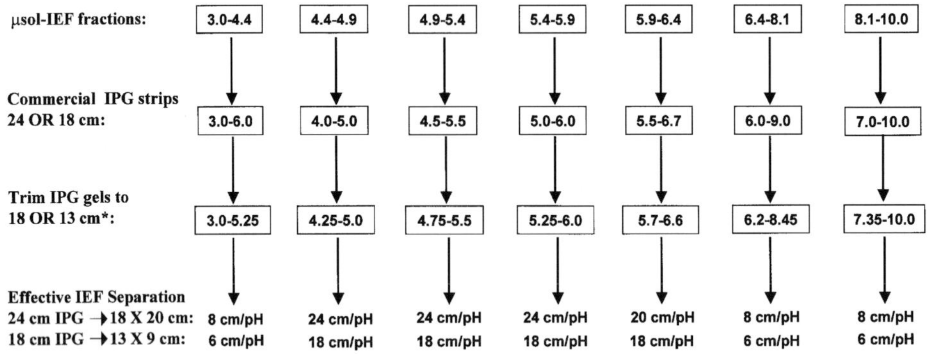

Fig. 6. A strategy for optimizing 2D PAGE resolution and throughput when systematically analyzing all μsol-IEF fractions. This study is based upon the seven pH range separation illustrated in Fig. 1A. Fractionated proteins are focused on the most suitable commercial IPG gels; i.e. IPG gels that maximize separation distances and are at least ±0.1 pH wider than the fraction pH ranges of the μsol-IEF fractions. After focusing, the IPG gels are trimmed from one or both ends to minimize the 'unused' separation area and to allow the trimmed IPG gels to fit into smaller second dimension gels. Reducing 24 cm IPG gels to 18 cm second dimension gels provides a throughput and expense advantage without sacrificing resolution. Similar advantages are obtained if 18 cm IPG strips are trimmed to 13 cm prior to the second dimension. *The pH ranges for trimmed gels are for 24 cm commercial IPG strips that have been cut down to 18 cm; the pH values are slightly different for the 18 cm IPG strips that are cut down to13 cm.

pools are focused on the narrowest available 24- or 18-cm commercial IPG strips that are at least ±0.1 pH wider than the pH ranges of the μsol-IEF fractions. The IPG gels are then trimmed by removing one or both ends of the strips if their pH ranges are greater than ±0.1 pH units wider than the fraction pH ranges to minimize the unused area and to allow them to fit on smaller second dimension gels. Because time and difficulty of running second dimension gels increases as the gel size increases, this down-sizing of the second dimension gel increases throughput without sacrificing resolution. In addition, smaller gels provide cost savings for gel reagents and expensive high-sensitivity stains. More importantly, for the scheme illustrated in Fig. 6, the effective IEF separation length of proteins with p*I*s from 3.0 to 10.0 is 88 cm when commercial 24-cm IPG strips are used, and as shown earlier, protein loads and quality of separation is much improved when prefractionated samples are used compared with running unfractionated samples on a series of narrow range 2D gels.

4.2. 1D PAGE

As described earlier, a powerful downstream analysis method after solution IEF prefractionation is 2D PAGE. However, although prefractionation improves sample load capacities and resolution, some protein groups can still not be reliably detected by 2D gels. Analysis of fractions on 1D gels in parallel with the 2D gels provides a number of advantages, including: (i) 1D PAGE is a high-throughput method for initial evaluation of solution IEF prefractionation efficiency; (ii) several types of proteins such as very large, very basic, and very acidic proteins are missed or poorly detected when analyzed on 2D gels, but these proteins can be readily detected on 1D gels; (iii) multiple fractionated samples can be rapidly separated and compared on a single 1D gel; and (iv) 1D gel lanes can be cut into many slices and all slices can be analyzed by in situ trypsin digestion followed by LC–MS/MS analysis (Section 4.4.3). Typically, after completing a μsol-IEF prefractionation of a complex proteome IEF focusing buffers are used to extract proteins adsorbed to or trapped within partition membranes and small aliquots of these extracts as well as each soluble pool are analyzed on 10% Tris Tricine gels. This 1D separation combined with quantitative analysis using densitometry is a simple, rapid method to check overall protein recoveries, distribution of proteins among fractions, as well as reproducibility resulting from different μsol-IEF separation conditions. Also, for many proteomes, the most acidic and most basic fractions may be sufficiently simple so that 1D gel densitometric image comparisons may be adequate for detection of quantitative changes between experimental samples. Most important, fractionated samples after solution IEF separations can be readily analyzed using large pore size 1D SDS gels for large proteins (>100 kDa) that are revealed poorly and inconsistently on 2D gels. This is an important complement to the more difficult, time consuming narrow pH range 2D gel analyses because large proteins are not reliably recovered on 2D gels, but reliable quantitative differences of large proteins can be identified on the 1D gels [7,8].

A comprehensive strategy for protein profiling and quantitative comparison using μsol-IEF prefractionation and a combination of 2D and 1D PAGE is shown in Fig. 7. Large pore 1D gels and slightly overlapping narrow range 2D gels are used to detect soluble proteins, and large pore as well as 10% Tris Tricine 1D gels are used to analyze insoluble proteins. Using this strategy, much larger numbers of soluble and insoluble proteins including low abundance proteins can be detected, quantified and compared than by alternative gel-based methods.

4.3. 2D DIGE

Two-dimensional differential gel electrophoresis (2D DIGE) is a relatively new approach to quantitative proteome analysis using 2D gels ([38] and Chapter 3). It involves separately labeling two or three different protein extracts with distinct fluorophores (cyanine dyes) that have similar charge, size, and reactivity toward lysine residues in the proteins. Because different cyanine dyes have distinguishable spectra, the individual fluorescently labeled protein samples can be combined prior to gel electrophoresis. The proteins from each sample are visualized separately by imaging with appropriate excitation and emission wavelengths for

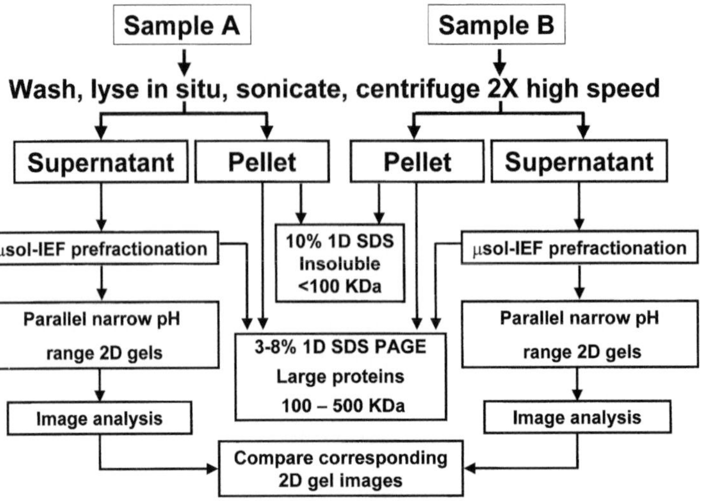

Fig. 7. A comprehensive protein profiling strategy utilizing μsol-IEF combined with 1D and 2D gels. Insoluble proteins are analyzed using a combination of large and small pore size 1D SDS gels followed by image analysis to detect differences. Soluble proteins are initially prefractionated using μsol-IEF and resulting fractions are analyzed using a combination of parallel slightly overlapping narrow pH range 2D gels for < 100 kDa proteins and large pore size 1D gels for large proteins (100–500$^+$ kDa).

the different dyes to obtain quantitative data. Currently there are three spectrally distinct fluorescent dyes commercially available (Amersham Biosciences). The ability to combine two or three labeled samples on a single 2D gel eliminates problems associated with gel-to-gel variation although other problems such as matching fluorescent spots with major stained bands for identification of proteins by LC–MS/MS can arise.

As described earlier, using an effective sample prefractionation can greatly improve 1D and 2D gel detection capacity. All of the solution IEF fractionation methods discussed in this chapter should be compatible with 2D DIGE. The μsol-IEF method with its smaller sample and volume requirements is particularly well suited for use with 2D DIGE. In fact, combination of μsol-IEF prefractionation with 2D DIGE should have advantages over prefractionation and gel-based analysis without labeling because any minor variations between multiple μsol-IEF runs as well as normal 2D gel-to-gel variability would be eliminated. A strategy combining fluorescent labeling with μsol-IEF prefractionation for quantitative proteome comparisons is shown in Fig. 8.

4.4. LC–MS/MS and LC/LC–MS/MS methods

MS analysis of tryptic peptides has emerged as the predominant method for identifying proteins in most proteome studies including gel-based protein profiling

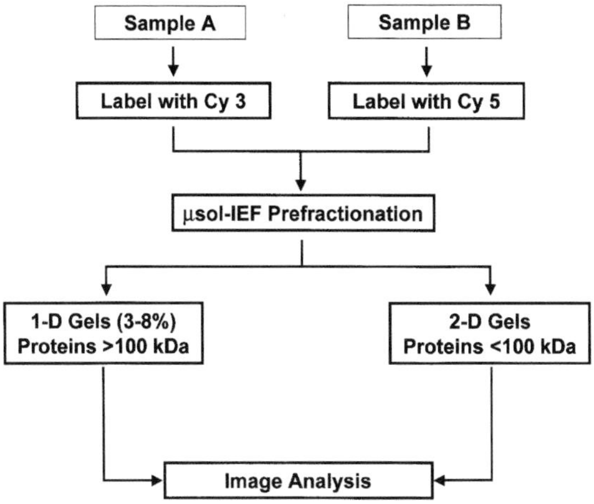

Fig. 8. A gel-based comprehensive protein profiling strategy utilizing 2D DIGE technology. Individual samples are initially labeled with distinct cyanine dyes and combined prior to μsol-IEF prefractionation. The fractionated samples are then separated on conventional 1D and 2D gels as described in Fig. 7, and images are compared using 2D DIGE software or analogous methods (see Chapter 3).

methods and non-gel methods [39]. Depending upon the complexity of the initial protein samples to be analyzed, non-gel approaches may use one, two, or three dimensions of chromatographic separations prior to MS and MS/MS analysis. The MS-based methods are interesting, promising alternatives to 2D gels because they can overcome some persistent limitations of 2D PAGE. Currently LC–MS/MS and LC/LC–MS/MS methods are not higher throughput and do not detect more proteins than the best 2D PAGE methods, although they do detect a different subset of proteins when complex proteomes are analyzed. However, these non-gel based approaches can be more readily automated than 2D gels, and detection dynamic ranges of about 10^4 can be obtained [40]. Two of the most common LC/LC–MS/MS approaches are the multi-dimensional protein identification technology (MudPIT) and isotope-coded affinity tag (ICAT) methods. Several overviews of MS-based approaches for comprehensive protein profiling and quantitative comparisons were recently published [41–43].

4.4.1. MudPIT

MudPIT was initially described by Yates and co-workers [44] several years ago. They demonstrated that a large number of proteins, e.g. ~1500 proteins from a yeast extract, could be identified from complex mixtures by digestion of crude extracts with trypsin followed by multidimensional nanocapillary chromatography interfaced directly with an ESI mass spectrometer (LC/LC–MS/MS). The MudPIT method, which was initially used only for qualitative analysis, could readily be made into a quantitative protein profile comparison method by using stable isotope labeling of one sample, which is mixed with an unlabeled reference sample prior to trypsin digest [45–47].

4.4.2. ICAT

The ICAT approach is a quantitative non-gel based method originally described by Aebersold and co-workers [48] for quantitative comparison of two samples. Two forms of a cysteine-specific labeling reagent coupled to a biotin affinity tag via a linker, which are known as heavy and light ICAT, are used. Heavy ICAT has eight deuterium atoms in the reagent's linker region, while light ICAT contains hydrogens instead. This creates a mass difference of 8 Da between the two ICAT forms. To compare two protein samples, each sample is labeled with a different form of the ICAT reagent under conditions where all cysteines are derivatized (end-point labeling), the samples are combined, subjected to trypsin digestion, cysteine containing peptides are affinity purified using the biotin tag, and these peptides are analyzed by LC–MS/MS. The relative abundance of proteins identified by MS/MS analysis of cysteine-containing peptides are estimated from the relative amounts of the light and heavy forms of each peptide detected.

Recently, a cleavage version of the ICAT reagent became commercially available (Applied Biosystems, Foster City, CA, USA) which contains ^{13}C atoms rather than deuteriums. Hence, this second generation reagent has a number of advantages over the original reagent. Elimination of the deuteriums reduces potential chromatographic separation of the heavy and light forms of each peptide. Also, the cleavable linker allows higher yield peptide release from affinity columns and removal of the bulky biotin group improves mass analysis. Several studies verifying the advantages of using the cleavable ICAT for quantitative protein profiling have been recently published [49–52].

The ICAT method is an interesting alternative to 2D PAGE. In contrast to 2D PAGE, the amount of protein that can be processed is theoretically unlimited. This is an important strength of the approach because increasing the total amount of sample analyzed can improve detection of lower abundance proteins [53]. However, although the ICAT method is quite promising and can be automated, ICAT and other MS-based methods currently require at least several days of mass spectrometer and computer time, and in only a few cases have more proteins been quantitatively profiled than the ~ 1500 proteins that can be separated in a single high resolution 2D gel.

4.4.3. Three-dimensional protein profiling

Since eukaryotic proteomes are complex and typically contain more than 10,000 proteins, it is impossible to identify all the proteins present using MudPIT, ICAT, or similar approaches. When a trypsin digested unfractionated protein mixture is analyzed, the number of peptides produced far exceeds the number that current mass spectrometers can detect. Hence tryptic peptides present at relatively high concentrations and with good ionization properties are preferentially and repetitively detected. The number of proteins that can be identified using these methods can be increased by increasing the resolution in each chromatographic separation. However, if one starts with the highest resolution separation media that is available, further increases in resolution require that analysis time be increased. A point is quickly reached where further increases in analysis times will result in only marginally greater numbers of proteins detected. An attractive method for further increasing the number of detected proteins in MudPIT, ICAT, or similar LC/LC–MS/MS experiments is to use μsol-IEF of mixed protein samples prior to trypsin digestion and subsequent LC/LC–MS/MS analysis (Fig. 9). The reduced protein complexity in μsol-IEF fractions should allow identification of more

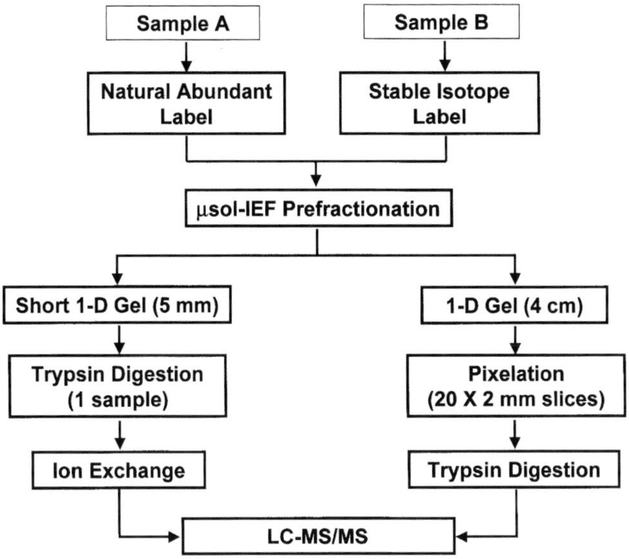

Fig. 9. Two '3D protein profiling' strategies utilizing μsol-IEF. For both 3D protein profiling methods, two proteome samples are labeled with natural abundance and stable isotope forms of a cysteine-specific reagent that does not have an affinity tag. The samples are then mixed and prefractionated using μsol-IEF into 5–7 or more narrow pH range pools. One 3D separation, IEF/SDS/LC–MS/MS, involves separation of μsol-IEF pools on 4-cm 1D SDS gels, the entire lanes are cut into uniform 1 or 2 mm slices, and all slices are subjected to trypsin digestion and analyzed by LC–MS/MS. The other 3D strategy, IEF/LC/LC–MS/MS, utilizes μsol-IEF fractionation prior to a MudPIT or type of MS/MS analysis. After stable isotope labeling and μsol-IEF separation each fraction is electrophoresed into a mini-gel for a short distance to remove interfering reagents (urea, ampholytes, detergents) prior to in-gel digestion of the entire sample in a single reaction followed by conventional MudPIT type analyses of each μsol-IEF pool.

proteins and/or yield better sequence coverage of identified proteins. In addition, this high-resolution front-end separation may require less complex chromatography in subsequent steps, which can provide at least partially compensatory analysis time savings. For example, when analyzing mammalian cell extracts or similarly complex proteomes, it will probably be more productive to digest five μsol-IEF fractions followed by 10 salt steps on a cation exchange rather than to simply perform 50 salt steps on a trypsin digestion of the unfractionated extract. Both strategies would require 50 LC–MS/MS runs, but the IEF/LC/LC–MS/MS approach should typically identify more proteins. A strategy where μsol-IEF prefractionation is introduced between the mixing of differentially labeled samples and the trypsin digestion steps is shown in Fig. 9.

An alternative 3D method produces a 'batch 2D' separation by combining μsol-IEF and subsequent short 1D SDS gels prior to LC–MS/MS analysis (IEF/SDS/LC–MS/MS). Briefly, two protein samples are initially labeled with different isotopic forms of a cysteine-specific reagent without an affinity tag. The samples are mixed and fractionated using μsol-IEF into a number of narrow pH range pools. Fractions are then separated on a mini 1D SDS gel by allowing the tracking dye to migrate 4 cm. Individual lanes are excised from the 4 cm long separation gel and sliced into uniform pieces (each 1 or 2 mm) using a gel slicer prior to digestion with trypsin and LC–MS/MS. This method is more scaleable than multi-dimensional separations of tryptic peptides.

5. Summary

Due to the limited capacities of both gel-based and non-gel-based protein profiling methods, it has become apparent that more powerful and reliable methods are needed for prefractionation of complex proteomes prior to 2D gels and/or alternative LC–MS analysis. Although preparative IEF prefractionation methods are not orthogonal to 2D gels, they show the most promise due to the very high resolution that can be obtained. Alternative lower resolution prefractionation methods severely compromise the ability to perform comprehensive quantitative comparisons due to variable cross-contamination between adjacent fractions and greater fraction complexity. A number of preparative solution-based IEF methods have been productively integrated into quantitative protein profiling strategies including the Rotofor, FFE, IsoPrime, the MCE and μsol-IEF (Zoom IEF fractionation). However, some of these methods require large sample amounts and result in large dilute fractions that are not compatible with direct analysis using downstream protein profiling methods. Of the methods listed earlier, the recently developed μsol-IEF method is the most compatible with samples that are difficult to obtain in large quantities. The μsol-IEF prefractionation method is capable of slicing complex proteomes into well-resolved fractions based on the protein's p*I*s on a small volume scale (less than 0.7 ml per fraction) that is compatible with subsequent direct transfer to downstream analytical analyses including 1D PAGE, narrow pH range 2D PAGE, LC–MS/MS or LC/LC–MS/MS.

Several alternative protein profiling strategies using the μsol-IEF prefractionation are presented. When μsol-IEF is combined with narrow pH range 2D gels and complementary large pore 1D gels, far more proteins can be quantitatively compared than without prefractionation. The high protein loads on narrow range gels that are feasible after prefractionation, combined with sensitive stains such as Sypro Ruby or silver stain, enables the detection of low abundance proteins. This gel-based comprehensive protein profiling strategy is also compatible with the 2D-DIGE multiple fluorescent tag labeling technology. Alternatively, solution IEF prefractionation methods can be combined with MudPIT and similar stable isotope labeling and LC/LC–MS/MS analysis methods to produce a powerful 3D protein profiling method (IEF/LC/LC–MS/MS). An alternative 3D strategy produces a facile high-throughput 'batch 2D' separation by following μsol-IEF with short 1D SDS mini-gels. Each section of this batch 2D separation is then subjected to LC–MS/MS.

Acknowledgments

The authors gratefully acknowledge Peter Hembach, Lynn Echan, Nadeem Ai-Khan, Kaye Speicher, Hsin-Yao Tang, and Sandra Harper for their assistance throughout the project of development of μsol-IEF prefractionation technology and integration of downstream analysis methods. We also thank Dr M.M. Mareel (Ghent, Belgium) for providing us with the human breast cancer cell lines (MCF-7/6 and MCF-7/AZ). This work was supported by NIH grants CA77048 and CA92725 as well as an NCI cancer center grant (CA10815), and the Commonwealth Universal Research Enhancement Program, Pennsylvania Department of Health.

References

1. Miklos, G.L.G. and Rubin, G.M., The role of the genome project in determining gene function: insights from model organisms. *Cell*, **86**, 521–529 (1996).
2. Godley, A.A. and Packer, N.H., The importance of protein co- and post-translational modifications in proteome projects, In: Wilkins, M.R., Williams, K.L., Appel, R.D. and Hochestrasser, D.F. (Eds.), *Proteome Research: New Frontiers in Functional Genomics*. Springer, Berlin, 1997, pp. 65–91.
3. Corthals, G.L., Wasinger, V.C., Hochstrasser, D.F. and Sanchez, J., The dynamic range of protein expression: a challenge for proteome research. *Electrophoresis*, **21**, 1104–1115 (2000).
4. Herbert, B.R., Sanchez, J.C. and Bini, L., Two-dimensional electrophoresis: the state of the art and future directions, In: Wilkins, M.R., Williams, K.L., Appel, R.D. and

Hochestrasser, D.F. (Eds.), *Proteome Research: New Frontiers in Functional Genomics*. Springer, Berlin, 1997, pp. 13–33.
5. Wasinger, V.C., Bjellqvist, B. and Humphery-Smith, I., Proteomic 'contigs' of *Ochrobactrum anthropi*, application of extensive pH gradients. *Electrophoresis*, **18**, 1373–1383 (1997).
6. Hoving, S., Voshol, H. and Oostrum, J.V., Towards high performance two-dimensional gel electrophoresis using ultrazoom gels. *Electrophoresis*, **21**, 2617–2621 (2000).
7. Zuo, X., Echan, L., Hembach, P., Tang, H.Y., Speicher, K.D., Santoli, D. and Speicher, D.W., Towards global analysis of mammalian proteomes using sample prefractionation prior to narrow pH range two-dimensional gels and using one-dimensional gels for insoluble and large proteins. *Electrophoresis*, **22**, 1603–1615 (2001).
8. Zuo, X., Hembach, P., Echan, L. and Speicher, D.W., Enhanced analysis of human breast cancer proteomes using micro-scale solution isoelectrofocusing combined with high resolution 1-D and 2-D gels. *J. Chromatogr. B*, **782**, 253–265 (2002).
9. Williams, K.L., Genomes and proteomes: towards a multidimensional view of biology. *Electrophoresis*, **20**, 678–688 (1999).
10. Righetti, P.G., Castagna, A. and Herbert, B., Prefractionation techniques in proteome analysis: a new approach identifies more low-abundance proteins. *Anal. Chem.*, **73**, 320–326 (2001).
11. Molloy, M.P., Herbert, B.R., Walsh, B.J., Tyler, M.I., Traini, M., Sanchez, J.C., Hochstrasser, D.F., Willams, K.L. and Gooley, A.A., Extraction of membrane proteins by differential solubilization for separation using two-dimensional gel electrophoresis. *Electrophoresis*, **19**, 837–844 (1998).
12. Huber, L.A., Pasquali, C., Gagescu, R., Zuk, A., Gruenber, G.J. and Matlin, K.S., Endosomal fractions from viral K-ras-transformed MDCK cell reveal transformation specific changes on two-dimensional gel maps. *Electrophoresis*, **17**, 1734–1740 (1996).
13. Lollo, B.A., Harvey, S., Liao, J., Stevens, A.C., Wagenknecht, R., Sayen, R., Whaley, J. and Sajjadi, F.G., Improved two-dimensional gel electrophoresis representation of serum proteins by using ProtoClear™. *Electrophoresis*, **20**, 854 (1999).
14. Fountoulakis, M., Langen, H., Gray, C. and Takacs, B., Enrichment and purification of proteins of *Haemophilus influenzae* by chromatofocusing. *J. Chromtogr.*, **806**, 279–291 (1998).
15. Fountoulakis, M., Takacs, M.F. and Takacs, B., Enrichment of low-copy-number gene products by hydrophobic interaction chromatography. *J. Chromtogr.*, **833**, 157–168 (1999).
16. Fountoulakis, M., Takacs, M.F., Berndt, P., Langen, H. and Takacs, B., Enrichment of low abundance proteins of *Escherichia coli* by hydroxyapatite chromatography. *Electrophoresis*, **20**, 2181–2195 (1999).
17. Zuo, X. and Speicher, D.W., A method for global analysis of complex proteomes using sample prefractionation by solution isofocusing prior to two-dimensional electrophoresis. *Anal. Biochem.*, **284**, 266–278 (2000).
18. Bier, M., Egen, N.B., Allgyer, T.T., Twitty, G.E. and Mosher, R.A., New developments in isoelectric focusing, In: Gross, E. and Meienhofer, J. (Eds.), *Peptides: Structure and Biological Functions*. Pierce Chemical Co Rockford, Illinois, 1979, pp. 79–89.
19. Egen, N.B., Thormann, W., Twitty, G.E. and Bier, M., A new preparative isoelectric focusing apparatus, In: Hirai, H. (Ed.), *Electrophoresis*. de Gruyter, Berlin, 1984, pp. 547–549.
20. Nilsson, C.L., Larsson, T., Gustafsson, E., Karlsson, K.A. and Davidsson, P., Identification of protein vaccine candidates from *Helicobacter pylori* using a preparative two-dimensional electrophoretic procedure and mass spectrometry. *Anal. Chem.*, **72**, 2148–2153 (2000).
21. Hesse, C., Nilsson, C.L., Blennow, K. and Davidsson, P., Identification of the apolipoprotein E4 isoform in cerebrospinal fluid with preparative two-dimensional electrophoresis and matrix

assisted laser desorption/ionization-time of flight-mass spectrometry. *Electrophoresis*, **22**, 1834–1837 (2001).

22. Wall, D.B., Kachman, M.T., Gong, S., Hinderer, R., Parus, S., Misek, D.E., Hanash, S.M. and Lubman, D.M., Isoelectric focusing nonporous RP HPLC: a two-dimensional liquid-phase separation method for mapping of cellular proteins with identification using MALDI-TOP mass spectrometry. *Anal. Chem.*, **72**, 1099–1111 (2000).
23. Davidsson, P., Puchades, M. and Blennow, K., Identification of synaptic vesicle, pre- and postsynaptic proteins in human cerebrospinal fluid using liquid-phase isoelectric focusing. *Electrophoresis*, **20**, 431–437 (1999).
24. Lubman, D.M., Kachman, M.T., Wang, H., Gong, S., Yan, F., Hamler, R.L., O'Neil, K.A., Zhu, K., Buchanan, N.S. and Barder, T.J., Two-dimensional liquid separations-mass mapping of proteins from human cancer cell lysates. *J. Chromatogr. B*, **782**, 183–196 (2002).
25. Hannig, K., New aspects in preparative and analytical continuous free-flow cell electrophoresis. *Electrophoresis*, **3**, 235–243 (1982).
26. Krivankova, L. and Bocek, P., Continuous free-flow electrophoresis. *Electrophoresis*, **19**, 1064–1074 (1998).
27. Bauer, J., Advances in cell separation: recent developments in counterflow centrifugal elutriation and continuous flow cell separation. *J. Chromatogr. B*, **722**, 55–69 (1999).
28. Weber, G. and Bocek, P., Recent developments in preparative free flow isoelectric focusing. *Electrophoresis*, **19**, 1649–1653 (1998).
29. Weber, G. and Bocek, P., Stability of continuous flow electrophoresis. *Electrophoresis*, **19**, 3094–3095 (1998).
30. Weber, G., Grimm, D. and Bauer, J., Application of binary buffer systems to free flow cell electrophoresis. *Electrophoresis*, **21**, 325–328 (2000).
31. Hoffmann, P., Hong, J., Moritz, R.L., Connolly, L.M., Frecklington, D.F., Layton, M.J., Eddes, J.S. and Simpson, R.J., Continuous free-flow electrophoresis separation of cytosolic proteins from the human colon carcinoma cell line LIM 1215: a non two-dimensional gel electrophoresis-based proteome analysis strategy. *Proteomics*, **1**, 807–818 (2001).
32. Righetti, P.G., Wenisch, E. and Faupel, M., Preparative protein purification in a multi-compartment electrolyser with immobiline membranes. *J. Chromatogr.*, **475**, 293–309 (1989).
33. Righetti, P.G., Wenisch, E., Jungbauer, A., Katinger, H. and Faupel, M., Preparative purification of human monoclonal antibody isoforms in a multi-compartment electrolyser with immobiline membranes. *J. Chromatogr.*, **500**, 681–696 (1990).
34. Wenisch, E., Righetti, P.G. and Weber, W., Purification to single isoforms of a secreted epidermal growth factor receptor in a multicompartment electrolyzer with isoelectric membranes. *Electrophoresis*, **13**, 668–673 (1992).
35. Herbert, B. and Righetti, P.G., A turning point in proteome analysis: sample prefractionation via multicompartment electrolyzers with isoelectric membranes. *Electrophoresis*, **21**, 3639–3648 (2000).
36. Zuo, X. and Speicher, D.W., Comprehensive analysis of complex proteomes using microscale solution isoelectrofocusing prior to narrow pH range two-dimensional electrophoresis. *Proteomics*, **2**, 58–68 (2002).
37. Righetti, P.G., Immobilized pH Gradients: Theory and Methodology, Elsevier, Amsterdam, 1990.
38. Unlu, M., Morgan, M.E. and Minden, J.S., Difference gel electrophoresis: a single gel method for detecting changes in protein extracts. *Electrophoresis*, **18**, 2071–2077 (1997).
39. Aebersold, R. and Mann, M., Mass spectrometry-based proteomics. *Nature*, **422**, 198–207 (2003).

40. Wolters, D.A., Washburn, M.P. and Yates, J.R. III, An automated multidimensional protein identification technology for short gun proteomics. *Anal. Chem.*, **73**, 5683–5690 (2001).
41. McDonald, W.H. and Yates, J.R. III, Shotgun proteomics and biomarker discovery. *Dis. Markers*, **18**, 99–105 (2002).
42. Lill, J., Proteomic tools for quantitation by mass spectrometry. *Mass Spectrom. Rev.*, **22**, 182–194 (2003).
43. Tao, W.A. and Aebersold, R., Advances in quantitative proteomics via stable isotope tagging and mass spectrometry. *Curr. Opin. Biotechnol.*, **14**, 110–118 (2003).
44. Washburn, M.P., Wolters, D. and Yates, J.R. III, Large-scale analysis of the yeast proteome by multidimensional protein identification technology. *Nat. Biotechnol.*, **19**, 242–247 (2001).
45. Washburn, M.P., Ulaszek, R., Deciu, C., Schieltz, D.M. and Yates, J.R. III, Analysis of quantitative proteome data generated via multidimensional protein identification technology. *Anal. Chem.*, **74**, 1650–1657 (2002).
46. Peng, J., Elias, J.E., Thoreen, C.C., Licklider, L.J. and Gygi, S.P., Evaluation of multidimensional chromatography coupled with tandem mass spectrometry (LC/LC–MS/MS) for large-scale protein analysis: the yeast proteome. *J. Proteome Res.*, **2**, 43–50 (2003).
47. Kubota, K., Wakabayashi, K. and Matsuoka, T., Proteome analysis of secreted proteins during osteoclast differentiation using two different methods: two-dimensional electrophoresis and isotope-coded affinity tags analysis with two-dimensional chromatography. *Proteomics*, **3**, 616–626 (2003).
48. Gygi, S.P., Rist, B., Gerber, S.A., Turecek, F., Gelb, M.H. and Aebersold, R., Quantitative analysis of complex protein mixtures using isotope-coded affinity tags. *Nat. Biotechnol.*, **17**, 994–999 (1999).
49. Smolka, M., Zhou, H. and Aebersold, R., Quantitative protein profiling using two-dimensional gel electrophoresis, isotope-coded affinity tag labeling, and mass spectrometry. *Mol. Cell. Proteom.*, **1**, 19–29 (2002).
50. Hansen, K.C., Schmitt-Ulms, G., Chalkley, R.J., Hirsch, J., Baldwin, M.A. and Burlingame, A.L., Mass spectrometric analysis of protein mixtures at low levels using cleavable ^{13}C-ICAT and multi-dimensional chromatography. *Mol. Cell. Proteom.*, **2**, 299–314 (2003).
51. Griffin, T.J., Lock, C.M., Li, X.J., Patel, A., Chervetsova, I., Lee, H., Wright, M.E., Ranish, J.A., Chen, S.S. and Aebersold, R., Abundance ratio-dependent proteomic analysis by mass spectrometry. *Anal. Chem.*, **75**, 867–874 (2003).
52. Ranish, J.A., Yi, E.C., Leslie, D.M., Purvine, S.O., Goodlett, D.R., Eng, J. and Aebersold, R., The study of macromolecular complexes by quantitative proteomics. *Nat. Genet.*, **33**, 349–355 (2003).
53. Gygi, S.P., Rist, B., Griffin, T.J., Eng, J. and Aebersold, R., Proteome analysis of low-abundance proteins using multidimensional chromatography and isotope-coded affinity tags. *J. Proteome Res.*, **1**, 47–54 (2002).

Proteome Analysis. Interpreting the Genome.
D.W. Speicher (editor)

Chapter 5

Modification specific proteomics applied to protein glycosylation and nitration

JUDITH JEBANATHIRAJAH[1] and PETER ROEPSTORFF*

Department of Biochemistry and Molecular Biology, University of Southern Denmark, Camusvej 55, DK-5230 Odense M, Denmark

1. Introduction

The term proteomics was introduced in 1995 for analysis of the complete protein complements expressed by a genome in a cell or a tissue type [1]. The ability to perform proteome analysis was the result of a number of independent scientific achievements through the past decade, the most important being the large genome sequencing projects resulting in the availability of the genetic blueprint for a number of organisms beginning with the yeast genome in 1997 [2], and the capacity of mass spectrometry to analyze minute amounts of proteins and peptides. The information obtained by digestion of proteins with specific endoproteases followed by either determination of the resulting peptide masses [3–5] or by generation of sequence information by tandem mass spectrometry of selected peptides [6,7] can be used to identify proteins by comparison with their

* Corresponding author. Tel.: +45-65502404; fax: +45-65502467. E-mail: roe@bmb.sdu.dk (P. Roepstorff).
[1] Present address: Department of Cell Biology, Harvard Medical School, Boston, MA 02115, USA.

sequences listed in protein sequence databases. The genomes of higher eukaryote organisms, to the surprise of most biologists, did not contain as many genes as predicted in spite of much larger genomes. Thus, for example, the yeast genome with a size of approximately 13 mbases contains an estimate of 6183 genes, whereas the human genome (approximately 3400 mbases or 250 times the size of the yeast genome) at present is expected to contain between 30,000 and 40,000 genes, the most recent estimate being as low as 28,000 [8], which is only approximately five times as many as that of yeast and twice as many as that of fly or worm [9]. This strongly implies that protein complexity and regulation in higher eukaryotes are governed by protein modifications and variable mRNA splicing. The combination of co- and post-translational modification and splicing events might increase the number of different molecular forms of proteins by a factor of 10–30 or maybe even more. A general goal of most present proteomics studies is expression proteomics, i.e. to establish which proteins are expressed in a given biological state and to establish quantitative differences in expression levels between different states. However, analysis of all different modified forms of proteins must be considered in order to understand the function of proteins in the living organism, and such studies can only be performed at the protein level.

Several hundred post-translational protein modifications (PTMs) have been reported in the literature [10], many of which are rare and are found to occur only in one or a few proteins under specific physiological conditions. However, some PTMs are ubiquitous. These include protein phosphorylation, glycosylation, acylation and processing such as proteolytic cleavage. Determination of PTMs represents a major challenge because only a limited portion of the molecules of a given protein in the cell might be modified and the modifications will often be located on different positions of the molecule depending on its actual functional state. This results in a heterogeneous population of the given protein, e.g. the typical situation for phosphorylation and O-glycosylation. The modifying group may also be highly heterogeneous as is typically seen for glycosylation where a given site may be occupied with more than 20 different glycan structures [11]. As a consequence, highly sensitive and selective analytical methodology is required for complete analysis of all the post-translationally modified forms of proteins present in a cell or a tissue.

A concept termed 'modification specific proteomics' with the goal of determining which proteins and which specific sites in these proteins contain particular types of PTMs has been developed in the authors' research group. Modification specific proteomics takes advantage of characteristics of the modifying groups for either visualization in gels, specific pull-downs from complex protein or peptide mixtures, or specific mass spectrometric detection [12]. This excludes the commonly occurring protein processing through N- or C-terminal truncation because such modified proteins do not have any characteristics different from other non-modified proteins. These efforts have concentrated on some of the ubiquitous types of PTMs, i.e. phosphorylation (described in Chapter 6), glycosylation, and the less common tyrosine nitration. This chapter will focus on glyco-specific and nitrotyrosine-specific proteomics.

2. Glycosylation

2.1. Why study protein glycosylation?

The most prevalent protein modification is glycosylation and 50% of all proteins are thought to be glycosylated based on bioinformatic analysis of genomic information [13,14]. The two major types of protein glycosylation found are O- and N-glycosylation (see Fig. 1). The N-linked glycans are attached to the protein via an amide bond to an asparagine residue and the O-linked glycans are attached via a glycosidic bond to either a serine or threonine. Proteins can also be 'glycated'; this is caused by an Amadori reaction of proteins in the presence of a physiologically high concentration of sugar. This is a non-enzymatic Schiff-base reaction between reducing sugars such as glucose and the α-amino group of N-terminal amino acid residues and/or other amino groups on the proteins. Excessive glycation such as glycated hemoglobin is observed in diabetic patients and is one of the diagnostic indications for diabetes [15]. The need to develop methods to study glycosylation is especially clear in the context of the importance of glycosylation in biology. Both N- and O-linked glycosylations are ubiquitous in prokaryotic as well as in eukaryotic organisms, however, vast differences in glycosylation patterns are found between species. It was previously thought that prokaryotes were unable to synthesize glycoproteins, however, it is increasingly evident that the variety of glycoproteins found in bacteria exceeds that found in

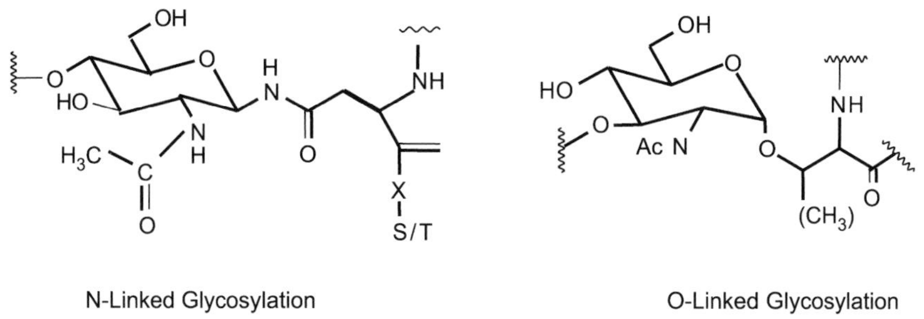

Fig. 1. N-linked glycosylation is shown on the left, where an amide bond links the sugar to the protein via an asparagyl residue. The N-linked consensus sequence is also shown, NXS/T. The O-linked glycosylation is shown to the right where a glycosidic bond links the sugar to the hydroxyl groups of Ser or Thr. The N-glycan can be very complex whereas the O-linked glycans can be small with only one sugar residue being attached.

eukaryotes [16,17]. These differences in glycoproteins and glycans found in prokaryotes are being exploited in the design of new vaccines.

In eukaryotes, the process of glycosylation can either be co-translational in the case of initial N-glycosylation or post-translational in the case of some O-glycosylations. The N-linked glycosylation occurs in the endoplasmic reticulum–golgi complex (ER–Golgi) during translation. Several glycosylation processing enzymes exist in the ER and these enzymes either add sugars to the initial structure or trim them back. This trimming and adding process continues in the Golgi apparatus [18]. Complex O-glycosylation takes place in the Golgi, whereas the addition of a simple O-linked *N*-acetyl-glucosamine (GlcNAc), which is a post-translational modification, occurs in the nucleus and cytosol. This simple modification is dynamic and has been found at sites that are identical to those used by serine/threonine kinases. The O-GlcNAc modification appears to play a pivotal role in signaling [19].

2.1.1. The effect of glycosylation on protein structure and function

Glycosylations not only determine the final three-dimensional (3D) structure of proteins but are also thought to act as scaffolds for the initial folding of the proteins [20]. Different organisms have diverse glycosylation patterns, and although the details of oligosaccharide processing reactions vary, similar functions can be deduced for specific patterns of glycosylation. An example of this is that in both yeast and mammals 'tether' regions separating two functionally different domains are often highly glycosylated by short O-linked glycans. These highly glycosylated regions are thought to maintain the 'tether' region in an open conformation. Glycosylation can also have dramatic effects on the properties of proteins such as in vivo protein stability, thermal stability, proteolytic susceptibility and solubility [21].

Differences in the types of sugars found in protein glycosylation can have dramatic evolutionary effects. For example, an interesting fact is that humans do not have a gene that codes for CMP-sialic acid hydroxylase (catalyzes the conversion of *N*-acetylneuraminic acid to *N*-glycolylneuraminic acid (Neu5Gc)), whereas other mammals have this enzyme. Intriguingly, this enzyme is less expressed in the brain relative to other organs in all species examined. This observation has caused speculations that the deletion of this enzyme in humans could have resulted in major differences of human brain development [22].

2.1.2. Glycosylation and diseases

The importance of glycosylation is underscored by the fact that several diseases are caused by defects in glycosylation mechanisms, these include various congenital disorders of glycosylation diseases (CDGD) which can have lethal or fatal effects such as death, chronic gastrointestinal problems, or severe neurological problems. These diseases are caused by genetic defects in genes coding for enzymes involved in the processing of the glycans on proteins. An example of this is CDGD1b where individuals lack an enzyme that converts fructose-6-phosphate into mannose-6-phosphate. The disease has serious pleiotropic effects and can be completely alleviated by supplementing the diet with mannose. Changes to glycosylation patterns of cell surface proteins are often observed in cancer cells and these changes facilitate evasion of the immune system. The role of glycosylation in the immune system cannot be overstated. Specific glycoforms of proteins are involved in the folding, quality assessment, and assembly of the peptide loaded major histocompatability complex (MHC) antigens and the T cell receptor complex.

Since glycosylation adds to functional properties of proteins and is involved in numerous diseases it is extremely important to characterize this modification. So what does this entail? First, it is important to identify which proteins are glycosylated under certain biological conditions. Secondly, the site(s) of glycosylation must be identified and thirdly the structure of the glycan determined. Some proteins may be glycosylated under one condition and not in another. The modification sites as well as glycan structures might change depending on the state of the studied system and the specific conditions addressed in the study. The resulting population of glycoproteins will be highly heterogeneous making heavy demands on the analytical procedures used.

2.2. Strategies for studying glycosylation

Glycosylation analysis by mass spectrometry poses some particular challenges. In order to fully characterize glycosylation status of a protein both the structure of the glycan and the site of glycosylation have to be defined. Once a protein is known to be glycosylated, methods are available to determine the structure of the glycan and the site of glycosylation [23]. These methods are extremely useful for the detailed characterization of proteins known to be glycosylated, but they require large amounts of sample and are not applicable to high throughput proteomics where the aim is to find unknown modifications. A further complication is that, although consensus sites for N-glycosylation are well defined, not all N-glycosylation consensus sites are occupied. Out of approximately 75,000 entries in SWISS-PROT 2/3 include a potential N-glycosylation consensus sequon: NXS/T. Only 10.6% of these are filed as glycoproteins and only 1% has been characterized with respect to site of attachment and number of carbohydrate units [12]. O-glycosylation sites have an additional setback of not being well defined and although attempts have been made to predict O-glycosylation sites, these predictions are not reliable. In addition, glycosylation and especially O-glycosylation are dynamic and change as a function of physiological conditions. Thus a specific site might be occupied under certain conditions and not under other conditions, resulting in a heterogeneous protein population in terms of occupied sites. In order to study the dynamics there is a need to be able to quantify the degree of occupancy at any given site under different conditions.

A specific interest of this laboratory is to develop methods compatible with high throughput proteomics to determine unknown glycan modifications of proteins. As a first attempt, efforts have been concentrated on development of methods compatible with sensitivity requirements in proteomics that allow determination of which proteins are glycosylated and which sites are occupied in these proteins. An overview of the strategy considerations prior to development of such methods is outlined in Fig. 2. The possible approaches encompass specific detection of glycoproteins in electrophoretic gels, selective enrichment of glycoproteins and

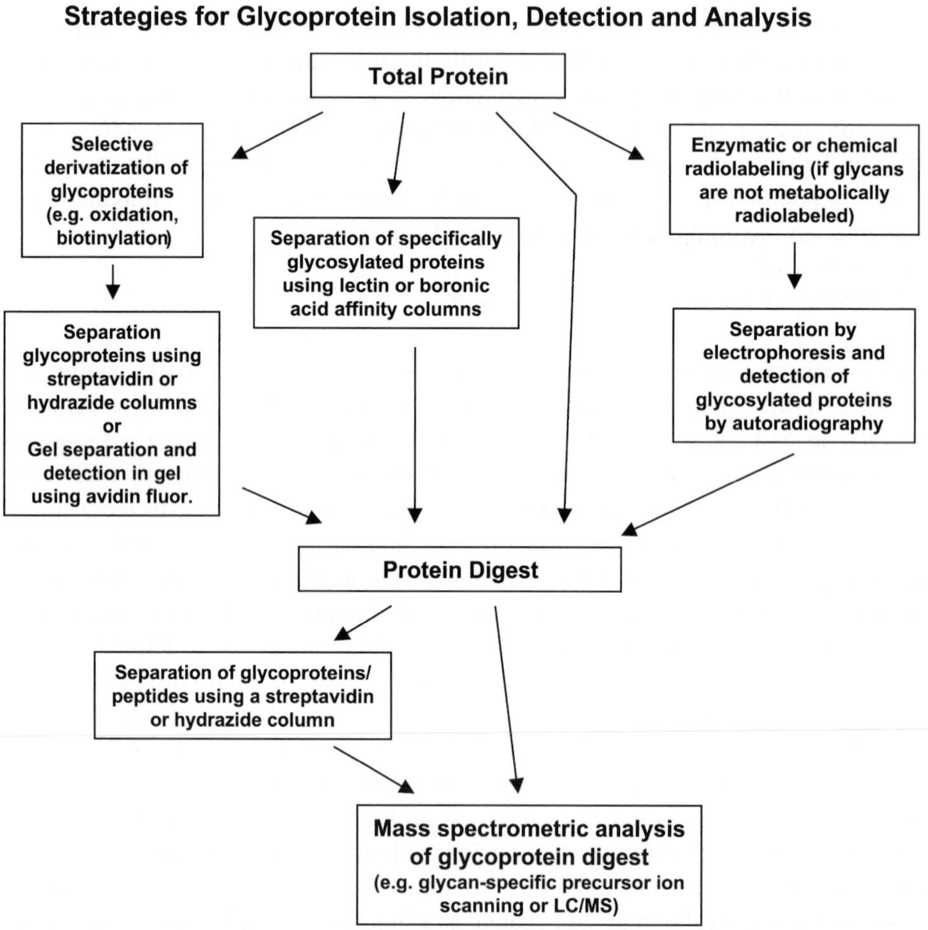

Fig. 2. Strategies for the specific detection, isolation and analysis of glycoproteins and glycopeptides.

glycopeptides and specific mass spectrometric detection of glycopeptides. A long term goal is to be able to determine and quantify all the possible glycan structures on each site as well as the degree of occupancy on each site.

2.2.1. Identification of glycosylated proteins by gel electrophoresis

Several different strategies can be followed to investigate which proteins are glycosylated. One strategy, which has not been used much, is in vivo metabolic labeling of glycans. This strategy can be used if cell cultures are utilized for pulse chase experiments using ^{14}C labeled sugars or other labeled sugars such as ^{35}S thiolated sugars. Radioactivity can be measured or SDS-PAGE followed by autoradiography can be used to analyze differential glycosylation under control and experimental conditions. Another labeling strategy that can be employed is non-metabolic labeling whereby enzymes such as galactosyl transferase add a stable or radioactive labeled sugar to an already present glycan [47]. This labeling can allow the protein to be detected using autoradiography or mass spectrometry.

A protein can often be identified as being glycosylated when it exhibits a diffuse banding pattern on 1D SDS-PAGE or when a train of multiple spots in the second dimension of a 2D gel can be observed. In order to verify this as being a glycosylated protein, the gel can be stained using a periodic-acid Schiff (PAS) stain (see Fig. 5). Alternatively, several commercially available stains and Western blotting kits with digoxigenin or biotin/streptavidin probes can be purchased from companies such as Molecular Probes and BioRad. The gel separation and subsequent glycoprotein detection methods are described in greater detail in a review by Packer et al. [24]. The proteins identified as glycoproteins using gel-based methods can then be subjected to analysis by liquid chromatography–mass spectrometry (LC–MS) in order to identify the protein in question. Unfortunately most labeling and staining methods such as PAS modify the glycan attached and therefore glycan structural information cannot always be derived. Once a protein is identified as being glycosylated, other strategies that were previously described can be followed; see Kuster et al. for a recent review [23]. The major disadvantage associated with the mass spectrometric methods described in the review is that they require larger amounts of protein than are usually available in proteomics studies. Therefore development of methods that allow selective enrichment and/or detection of the glycopeptides is needed.

2.2.2. Enrichment analysis and LC–MS of glycopeptides

As suppression effects are a major problem in MALDI or ESI of glycopeptides, any type of separation and enrichment of glycopeptides significantly improves the identification of these peptides. Enrichment analysis, whereby a specific type of modified peptide is selectively purified from a complex mixture of peptides has proved to be a successful strategy for analyzing other modifications such as phosphorylation [25]. In contrast, glycosylation analysis by mass spectrometry poses particular challenges. One of the problems experienced with glycosylated peptides is that ionization of glycosylated peptides is often suppressed in the presence of non-glycosylated peptides. When using nanoelectrospray, hydrophilic peptides, such as glycosylated peptides tend to partition into the aqueous phase, which is found in the center of the droplet and therefore the ionization efficiency of these peptides is lower than more hydrophobic peptides found at the surface of the nanodroplets where they are more easily desolvated and ionized. This problem becomes more severe when the mixture is extremely complex, e.g. whole cell lysate peptide mixtures. Glycopeptides also suffer ionization suppression when using MALDI; however, typically different glycopeptides are suppressed during MALDI ionization as compared with ESI, thus providing complimentary information. The ionization suppression experienced by glycopeptides is not the only factor contributing to the necessity of extremely sensitive methods while analyzing glycopeptides. Another factor is that a glycosylation site of one particular glycopeptide can be occupied by several different glycan groups with different masses. Thus, one peptide is distributed over a number of different *m/z* values, and therefore, the MS techniques used for the analysis of glycopeptides have to be extremely sensitive.

2.2.2.1. Lectin affinity chromatography. A number of separation methods can be used for the enrichment of glycopeptides. These include lectin affinity chromatography, boronic acid affinity chromatography, and PAS coupled affinity chromatography. There are several lectin affinity chromatography materials available (see Table 1) for the separation of glycoproteins from a complex protein lysate. Some lectin affinity columns have also been found to work with glycopeptides, although the majority of lectins only recognize glycoprotein epitopes. Listed in Table 1 are the sugars or glycan types that are recognized by the respective lectin affinity columns. Several lectins have been described as being effective for proteomics studies. One example is described for a *Caenorhabditis. elegans* study [26]. However, lectins are highly specific for certain types of glycans and only specific types of glycosylated peptides will be enriched. The other drawback encountered when working with lectins is that they often recognize specific protein/glycan epitopes and, as mentioned earlier, may not be as efficient at enriching for glycosylated peptides.

2.2.2.2. Boronic acid affinity chromatography. One major goal of this laboratory is to characterize glycosylated proteins from complex mixtures. In order to isolate a complete set of glycans from complex protein mixtures, the use of non-lectin affinity chromatography has been investigated. Enrichment analysis of glycopeptides from a complex mixture such as a whole cell lysate can greatly overcome

Table 1
Lectin types and affinities

Lectin	Specificity
ConA	Mannose, glucose, GlcNAc
AAA	Fucose
WGA	β-GlcNAc, sialic
DSA	GlcNAc
GNA	Mannose
MAA	Sialic
PNA	β-Gal
SNA	β-Gal, sialic
ECA	α and β GalNAc
LCH	α-Mannose, α-glucose, α-GlcNAc
SBA	α and β GalNAc,
Allo S	β-Gal
DBA	α GalNAc
Lotus	α-Fucose
UEA-1	α-Fucose

Selectivity of lectin columns: lectins are proteins which bind specific configurations of sugars. This specificity may be used for a variety of purposes, i.e. for affinity chromatography, to localize glycoproteins in gels and on Western blots and also to precipitate glycoproteins.

the problems encountered in glycosylation analyses that were described earlier. The ideal enrichment method would allow selective purification of all types of both O- and N-linked glycosylated peptides from a complex mixture of peptides. Phenylboronic affinity columns only require that the binding partner has a diol moiety and thus can be used for enrichment of various glycosylated peptides from a complex mixture. Most, if not all, glycosylated peptides contain a *cis*-diol that can bind to boronic acid affinity beads reversibly in a pH dependent manner. At a high pH boronic acid forms a cyclic ester with diols and at a low pH the diol is released. Affinity chromatography using an *m*-aminophenylborate polyacrylamide gel has been used for unbiased glycosylated protein enrichment [27] and it can be used for enrichment of glycosylated peptides as well. This method offers a universal glycosylated peptide enrichment, which is especially compatible with mass spectrometry because conditions for the binding and elution of glycoproteins and glycopeptides are gentle. For example, glycopeptides can be eluted from these columns using 5% formic acid and the eluate can be directly loaded onto a C18 reverse phase column or used for nanospray analysis if the peptide mixture is simple. The formation of the cyclic ester with the *cis*-diol is shown in Fig. 3. The boronic acid enrichment of glycopeptides is shown in Fig. 4 for urokinase plasminogen activated receptor kinase, a glycosylated GPI anchor protein [28].

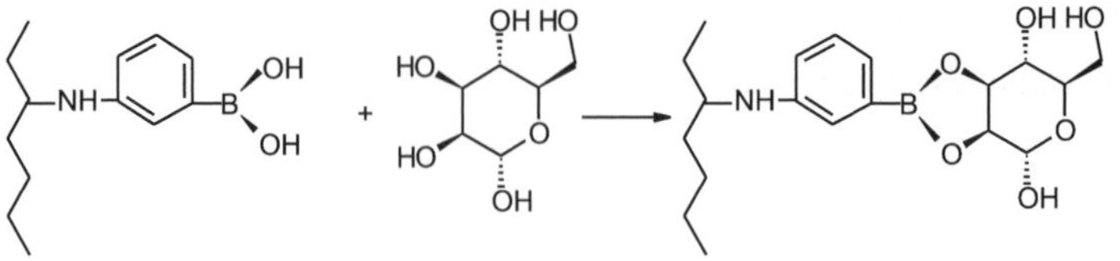

Fig. 3. The principle of boronic acid affinity purification of carbohydrates and glycoproteins/peptides. Glycosylated peptides contain a *cis*-diol that can bind to boronic acid affinity beads reversibly in a pH dependent manner. When glycosylated proteins or peptides are incubated with an *m*-aminophenylborate beads at pH boronic acid, a cyclic ester is produced with diols. On lowering the low pH the diol is released.

If a complex glycosylated peptide mixture is being analyzed, e.g. a boronic acid enriched fraction from a whole cell lysate, it is useful to use liquid chromatography tandem mass spectrometry (LC–MS/MS) for the analysis. During the LC–MS/MS analysis, conditions for fragmentation of the glycosylated peptides should be

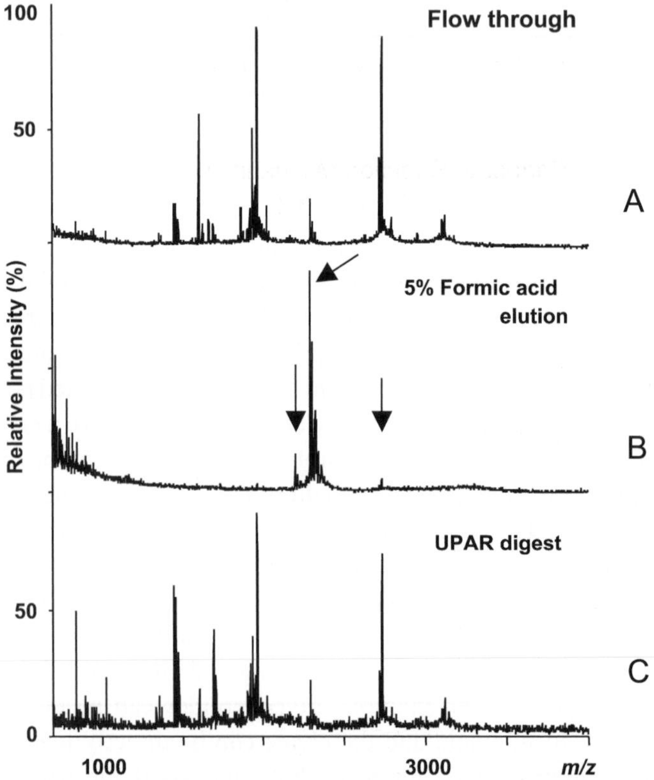

Fig. 4. MALDI spectra obtained during selective isolation of the glycopeptides from a tryptic digest of recombinant UPAR by affinity chromatography on a boronic acid microcolumn. (A) Flow through. (B) The glycopeptides (marked with arrows) eluted with formic acid solution. (C) The digest prior to affinity chromatography.

adjusted so that good fragmentation of the peptide backbone occurs to provide sufficient fragment ions for identification purposes using standard database search programs.

2.2.2.3. Periodic-acid-Schiff coupled affinity (PAS) chromatography

Affinity tags to selectively enrich glycoproteins and glycopeptides such as biotin or digoxigenin can be used. These affinity tags should be attached to the glycopeptide by utilizing a property unique to the glycan and not to the peptide backbone. This can be achieved by using the unique chemistry provided by the *cis*-diol groups found on sugars. A mild periodate reaction oxidizes the vicinal diols to aldehydes without affecting the peptide. A hydrazone coupling reaction can then be used to attach a tag such as iminobiotin to the glycan or to directly couple the glycosylated proteins to a hydrazine column. Using an iminobiotin tag is beneficial for a number of reasons: (i) the imino group added a positive charge to the glycopeptide; (ii) this group was used to provide signature precursor ions (*m/z* 226.10); (iii) the iminobiotin had a higher dissociation rate than biotin. The schematic for the periodate reaction is shown in Fig. 5 and

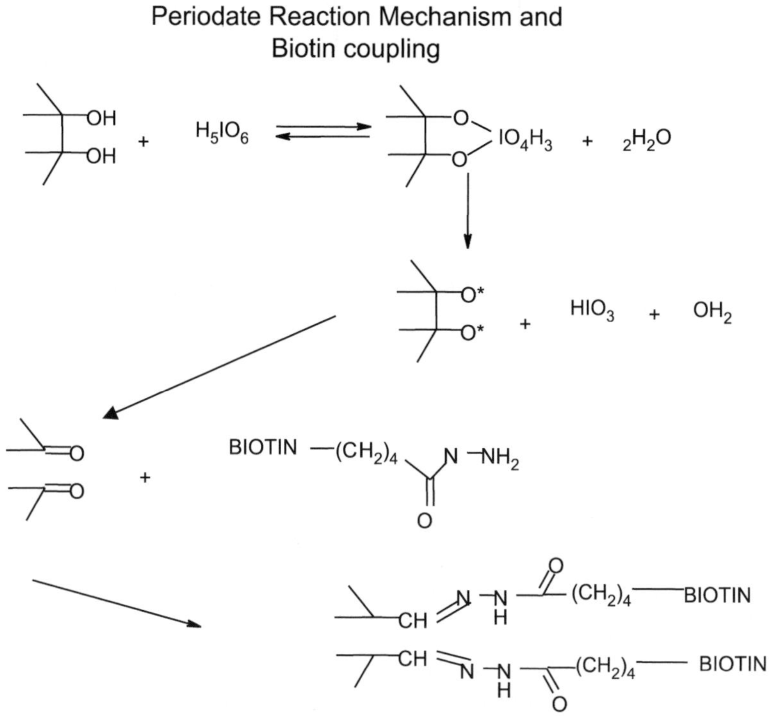

Fig. 5. The reactions involved in selective biotinylation of glycans.

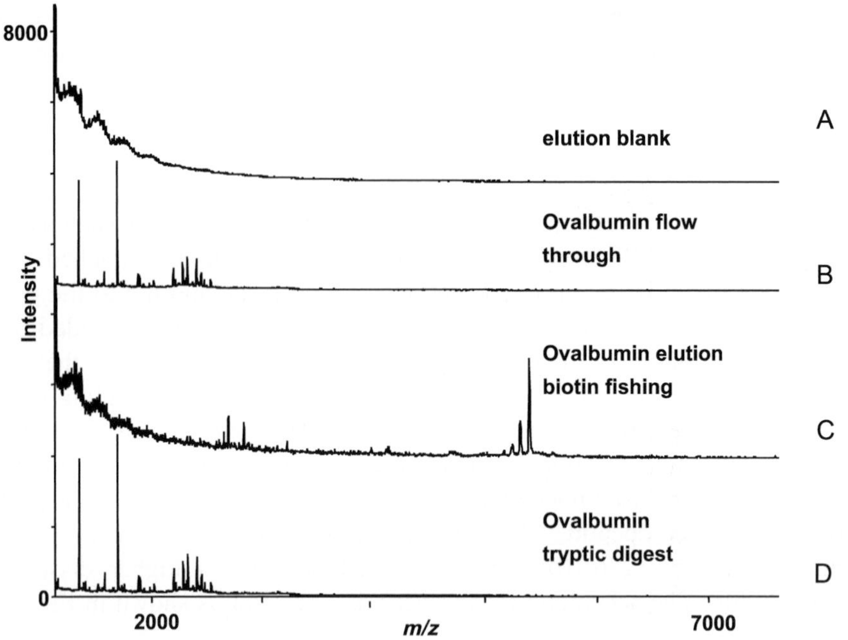

Fig. 6. MALDI spectra obtained in selective isolation of the glycopeptide from a tryptic digest of ovalbumin by affinity purification on a streptavidin microcolumn after iminobiotinylation of the glycan. (A) MALDI spectrum of a control blank elution. (B) Flow through after sample application. (C) Elution of the iminobiotinylated peptide with acid. (D) The ovalbumin digest prior to affinity chromatography.

an example of the derivatization and enrichment for ovalbumin is shown in Fig. 6. Coupling the glycoprotein/glycopeptide directly to a hydrazine column is useful if the only information required from a sample is site occupancy. The bound glycosylated moiety can be eluted from the column using PNGase F/H [28].

2.2.3. Precursor ion scanning

One approach that can be used to identify glycosylated peptides within complex peptide mixtures such as protein digests is precursor ion scanning for glycan derived fragment ions, i.e. the selective detection of precursor ions that gives rise to glycosylation-specific fragment ions. The main advantage of this approach is that no prior derivatization or enrichment strategies are absolutely necessary for its success. The characteristic fragment ions used for this purpose are normally the 'reporter' oxonium ions of hexose at *m/z* 163.06, of *N*-acetylhexosamines at *m/z* 204.08, and of hexoylhexosamine at *m/z* 366.14 [29], but much larger oxonium ions have also recently been used [30]. This can be performed in discovery type

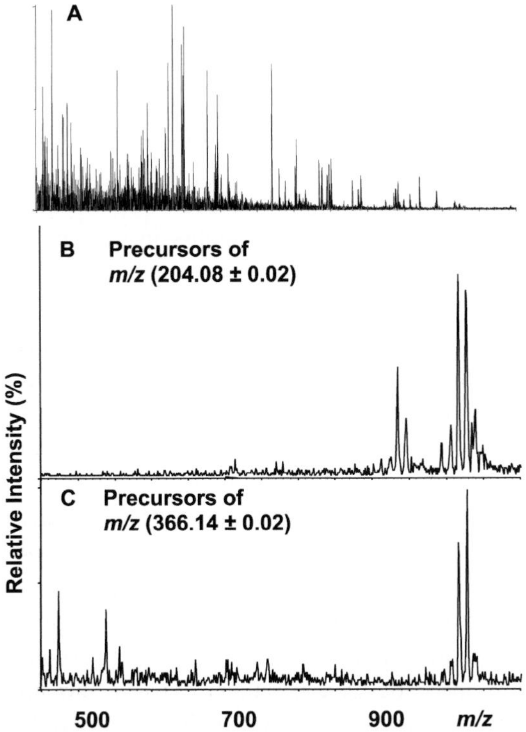

Fig. 7. Identification of a glycopeptide in a complex protein digest by high-resolution precursor ion scanning. (A) ESI spectrum of a protein digest doped with a small glycopeptide. (B) Precursor ion scan for the presence of HexNAc. (C) Precursor ion scan for the presence of HexNAc-Hex structures.

proteomics where selected ion traces for signature glycan fragments can be extracted from an LC–MS/MS run in order to identify the fractions containing glycosylated species [29]. Additional experiments are necessary in order to confirm the presence of glycopeptides and to obtain peptide and glycan information. The reporter fragment ions can also be used for (nano-) electrospray and precursor ion scanning in order to find the glycosylated peptides in the protein digest once it is established that the protein is glycosylated [31]. These precursor ion methods were found to be highly non-specific because interfering ions derived from the peptide backbone resulted in non-specificity. This limited the ability of precursor ion scanning from glycan identification. However, recent experiments showed that high resolution, high accuracy fragment ion selection on quadrupole TOF type instruments increases the specificity of this type of experiment as illustrated in Fig. 7 [49]. Using off-line nano-electrospray precursor ion scanning has the advantage that further MS/MS studies of the identified glycopeptides can be performed immediately after their identification, using appropriate conditions that allow one to obtain peptide sequence information, the glycosylation site and possibly glycan structure information.

2.2.4. Site-specific determination of O-glycosylation

The site-specific determination of O-glycosylation is particularly difficult for two reasons. First, O-glycosylation frequently occurs in regions very rich in serine, threonine, and proline (so-called mucin type sequences). In these regions there are many potential glycosylation sites of which only a few may be occupied and due to the nature of the amino acid sequences, proteolytic cleavage between the potential sites to generate appropriate peptides is often not possible. Second, attempts to identify the glycosylated sites by tandem mass spectrometry using collision-induced dissociation often fail because the O-glycosidic bond is fragmented prior to peptide bond fragmentation. A number of strategies have been developed to overcome these limitations. One takes advantage of the fact that the O-glycosidic bond is less labile than peptide bonds under acid hydrolysis conditions. It has been demonstrated that gas phase hydrolysis of the glycopeptides with pentafluoropropionic acid results in generation of a series of peptides forming a sequence ladder with partial conservation of the O-linked glycans. The resulting mass spectra are rather complex, but since the sequence is normally known, it has been demonstrated that it is possible to assign the O-glycosylated residues in peptides with up to more than 20 residues of which three contained an O-linked *N*-acetyl galactosamine. Partial loss of the acetyl group further facilitated the assignment of the number of glycosylated residues in each peptide [32]. Beta elimination of the O-linked glycans with methylamine in the gas phase results in generation of a new amino acid residue with a residue weight 13 Da above that of Thr or Ser for formerly O-glycosylated Thr and Ser residues. This modification is stable under CID conditions and thus allows identification of the O-glycosylated residues by tandem mass spectrometry [33]. Unfortunately, the structural information about the glycan is lost when this procedure is used. This can be obtained by analysis of the removed glycans [23], but with loss of the site specificity. Similar strategies can be used to identify single O-GlcNAc sites using beta elimination followed by derivatization with affinity tags [48]. Electron capture dissociation (ECD) is a new peptide fragmentation technique, which can be applied to multiply charged peptide ions in Fourier transform mass spectrometry (FTMS). It is selective for peptide bonds and does not affect the O-glycosidic bond, thus allowing the site-specific assignment of the O-linked glycan structures [34]. The specific assignment of O-glycosylated sites in 60-residues-long polypeptides has recently been demonstrated by FTMS combined with ECD [35]. Thus a number of methods are available for the site-specific assignment of O-linked glycan structures. However, none of these methods are presently suitable for a high throughput modification specific proteomics approach.

3. Tyrosine nitration

Nitration of tyrosine residues in proteins has been reported to take place in cells during oxidative stress and inflammation by the generation of peroxynitrite produced by the reaction of the superoxide anion in the presence of excess nitric oxide [36]. Based upon Western blotting, tyrosine nitration has been suggested to be involved in more than 60 human disorders [37]. A recent proteomics study using Western blotting with antibodies directed against 3-nitrotyrosine identified more than 40 proteins as being potentially tyrosine nitrated as a consequence of in vivo inflammatory response [38]. The proteins, most of which were not previously known to be nitrated, were identified by MALDI mass mapping. Unfortunately, attempts were not made to identify the peptides. Tyrosine nitration would probably not affect the electrophoretic properties of the proteins. Therefore, the nitrated proteins will probably not be separated from the non-nitrated proteins. They will also constitute only a minor fraction of the molecules in the protein spot in the gel. In addition, it is not known if in vivo tyrosine nitration is a residue specific or a random process. In the latter case, identification of the nitrated tyrosine residues will be extremely difficult at low levels of nitration. To investigate if in vivo nitration is random or limited to specific tyrosine residues, it is necessary to develop highly sensitive and selective methods to identify the specific nitration sites. Recent efforts in this laboratory have demonstrated that it is possible to localize nitrated tyrosine residues by precursor ion scanning analysis of the peptide mixture derived by in-gel digestion of chemically nitrated bovine serum albumin (BSA) [39]. The immonium ion for nitrotyrosine at *m/z* 181.06 was found to be a good reporter ion for nitrotyrosine and this allowed identification of the nitrotyrosine containing peptides. MS/MS spectra of the peptides giving rise to the ion at *m/z* 181.06 allowed identification of the nitrated tyrosine residues in cases where a peptide contained more than one tyrosine residue. In BSA only 4 out of 21 tyrosine residues were nitrated in vitro with tetranitromethane [39] in agreement with independent studies in which it was also observed that chemical nitration with a variety of nitrating reagents leads to limited site-specific nitration of proteins [40, 41]. An alternative procedure for determination of tyrosine nitration based upon conversion of nitrotyrosine to aminotyrosine followed by MALDI MS or LC–MS/MS analysis has recently been published [42,43]. In these studies, nitration of specific tyrosine residues was also observed. Thus, the present evidence indicates that tyrosine nitration independently of being induced in vivo or in vitro is a site-specific event indicating the need for further studies by modification-specific proteomics to understand its possible biological consequences.

4. Summary

Modification-specific proteomics requires development of a number of highly specific tools that allows analysis of PTMs in biological material. The strategies described in this and other chapters are just the first steps in this direction. The high degree of heterogeneity with respect to site and structure observed for many types of PTMs, especially glycosylation, immediately raises the questions: (i) what percentage of the peptides is modified at a given site and (ii) in cases where the prosthetic group itself is heterogeneous, what is the percentage distribution of the different forms. This latter question will be most relevant for glycoproteins and it has been argued that the intensities of the peaks representing the different glycoforms in mass spectra also are representative of the relative amounts of the species with the exception of sialic acid containing and sulphated glycans [44,45]. However, the biological importance of glycan heterogeneity, if any, is not well understood. O-linked glycosylation also recently has been demonstrated to be a highly controlled process involving several O-glycosyl transferases with different specificities [46] where the site-specific glycosylation of one must be preceded by the correct action of another. Is that a regulatory event? Many questions concerning the few PTMs dealt with in this chapter are still open, e.g. what is the biological function of glycan heterogeneity, how does the highly regulated site-specific O-glycosylation affect protein function, and is tyrosine nitration a biological meaningful event or just an artifact. Much more information on protein modification at a variety of levels must be generated by systematic proteomics studies to answer these and many other questions. At present, proteomics has given rise to more new questions than answers, and there is sufficient work for many investigators for a long time.

References

1. Wilkins, M.R., Pasquali, C., Appel, R.D., Ou, K., Golaz, O., Sanchez, J.C., Yan, J.X., Gooley, A.A., Hughes, G., Humphery-Smith, I., Williams, K.L. and Hochstrasser, D.F., From proteins to proteomes: large scale protein identification by two-dimensional electrophoresis and amino acid analysis. *Biotechnology*, **14**, 61–65 (1996).
2. Mewes, H.W., Albermann, K., Bahr, M., Frishman, D., Gkeissner, A., Hani, J., Heumann, K., Kleine, K., Maierl, A., Olover, S.G., Pheiffer, F. and Zollner, A., Overview of the yeast genome. *Nature*, **387(6632 Suppl.)**, 7–65 (1997).
3. Mann, M., Højrup, P. and Roepstorff, P., Use of mass spectrometric molecular weight information to identify proteins in sequence databases. *Biol. Mass Spectrom.*, **22**, 338–345 (1993).
4. Henzel, W.J., Billeci, T.M., Stults, J.T. and Wong, S.C., Identifying proteins from two-dimensional gels by molecular mass searching of peptide fragments in protein sequence databases. *Proc. Natl Acad. Sci. USA*, **90**, 5011–5015 (1992).

5. James, P., Quadroni, M., Carafoli, E. and Gonnet, G., Protein identification by mass profile fingerprinting. *Biochem. Biophys. Res. Commun.*, **195**, 58–64 (1993).
6. Eng, J.K., McCormack, A.L. and Yates, J.R., An approach to correlate tandem mass spectral data of peptides with amino acid sequences in a protein database. *J. Am. Soc. Mass Spectrom.*, **5**, 976–989 (1994).
7. Mann, M. and Wilm, M., Error tolerant identification of peptides in sequence databases by peptide sequence tags. *Anal. Chem.*, **66**, 4390–4399 (1994).
8. Lander, E. Whitehead Institute, MIT, personal communication.
9. International Human Genome Consortium, Lander, E.S., et al., Initial sequencing and analysis of the human genome. *Nature*, **409**, 860–921 (2001).
10. Krishna, R.G. and Wold, F., Post-translational modification of proteins. *Adv. Enzymol. Relat. Areas Mol. Biol.*, **67**, 265–298 (1993).
11. Mortz, E., Sareneva, T., Haebel, S., Julkunen, I. and Roepstorff, P., Mass spectrometric characterization of glycosylated interferon-gamma variants separated by gel electrophoresis. *Electrophoresis*, **17(5)**, 925–931 (1996).
12. Jensen, O.N., Modification-specific proteomics: strategies for systematic studies of post-translationally modified proteins, In: Blackstock, W. and Mann, M. (Eds.), *Proteomics: A Trends Guide*. Elsevier, London, 2000, pp. 36–42.
13. Apweiler, R., Hermjakob, H. and Sharon, N., On the frequency of protein glycosylation, as deduced from analysis of the SWISS-PROT database. *Biochim. Biophys. Acta*, **1473**, 4–8 (1999).
14. Hansen, J.E., Lund, O., Engelbrecht, J., Bohr, H., Nielsen, J.O. and Hansen, J.E., Prediction of O-glycosylation of mammalian proteins: specificity patterns of UDP-GalNAc:polypeptide *N*-acetylgalactosaminyltransferase. *Biochem. J.*, **308**, 801–813 (1995).
15. Bunn, H.F., Gabbay, K.H. and Gallop, P.M., The glycosylation of hemoglobin: relevance to diabetes mellitus. *Science*, **200**, 21–27 (1978).
16. Schaffer, C., Graninger, M. and Messner, P., Prokaryotic glycosylation. *Proteomics*, **1**, 248–261 (2001).
17. Benz, I. and Schmidt, M.A., Never say never again: protein glycosylation in pathogenic bacteria. *Mol. Microbiol.*, **45**, 267–276 (2002).
18. Schleip, I., Heiss, E. and Lehle, L., The yeast SEC20 gene is required for N- and O-glycosylation in the Golgi. Evidence that impaired glycosylation does not correlate with the secretory defect. *J. Biol. Chem.*, **276**, 28751–28758 (2001).
19. Wells, L., Vosseller, K. and Hart, G.W., Glycosylation of nucleocytoplasmic proteins: signal transduction and O- GlcNAc. *Science*, **291**, 2376–2378 (2001).
20. Dean, N., Asparagine-linked glycosylation in the yeast Golgi. *Biochim. Biophys. Acta*, **1426**, 309–322 (1999).
21. Van den Steen, P., Rudd, P.M., Dwek, R.A. and Opdenakker, G., Concepts and principles of O-linked glycosylation. *Crit. Rev. Biochem. Mol. Biol.*, **33**, 151–208 (1998).
22. Chou, H.H., Hayakawa, T., Diaz, S., Krings, M., Indriati, E., Leakey, M., Paabo, S., Satta, Y., Takahata, N. and Varki, A., Inactivation of CMP-*N*-acetylneuraminic acid hydroxylase occurred prior to brain expansion during human evolution. *Proc. Natl Acad. Sci. USA*, **99**, 11736–11741 (2002).
23. Kuster, B., Krogh, T.N., Mortz, E. and Harvey, D.J., Glycosylation analysis of gel-separated proteins. *Proteomics*, **1**, 350–361 (2001).
24. Packer, N.H., Ball, M.S. and Devine, P.L., Glycoprotein detection of 2-D separated proteins. *Methods Mol. Biol.*, **112**, 341–352 (1999).
25. Stensballe, A., Andersen, S. and Jensen, O.N., Characterization of phosphoproteins from electrophoretic gels by nanoscale Fe(III) affinity chromatography with off-line mass spectrometry analysis. *Proteomics*, **1**, 207–222 (2001).

26. Hirabayashi, J., Arata, Y. and Kasai, K., Glycome project: concept, strategy and preliminary application to *Caenorhabditis elegans*. *Proteomics*, **1**, 295–303 (2001).
27. Li, Y., Larsson, E.L., Jungvid, H., Galaev, I.Y.U. and Mattiasson, B., Affinity chromatography of neoglycoproteins. *Bioseparation*, **9**, 315–323 (2000).
28. Jebanathirajah, J., Steen, H., Borch, J., Stensballe, A., Jensen, O.N. and Roepstorff, P., *Integrated proteomics strategy for studying glycosylation and phosphorylation*, 50th ASMS Conference on Mass Spectrometry and Allied Topics, ASMS, Orlando, FL, 2002.
29. Carr, S.A. and Bean, M.F., Selective identification and differentiation of N- and O-linked oligosaccharides in glycoproteins by liquid chromatography–mass spectrometry. *Protein Sci.*, **2**, 183–196 (1993).
30. Ritchie, M.A., Gill, A.C., Deery, M.J. and Lilley, K., Precursor ion scanning for detection and structural characterization of heterogeneous glycopeptide mixtures. *J. Am. Soc. Mass Spectrom.*, **13**, 1065–1077 (2002).
31. Annan, R.S. and Carr, S.A., The essential role of mass spectrometry in characterizing protein structure: mapping posttranslational modifications. *J. Protein Chem.*, **16**, 391–402 (1997).
32. Mirgorodskaya, E., Hassan, H., Wandall, H.H., Clausen, H. and Roepstorff, P., Partial vapor-phase hydrolysis of peptide bonds: a method for mass spectrometric determination of O-glycosylated sites in glycopeptides. *Anal. Biochem.*, **269**, 54–65 (1999).
33. Mirgorodskaya, E., Hassan, H., Clausen, H. and Roepstorff, P., Mass spectrometric determination of O-glycosylation sites using beta-elimination and partial acid hydrolysis. *Anal. Chem.*, **73**, 1263–1269 (2001).
34. Mirgorodskaya, E., Roepstorff, P. and Zubarev, R.A., Localization of O-glycosylation sites in peptides by electron capture dissociation in a Fourier transform mass spectrometer. *Anal. Chem.*, **71**, 4431–4436 (1999).
35. Schwientek, T., Bennett, E.P., Flores, C., Thacker, J., Hollmann, M., Reis, C.A., Behrens, J., Mandel, U., Keck, B., Schäfer, M.A., Haselmann, K., Zubarev, R., Roepstorff, P., Burchell, J.M., Taylor-Papadimitriou, J., Hollingsworth, M.A. and Clausen, H., Functional conservation of subfamilies of putative UDP-*N*-acetylgalactosamine: polypeptide *N*-acetylgalactosaminyl-transferases in *Drosophila*, *Caenorhabditis elegans*, and mammals. *J. Biol. Chem.*, **277**, 22623–22638 (2002).
36. Beckman, J.S., Beckman, T.W., Chen, J., Marshall, P.A. and Freeman, B.A., Apparent hydroxyl radical production by peroxynitrite: implications for endothelial injury from nitric oxide and superoxide. *Proc. Natl Acad. Sci. USA*, **87**, 1620–1624 (1990).
37. Ischiropoulus, H., Biological tyrosine nitration: a pathophysiological function of nitric oxide and reactive oxygen species. *Arch. Biochem. Biophys.*, **356**, 1–45 (1998).
38. Aulak, K.S., Miyagi, M., Yan, L., West, K.A., Massillon, D., Crabb, J.W. and Stuehr, D.J., Proteomic method identifies proteins nitrated in vivo during inflammatory challenge. *Proc. Natl Acad. Sci. USA*, **98**, 12056–12061 (2001).
39. Petersson, A.S., Steen, H., Kalume, D.E., Caidahl, K. and Roepstorff, P., Investigation of tyrosine nitration in proteins by mass spectrometry. *J. Mass Spectrom.*, **36**, 616–625 (2001).
40. Yamakura, F., Taka, H., Fujimura, T. and Murayama, K., Inactivation of human manganese-superoxide dismutase by peroxynitrite is caused by exclusive nitration of tyrosine 34 to 3-nitrotyrosine. *J. Biol. Chem.*, **273**, 14085–14089 (1998).
41. Viner, R.I., Ferrington, D.A., Williams, T.D., Bigelow, D.J. and Schoneich, C., Protein modification during biological aging: selective tyrosine nitration of the SERCA2a isoform of the sarcoplasmic reticulum Ca^{2+}-ATPase in skeletal muscle. *Biochem. J.*, **340**, 657–669 (1999).
42. Sarver, A., Scheffler, N.K., Shetlar, M.D. and Gibson, B.W., Analysis of peptides and proteins containing nitrotyrosine by matrix-assisted laser desorption/ionization mass spectrometry. *J. Am. Soc. Mass Spectrom.*, **12**, 439–448 (2001).

43. Miyagi, M., Sakaguchi, H., Darrow, R.M., Yan, L., West, K.A., Aulak, K.S., Stuehr, D.J., Hollyfield, J.G., Organisciak, D.T. and Crabb, J.W., Evidence that light modulates protein nitration in rat retina. *Mol. Cell. Proteom.*, **1**, 293–303 (2002).
44. Mørtz, E., Sareneva, T., Julkunen, I. and Roepstorff, P., Does matrix-assisted laser desorption/ionization mass spectrometry allow analysis of carbohydrate heterogeneity in glyco-proteins? A study of natural human interferon-γ. *J. Mass Spectrom.*, **31**, 1109–1118 (1996).
45. Talbo, G. and Roepstorff, P., Determination of sulfated peptides via prompt fragmentation by UV matrix assisted laser desorption/ionization mass spectrometry. *Rapid Commun. Mass Spectrom.*, **7**, 201–204 (1993).
46. Wandall, H.H., Hassan, H., Mirgorodskaya, E., Kristensen, A.K., Roepstorff, P., Bennett, E.P., Nielsen, P.A., Hollingsworth, M.A., Burchell, J., Taylor-Papadimitriou, J. and Clausen, H., Substrate specificities of three members of the human UDP-*N*-acetyl-α-D-galactosamine: polypeptide *N*-acetylgalactosaminyltransferase family, GalNAc-T1, -T2, and -T3. *J. Biol. Chem.*, **272**, 23503–23514 (1997).
47. Roquemore, E.P., Chou, T.Y. and Hart, G.W., Detection of O-linked *N*-acetylglycosamine (O-GlcNAc) on cytoplasmic and nuclear proteins. *Methods Enzymol.*, **230**, 443–460 (1994).
48. Wells, L., Vosseller, K., Cole, R.N., Cronshaw, J.M., Matunis, M.J. and Hart, G.W., Mapping sites of O-GlcNAc modification using affinity tags for serine and threonine post-translational modifications. *Mol. Cell. Proteom.*, **1**, 791–804 (2002).
49. Jebanathirajah, J.A., Steen, H. and Roepstorff, P., Using optimized collision energies, high resolution, high accuracy fragment ion selection to improve glycopeptide detection by precursor ion scanning. *J. Am. Soc. Mass Specrom.*, **14**, 777–784 (2003).

Proteome Analysis. Interpreting the Genome.
D.W. Speicher (editor)

Chapter 6

Phosphoproteomics: mass spectrometry based techniques for systematic phosphoprotein analysis

OLE NØRREGAARD JENSEN*,[1]

Department of Biochemistry and Molecular Biology, University of Southern Denmark, DK-5230 Odense M, Denmark

* Tel.: +45-6550-2368; fax: +45-6550-2467. E-mail: jenseno@bmb.sdu.dk (O.N. Jensen).
[1] www.protein.sdu.dk.

1. Introduction

All processes in living organisms are based on specific, spatially and temporally regulated interactions between biological molecules, including lipids, carbohydrates, proteins and nucleic acids. Proteins interact with each other and with other types of biomolecules in a controlled fashion to facilitate basic processes such as metabolism, reproduction, development and defense. A majority of these intricate processes are modulated and controlled by post-translational modifications of proteins.

The complete characterization of the proteome of a given organism, tissue or organelle must necessarily include identification and detailed structural analysis of the dynamically changing populations of post-translationally modified gene products. This ambitious task requires the development of sensitive and specific analytical techniques aimed at detection, identification and quantitation of modified amino acid residues in proteins. Mass spectrometry is the method of choice for detailed structural analysis of modified proteins because accurate mass determination at the protein, peptide or amino acid residue level reveals the presence of any post-translational modifications, which leads to a change in the molecular mass of the species in question. Combinations of genetic, biochemical and immunological techniques with mass spectrometry-based methods result in novel quantitative analytical strategies that may soon allow researchers to study the dynamics of complex cellular processes.

Reversible protein phosphorylation is among the best characterized of the hundreds of known post-translational modifications and it plays a pivotal role in cellular development, differentiation, growth and homeostasis. Phosphoproteins are involved in all biochemical processes in the cell, including transcriptional and translational regulation, metabolism, protein degradation, cellular signaling and communication, and in maintaining the structural integrity of cells.

Phosphorylation of a protein encompasses the covalent attachment of phosphate groups to amino acid residues. In a given phosphoprotein, only specific amino acid residues in defined sequence motifs or structural domains are phosphorylated. In eukaryotes, serine, threonine or tyrosine residues are the commonly phosphorylated residues, but other amino acids may also be modified. The phosphorylation process is catalyzed by specific and highly regulated protein kinases. The reverse process of removing phosphate groups from proteins is equally important and strictly controlled and is catalyzed by specific phosphoprotein phosphatases. The human genome contains on the order of 500 genes ($\sim 2\%$ of the genome) that encode protein kinases and about 170 genes that encode phosphoprotein phosphatases. ATP or GTP commonly donates the phosphate group in kinase-catalyzed phosphorylation reactions. The phosphorylation event may activate an enzyme, modulate interactions among protein units during the assembly of a functional molecular complex, or in the case of transcription factors affect the binding kinetics of proteins to specific nucleic acid sequences.

The crucial role of protein phosphorylation and dephosphorylation events is emphasized by the fact that protein kinases, phosphoprotein phosphatases and their respective substrates are implicated in a number of human diseases and ailments, including diabetes and many types of cancer [1,2]. A more detailed understanding of phosphorylation events is of major relevance for biomedical, biotechnological and pharmaceutical research. Thus, there is a growing interest in investigating at the level of the proteome the complex molecular networks underlying cellular metabolism and signaling. In such an endeavor there is an obvious need to study protein phosphorylation at a global scale, i.e. to investigate the dynamics of the phosphoproteome in living systems. Phosphoproteomics, i.e. the systematic identification as well as qualitative and quantitative characterization of phosphoproteins in cells, tissues or organisms is a rapidly growing field of research which requires parallel, yet integrated developments at the analytical, biological and informatics levels. This chapter outlines some of the current mass spectrometry driven strategies for phosphoprotein and phosphoproteomics research [3–11].

2. Modification-specific proteomics and phosphoproteomics

Protein modification reactions are rarely complete at any given amino acid residue in a protein meaning that a gene product may be present in the unmodified form and several different modified forms at the same time. The sub-stoichiometric quantities of modified protein calls for highly sensitive analytical techniques as well as for specific enrichment of these species. Thus, the approaches pursued for analysis of post-translationally modified proteins in this laboratory include selective and specific methods for detection and enrichment of these species. The general strategy for 'modification-specific proteomics' [10] includes: (i) detection and visualization of a given class of post-translationally modified proteins; (ii) enrichment/purification of this population of modified proteins or peptides; and (iii) detection, identification and characterization of the modified proteins and peptides by mass spectrometry.

In the case of phosphoproteins, combinations of physicochemical, biochemical and immunological techniques may be applied at all three strategy levels as depicted in Fig. 1. Phosphoproteomics experiments are conveniently performed in a 'differential display' setup, where a perturbed biological system is compared to a reference system. The number of proteins studied in a proteomic investigation may be in the range of just a few phosphoproteins present in purified macromolecular complexes to the whole complement of phosphoproteins present in a cell lysate or tissue homogenate. The analytical strategy may rely on

1. Detection/visualization of phosphoproteins

METHOD	*EXAMPLE*
Metabolic labeling	Radiolabeling(^{32}P, ^{33}P) and imaging
	Stable isotope labels (^{15}N,^{2}H,^{13}C) and MS
Antibodies	Western blotting (α-pY, α-pS, α-pT)
Chemical methods	Phosphate-specific reagents (β-elimination)
Enzyme-based methods	Kinases, phosphatases

2. Affinity-based enrichment of phosphoproteins and phosphopeptides

METHOD	*EXAMPLE*
Epitope binders	Immunoprecipitation (α-pY, α-pS, α-pT)
Affinity capture	Chelators (IMAC, Ga^{3+}, Fe^{3+})
Affinity tags	β-elimination chemistry
	Phosphoramidate chemistry
	Incorporation of affinity handle (e.g. biotin)

3. Phosphate-specific or -selective mass spectrometry techniques

METHOD	*EXAMPLE*
Product ion scan	Direct observation of pS, pT, pY residues in MS/MS spectra
Precursor ion scan	Negative ion. *m/z* 79, 63, 97 (pS, pT, pY)
	Positive ion mode: m/z 216.04 (pY)
Neutral loss scan	Loss of H_3PO_4, 98 Da
Quantitation:	
Stable isotope labeling	Mass signatures (^{15}N, ^{2}H, ^{13}C)

Fig. 1. Approaches for phosphoprotein analysis by mass spectrometry. The three conceptual levels of phosphoproteomics encompasses: (1) detection and visualization of intact phosphoproteins; (2) enrichment of phosphoproteins or phosphopeptides; and (3) specific and/or selective detection and identification of phosphorylated species by mass spectrometry techniques. A number of useful analytical techniques are listed.

'classical' 2D gel-based proteomics or the emerging mass spectrometry driven techniques such as multidimensional capillary liquid chromatography coupled to tandem mass spectrometry (LC–MS/MS) (Fig. 2).

The combination of technologies for detection, enrichment and structural characterization of phosphoproteins leads to improved sensitivity, specificity and selectivity in the phosphoproteome analysis. Robustness and dynamic range are also critical parameters of any analytical strategy.

2.1. Detection and visualization of phosphoproteins

The first analytical level includes the specific detection of intact phosphoproteins present in protein mixtures, such as cell lysates or partially purified protein fractions. A separation technique, such as SDS-PAGE or 2DE is often used in combination with either ^{32}P labeling and autoradiography/phosphorimaging or immunodetection by western blotting using phosphoamino acid specific antibodies (Figs. 1 and 2).

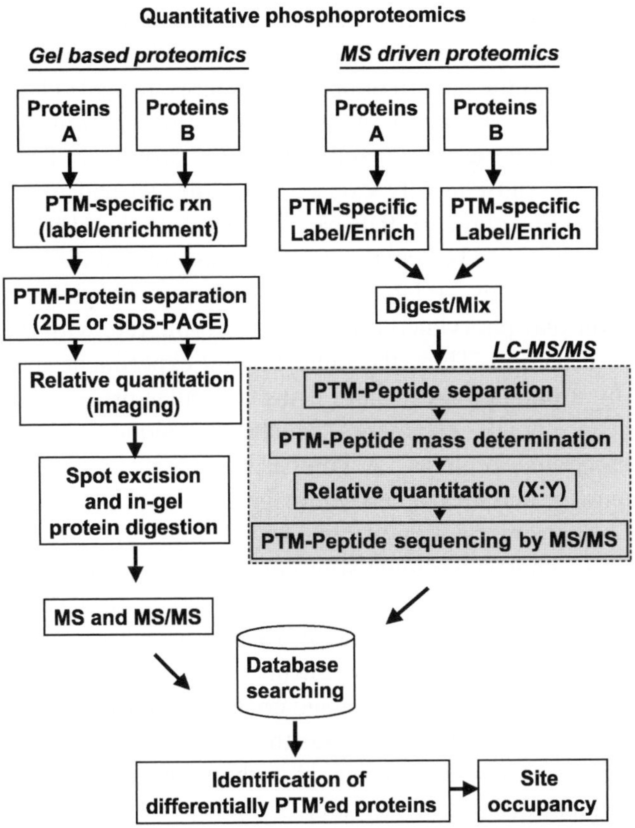

Fig. 2. Analytical strategies for phosphoprotein analysis. Left panel: proteomic analysis based on gel electrophoresis. Phosphoproteins are labeled with a phosphate-specific marker, e.g. 32-P. Protein samples are then separated on individual 2DE gels and the resulting phosphoprotein patterns are compared by image analysis to detect differentially expressed and differentially phosphorylated proteins. Proteins of interest are then identified by mass spectrometry and database searching. Detailed MS and MS/MS analysis may reveal phosphorylation sites. Right panel: proteomic analysis based on liquid chromatography interfaced to tandem mass spectrometry. Two protein populations are differentially labeled with a modification-specific mass tag or peptide mass tag, e.g. by beta-elimination/Michael addition (Fig. 3). The two protein populations are mixed and digested and then analyzed by LC–MS/MS. Due to the mass tag the relative amounts of differentially modified peptides may be determined. Peptides are sequenced by MS/MS and identified, allowing determination of modified amino acid residues.

Radiolabeling of proteins by ^{32}P phosphate is a highly sensitive and specific assay for selective detection of phosphoproteins. The radiolabeling experiment may be performed in a reconstituted system (in vitro) or in living cells (in vivo). ^{32}P or ^{33}P in the orthophosphate position is usually the phosphate donor in a phosphorylation reaction which may be catalyzed by recombinant kinases (in vitro), by a specific overexpressed kinase (in vivo) or by all endogenous kinase activities in the biological system in question (in vivo). Following separation of proteins by SDS-PAGE or 2DE, autoradiography or phosphorimaging is used to read out the pattern of phosphoproteins in a gel. These methods can also be applied to the detection of phosphoprotein containing fractions after centrifugal or chromatographic separation of proteins. The radiolabeled phosphoproteins are subsequently excised from the gel and identified by mass spectrometry. One advantage of 2DE-based methods is that different phosphorylated forms of proteins can be separated due to concomitant differences in charge states. One caveat of using ^{32}P labeling for visualizing phosphoproteins in gels is that abundant non-phosphorylated proteins may co-migrate with the phosphoprotein but escape detection. Thus, the radiolabeling experiment should always be accompanied by a parallel experiment using a regular protein staining method, such as fluorescence or silver staining. The latter techniques will reveal a majority of non-phosphorylated proteins. A comparison of the two patterns, i.e. the autoradiogram image and the stained gel image, allows an evaluation of the extent and efficiency of phosphorylation of proteins.

Visualization of phosphoproteins after gel electrophoretic separation can also be achieved by Western blotting using anti-phosphoamino acid antibodies [12]. Antiphosphotyrosine antibodies, such as 4G10 [13], have been found very efficient for this purpose. Western blotting experiments should be interpreted with care and always in the context of carefully designed control experiments to eliminate false positive and false negative results. As was the case for ^{32}P labeling, co-migrating proteins may obscure low-abundance phosphoproteins and interfere with or exclude correct identification of the latter.

Specific detection of phosphoproteins in electrophoretic gels using chemical approaches has been reported. For example, a commercial kit available from Pierce uses beta-elimination to release phosphate groups that in turn are visualized by complexation with a colored reagent. The efficiency and sensitivity of this method for phosphoproteome investigations remains to be established. It is only applicable to pSer and pThr containing proteins because pTyr is resistant to beta-elimination (see later). A biochemical approach based on differential display 2D gel electrophoresis (see Chapter 3) incorporates alkaline phosphatase treatment. Phosphoproteins typically exhibit a horizontal pearls-on-string pattern on 2DE gels due to the various states of phosphorylation. After phosphatase treatment the phosphoprotein spots will collapse into a single spot. Thus, a comparison of 2DE gels obtained from a phosphatase-treated sample and an untreated sample can reveal phosphoproteins [14]. An added advantage of this proteomic approach is that protein identification may be performed using the treated sample where the sample heterogeneity is minimized, thereby increasing sensitivity at the expense of details on the sites of phosphorylation.

These methods for visualization of phosphoproteins by SDS-PAGE or 2DE of whole cell lysates are appealing due to their sensitivity and relatively high specificity. However, they require experimental skills, special precautions (^{32}P) and have a number of inherent limitations. The limited resolution and dynamic range of 2DE gels for isolation of low-abundance phosphoproteins from cell lysates is an issue which can be improved upon by using narrow pH range gels [15] (see Chapter 2) or by enrichment of phosphoproteins prior to electrophoretic separation as described later.

2.2. Enrichment of phosphoproteins and phosphopeptides

The second analytical level of phosphoproteomics encompasses the enrichment of phosphoproteins from crude protein mixtures such as cell lysates (Fig. 1). The aim of this second tier of analysis is to secure sufficient protein material for identification of the cognate gene and preferably also characterization of the gene product. A number of research groups have pursued biochemical, immunological and chemical analytical strategies designed to target phosphoproteins in crude protein mixtures derived from cell lines or tissues.

2.2.1. Immunoprecipitation

Selective purification or enrichment of phosphoproteins by immunoprecipitation is a viable method for detailed investigation of cellular signaling processes. This approach relies on recognition of phospho-specific epitopes in the protein. Anti-phosphoamino acid antibodies are valuable reagents for immunoprecipitation of intact phosphotyrosine-containing proteins [13], and as was shown recently, phosphoserine/threonine containing proteins can be enriched by this approach [16]. Immunoprecipitation experiments should be performed in a differential display setup to compare a reference system to a perturbed system, e.g. growth factor stimulated cells vs. resting cells [13] or cells that are treated or not treated, respectively, with a phosphatase inhibitor, such as calyculin A [16]. The precipitated proteins are subsequently separated by SDS-PAGE or 2DE and identified by mass spectrometry methods.

Several groups have reported the use of phosphotyrosine antibodies for enrichment of phosphopeptides following proteolytic degradation of phosphoproteins [17,18]. The recovered phosphotyrosine-peptides were subsequently identified by mass spectrometry.

2.2.2. Immobilized metal affinity chromatography

Affinity-based enrichment of phosphopeptides is yet another method to isolate these species prior to their characterization by mass spectrometry. Phosphate groups exhibit a preference for binding to positively charged metal ions, such as Fe^{3+} and Ga^{3+} [19–25]. Thus, immobilized metal affinity chromatography

(IMAC) using Fe^{3+} or Ga^{3+} has proved to be a useful tool for phosphopeptide enrichment prior to, for example, tandem mass spectrometric analysis for phosphopeptide sequencing [23,24]. The IMAC approach is also amendable to isolation of certain proteins, but has not yet been useful for general enrichment of intact phosphoproteins from crude cell lysates (Gronborg, Pandey and Jensen, unpublished results).

Nevertheless, the sensitivity and selectivity of miniaturized IMAC procedures for phosphopeptide enrichment enables detailed analysis of phosphoproteins recovered from electrophoretic gels [23,24] or recovery of phosphopeptides from enzymatically digested protein preparations, including whole cell lysates [26] or cellular organelles [27]. The IMAC method should be used with some caution as the resin may bind significant amounts of non-phosphopeptides, either due to overloading or suboptimal loading conditions. It has been reported that O-methylesterification of free carboxyl groups can minimize binding of non-phosphopeptides to IMAC thereby improving the selectivity of this method [26]. The eluted phosphopeptide candidates should in all cases be analyzed by MS and MS/MS in an offline or online setup to confirm the presence of phosphorylated residues.

2.2.3. Chemical derivatization and tagging

The enrichment methods based on recognition of intact phosphate epitopes on proteins and peptides have been complemented by several methods that are based on chemical derivatization of the phosphoamino acid to a neutral or functional entity.

A widely used method involves conversion of phosphoserine and phosphothreonine to dehydroalanine and dehydroaminobutyric acid, respectively, by beta-elimination under alkaline conditions [28] (Fig. 3). The dehydroamino acid product generated by beta-elimination can be derivatized by a Michael addition reaction to generate molecules that are selectively tagged on phosphoamino acids [29,30]. In one case, the phosphopeptides were converted to acylated peptides, which were efficiently detected by their mass signatures in MALDI-TOF-MS [31]. An elegant implementation of a chemical derivatization procedure aimed at phosphoproteome analysis used beta-elimination and a biotin-derivative as the nucleophile for the Michael addition reaction [32,33] (Fig. 3). Phosphoproteins were tagged with biotin and then purified using avidin chromatography followed by identification by mass spectrometry. This approach facilitated recovery and identification of several abundant phosphoproteins from a yeast cell extract [33], but did not provide any information on phosphorylation sites. Further optimization of the chemistry for improving specificity and sensitivity should make this technique generally applicable to phosphoproteome analysis.

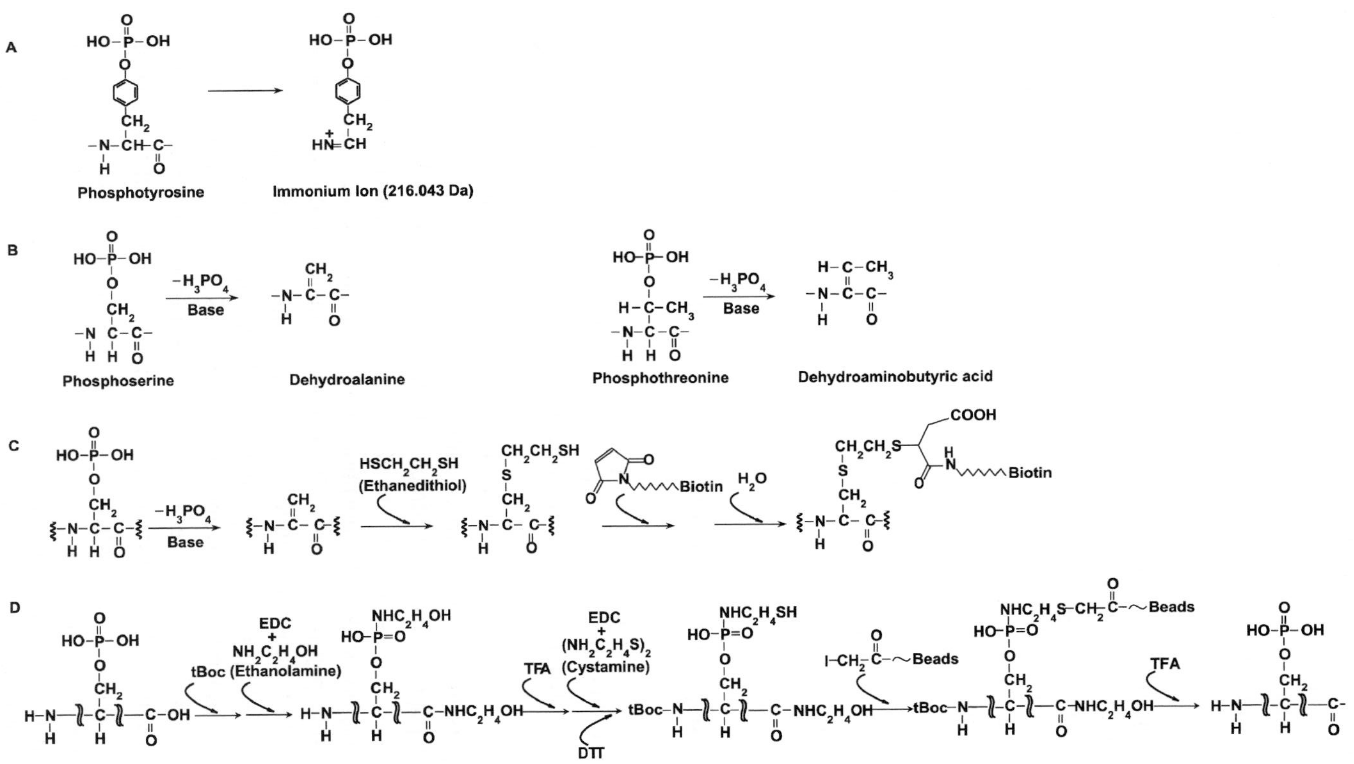

Fig. 3. Conversion or immobilization of phosphoamino acids and phosphopeptides by chemistry-based methods. (A) The phosphotyrosine immonium ion acts as reporter ion in MS/MS precursor ion scans for determination of phosphotyrosine containing peptides. (B) Base catalyzed beta-elimination of phosphoric acid from pSer and pThr residues. A similar reaction takes place in the gas phase in the mass spectrometer to generate the corresponding dehydroamino acid residues. (see also Fig. 4). (C) Biotinylation of phosphopeptides after beta-elimination and Michael addition of ethanedithiol. The biotinylated species are subsequently recovered by using immobilized avidin. (D) Immobilization of phosphopeptides/protein using phosphoramidate chemistry. The immobilized phopshopeptides are recovered by acid hydrolysis.

The derivatized phosphopeptide species are usually more amendable to tandem mass spectrometry relative to native phosphopeptides because of removal of the negatively charged and somewhat labile phosphate group. The beta-elimination/-Michael addition protocol is applicable to pSer and pThr and necessarily requires some sample handling and desalting/concentration steps prior to mass spectrometry. The sensitivity and selectivity of this method also relies on quantitative conversion of the phosphoamino acids to homogenous derivatives, which can be a challenge in the case of phosphothreonine, particularly when working at nanogram protein levels.

A potentially more versatile method was introduced by Aebersold and co-workers and includes conversion of phosphoamino acids into a phosphoramidate derivative [34] (Fig. 3). This method is, in principle, applicable to proteome wide screening of proteins containing pSer, pThr and pTyr residues as was demonstrated for a yeast whole cell lysate. One disadvantage of this approach is the need for protection of labile amino acid residues and requirements for chromatographic desalting of samples between reactions. As with alternative methods, further optimization and refinement is required before this technique can be applied as a general tool in phosphoproteomics.

3. Detection and sequencing of phosphopeptides by mass spectrometry

Although mass spectrometry is a highly specific and sensitive analytical technique for protein and peptide structure analysis, it is not a trivial task to analyze post-translationally modified peptides and proteins by this method. This is mainly because the inherent properties of the modifying moiety, in this case phosphorylation, affect the physicochemical characteristics of peptides. The phosphate group is negatively charged and therefore influences the ionization efficiency of the peptide and also affects the thermodynamic behavior of the peptide in solvents used in ESI MS and in the matrix used in MALDI MS. This often leads to 'suppression' of phosphopeptide ion signals in mass spectra obtained from crude peptide mixture. Phosphorylation is usually sub-stoichiometric meaning that phosphopeptides are less abundant than non-modified peptides from the same protein. Very small (<700 Da) phosphopeptides may be lost during sample preparation or escape detection by MS whereas large (>3 kDa) multiphosphorylated peptides are difficult to sequence by tandem mass spectrometry because of the lability of the phosphate groups and the complex fragmentation behavior. It is therefore usually advantageous to use several different proteases for generation of peptides, if sample amounts allow this approach. Problems associated with low stoichiometry and suppression effects of phosphopeptides present in complex peptide mixtures can be circumvented by applying enrichment techniques such as IMAC prior to MS analysis, and by using separation techniques like reversed phase capillary chromatography coupled on-line to ESI or offline to MALDI tandem mass spectrometry (LC–MS/MS). In most cases, it is necessary to use a combination of analytical methods, including various chromatographic methods and ESI and MALDI mass spectrometry, to identify all phosphorylation sites in a protein [24,35–37]. In the following sections, we will describe a series of such methods that can be applied to phosphoprotein and phosphoproteome analysis.

3.1. Phosphoprotein analysis by MALDI mass spectrometry

MALDI mass spectrometry is a sensitive method for peptide mass determination. Sample preparation encompasses enzymatic or chemical degradation of the phosphoprotein to peptides. Phosphoproteins isolated by SDS-PAGE or 2DE should be processed according to an established in-gel digestion protocol, for example, as reported by Shevchenko et al. [38]. Phosphopeptides are identified by MS because their mass is 80 Da or multiples of 80 Da higher than the mass of the unmodified peptide. MALDI MS usually enables direct characterization of phosphopeptides present in pure form or in simple mixtures when peptide amounts on the order of 0.5–5 pmol are available. It is substantially more challenging to characterize complex peptide mixtures derived by proteolytic digestion of low picomole to subpicomole levels of phosphoproteins by this method. Phosphopeptides may generate low intensity ion signals or be completely absent from MALDI mass spectra because of the suppression effects mentioned in previous section. The applied sample preparation method will determine the success of the experiment, and it is often necessary to try a series of different sample preparation techniques in order to identify and characterize phosphopeptides. Analysis of phosphopeptides in the negative ion mode may improve the ion detection efficiency [39,40]. Addition of ammonia reagents to the MALDI matrix has been reported to improve the ion signal for model phosphopeptides thereby improving sensitivity [41]. Sample concentration and desalting techniques, such as ZipTips or custom nanoscale columns are very useful and enable co-deposition of sample and matrix onto the MALDI probe [42,43]. Enrichment and recovery of phosphopeptides is possible by using IMAC resin in a nanoscale format [24] leading to improved selectivity and sensitivity in phosphoprotein characterization.

The observation of a neutral loss of H_3PO_4 (98 Da) from ions in MALDI reflector TOF instruments is a good indicator for the presence of phosphopeptides. However, it should be noted that the neutral loss will be detected as a metastable ion which does not follow the calibration curve of intact peptides. The neutral loss is therefore often observed as an unresolved ion signal with an apparent mass deficiency of 80–90 Da relative to the intact phosphopeptide. Only when the MALDI instrument is operated in the post-source-decay (PSD) mode will the neutral loss appear correctly as a 98 Da mass decrease. An additional advantage of this method is that the amino acid sequence information revealed in a PSD experiment may localize the exact phosphorylation site [44–47].

The presence of phosphopeptides in complex mixtures can also be investigated by using a combination of differential alkaline phosphatase treatment and MALDI MS [24,48–50]. Removal of phosphate groups by the phosphatase will lead to a mass change of the affected phosphopeptides relative to the spectrum obtained from the unprocessed sample. In addition, the dephosphorylated peptides usually produce more intense ion signals than the phosphorylated species, leading to improved sensitivity.

3.2. Phosphoprotein analysis by MALDI tandem mass spectrometry

A number of MALDI tandem mass spectrometry instruments have been developed over the past few years. Such instruments are very useful for phosphoprotein analysis because they combine the simplicity of MALDI sample preparation with peptide mass determination and peptide sequencing on the same instrument. Peptide fragmentation is achieved by either low or high-energy collision induced dissociation, depending on the system configuration. For example, MALDI ion trap, MALDI QTOF and MALDI TOF–TOF tandem mass spectrometers have been used for phosphopeptide analysis [51–54]. One advantage of these instruments is that the same sample can be reanalyzed multiple times because the sample consumption in MALDI MS is rather low. The mass determination can be temporarily separated from the amino acid sequencing experiments by minutes or hours because the MALDI sample deposit is stable and can be stored for days. Thus, it is possible to obtain MALDI MS data, interpret the data in due time and then re-measure candidate-modified peptides by MALDI tandem mass spectrometry. The combination of nanoscale separation techniques with MALDI MS/MS is a promising concept for characterization of post-translationally modified peptides. In this laboratory, nanoscale IMAC is used in combination with MALDI MS/MS for isolation, detection and sequencing of phosphopeptides from crude mixtures [52]. One advantage of MALDI MS/MS is that the neutral loss of phosphoric acid (98 Da, beta-elimination) from pSer and pThr can be induced by collision-induced dissociation in the instrument to identify phosphopeptide species in mixtures [52].

3.3. Phosphoprotein analysis by ESI tandem mass spectrometry

Electrospray ionization MS and MS/MS is applicable to phosphopeptide analysis using the same or similar strategies as described in the previous sections for MALDI MS and MS/MS. ESI differ from MALDI in several fundamental aspects. In ESI the peptide and protein ions are continuously produced whereas in MALDI the ions are generated as packages produced by a pulsed laser run at 10–200 Hz.

Both ionization techniques are used in tandem mass spectrometry product ion scan experiments, i.e. for peptide sequencing. The product ion scan mode in ESI MS/MS is very useful for phosphopeptide sequencing (Fig. 4). In addition to product ion scanning, the continuous ion beam in ESI facilitates two selective tandem mass spectrometry scan modes. The precursor ion scan and the neutral loss scan are excellent tools for selective detection of phosphopeptides in crude mixtures.

The precursor ion scan mode allows selective and specific detection of precursor ions that give rise to a specific reporter ion. In the case of phosphopeptides this method is used in the negative ion ESI mode for selective detection of ions that produce the *m/z* 79 (PO_3^-) ion and the *m/z* 97 ($H_2PO_4^-$) ion [55–57]. This precursor

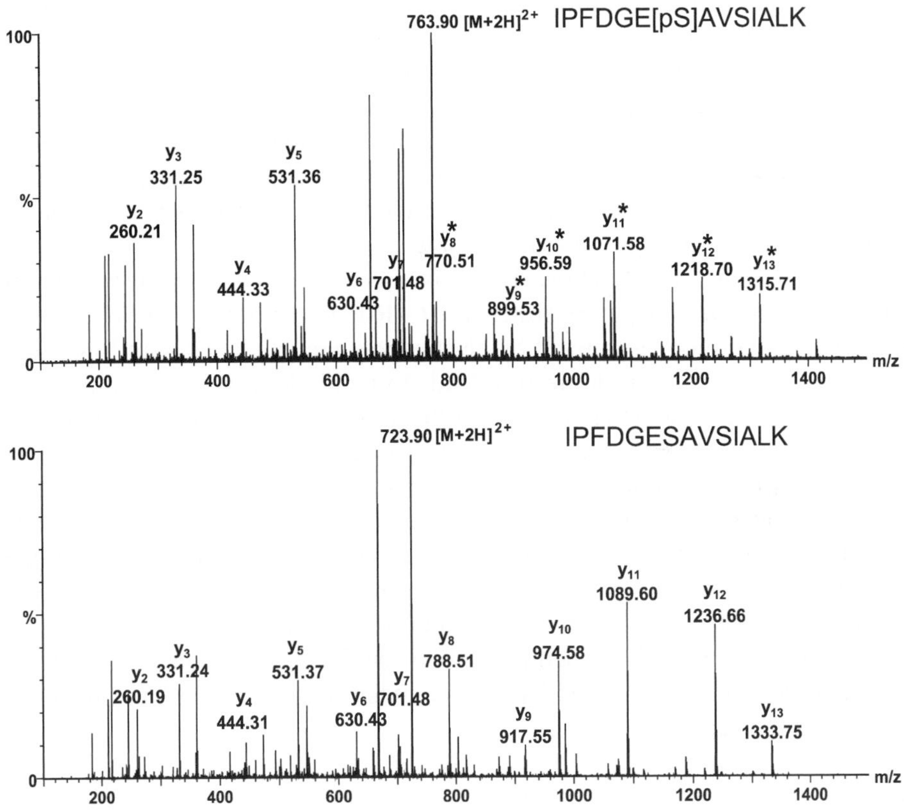

Fig. 4. Phosphopeptide sequencing by tandem mass spectrometry. Top panel: electrospray QTOF MS/MS spectrum of the doubly protonated phosphopeptide ion (IPFDGE[pS]AVSIALK, *m/z* 763.9) from *B. subtilis* PrkC protein. C-terminal peptide ion series is indicated (y and y^* ions). Note that this phosphopeptide generates a y_8^*–y_{13}^* ion series that reflects the loss of phosphoric acid from the y-ions (compare to lower panel). Lower panel: electrospray QTOF MS/MS spectrum of the unmodified peptide (IPFDGESAVSIALK, *m/z* 723.9). The spectrum exhibits a near-complete y-ion fragment ion series.

ion scan is applicable to pSer, pTyr and pThr containing peptides at the 0.5–1 pmol level when using nanoelectrospray MS/MS [58]. A specific phosphotyrosine scan is achieved by positive ion precursors of the phosphotyrosine immonium ion *m/z* 216.043 (Fig. 3) on high mass accuracy QTOF instruments equipped with a nanoelectrospray ion source [59] and is useful for the analysis of tyrosine receptor kinases and their associated proteins [60]. One drawback of this sensitive and specific technique is that it requires a rather long scantime which is only provided when using the nanoelectrospray ion source. It is therefore not readily compatible with LC–MS/MS experiments.

The neutral loss of phosphoric acid (H_3PO_4, 98 Da) by gas-phase beta-elimination is used for specific detection of phosphopeptides in the positive ion mode. This method is less efficient than the precursor ion scan techniques when using triple quadrupole instruments and is only applicable to pSer and pThr which readily undergo gas-phase beta-elimination of phosphoric acid in the mass spectrometer. It is a promising technique for performing real-time phosphopeptide screening during LC–MS/MS runs when using ion trap or QTOF type instruments with customized software for data dependent tandem mass spectrometry experiments [61]. In our laboratory this method is being used as a complement to LC–MS/MS product ion analysis for detailed characterization and sequencing of phosphopeptides present in crude samples.

Electron capture dissociation (ECD) is a recently developed technique for gentle fragmentation of polypeptides in FT–ICR instruments. Multiple charged peptide or protein ions generated by electrospray are irradiated with low-energy electrons. The capture of an electron leads to rapid cleavage of the imine bond between amino acid residues leading to the generation of c- and z-type peptide fragment ions [62]. The rapid nature of the dissociation event eliminates any loss of post-translational modification. This is a unique feature of ECD, which enables amino acid sequencing of labile modified peptides and small proteins. The ECD method has been successfully used for amino acid sequencing of multiple phosphorylated peptides in the 1.5–3.5 kDa range [63] and for sequencing an intact 23 kDa phosphoprotein, beta-casein [64]. On-going development and improvements to the ECD method and its implementation on other types of tandem mass spectrometers will probably make it a useful tool in proteomics research in the near future.

3.4. Integrated strategies for phosphoprotein and phosphoproteome analysis by mass spectrometry

Phosphoprotein and phosphoproteome analysis is orders of magnitude more complicated than 'simple' protein identification experiments. Phosphorylation of any given residue in a protein is usually sub-stoichiometric and it is therefore necessary to investigate every single residue in a protein to determine whether it is modified or not. Thus, high-sensitivity and high-sequence coverage as well as high specificity and selectivity are crucial parameters for any phosphoprotein or phosphoproteome experiment. It is therefore not surprising that most integrated strategies take advantage of multiple techniques to achieve complete characterization of phosphoproteins. As an example, Vihinen and Saarinen used MALDI MS and ESI MS/MS, ^{32}P labeling, IMAC and Edman degradation for determination of up to 20 phosphorylation sites in the Semliki forest virus protein Nsp3 [35].

Annan and co-workers have optimized a procedure based on nanoscale-chromatography interfaced to ESI tandem mass spectrometry with selected ion monitoring of *m/z* 79 produced upon CID of phosphopeptides. Putative phosphopeptide fractions are identified this way and collected for further analysis by MALDI MS and for sequencing by nanoelectrospray tandem mass spectrometry using precursor ion scans and product ion scans [65].

Lehmann and colleagues recently reported a method that takes advantage of a highly active enzyme, elastase, to produce small peptides from phosphoproteins. The phosphopeptides were detected by neutral loss scanning or recovered by IMAC prior to being sequenced by ESI tandem mass spectrometry [66,67]. The advantage of this method is that sets of overlapping peptides are generated leading to redundant sequence information to confirm phosphorylation sites. Yates and co-workers, who used several different enzymes to degrade protein mixtures, followed by LC/LC–MS/MS analysis for comprehensive peptide sequencing, also explored this idea. This 'shotgun sequencing' approach provides high sequence coverage as well as redundant sequence information for confident assignment of post-translational modification in proteins and protein complexes [68].

As mentioned previously, the combination of IMAC and mass spectrometry has proven very efficient for phosphoproteome analysis. Vener and co-workers used this method to identify in vivo phosphorylated proteins in *A. thaliana* thylakoid membranes [27]. Ficarro and colleagues used O-methylesterification as a sample preparation technique prior to IMAC based enrichment of phosphopeptides. The eluted peptides were then analyzed by LC–MS/MS to sequence the putative phosphopeptides. Several hundred phosphopeptides generated from a yeast cell lysate were determined by this approach [26].

3.5. Phosphoprotein characterization by Edman degradation and mass spectrometry

Metabolic labeling of proteins with radioactive phosphate allows monitoring of intact phosphoproteins or phosphopeptides derived from these phosphoproteins by HPLC or TLC combined with radiodetection or phosphorimaging. The recovered ^{32}P labeled peptides are subsequently analyzed by MALDI MS for accurate mass determination which, in combination with known protein cleavage criteria, leads to tentative identification of the phosphopeptide. This is then followed by Edman degradation to determine the position of the phosphorylated residue by counting the number of Edman cycles required to release the radioactive residue [69–71]. This principle has been further developed and extended to phosphoproteome analysis by using a combination of several complementary proteases for digestion, Edman degradation for 'counting residues' and computer software for combinatorial data analysis to determine which amino acid residues are phosphorylated in a protein [72].

4. Cellular dynamics: integrated methods for quantitative phosphoproteome analysis

Investigations of dynamic cellular processes obviously require scalable quantitative methods to study various classes of biomolecules. Furthermore, when the main goal is the determination of changes in post-translational modification of proteins then novel approaches are needed [11]. Mass spectrometry is a key technology for the determination of PTMs and therefore much effort is being devoted to the development of mass spectrometry driven methods for quantitative determination of protein modifications, including phosphorylation. A number of promising approaches have already been reported.

Metabolic labeling of biomolecules with stable isotopes is commonly used in pharmacological sciences and has now also entered the area of protein mass spectrometry. The idea is to differentially incorporate heavy and light isotope labeled compounds into protein populations. The reference system is labeled with the light reagent and the perturbed system is labeled with the heavy reagent, or vice versa. The two differentially labeled protein populations are then mixed and processed for mass spectrometry analysis. The mass spectrometry readout will distinguish between the heavy and light isotope labeled versions of peptides and the ratio between the measured signals for the light and heavy species will reflect the relative abundance of the cognate proteins. The ICAT method [73] is one such technique that unfortunately is not well suited for analysis of PTMs because it targets only cysteine containing peptides.

A general method for analysis of post-translationally modified proteins requires either incorporation of an isotope label in every single peptide or at least in all post-translationally modified peptides of interest. To achieve the latter goal, Goshe and colleagues used the beta-elimination/Michael addition chemistry for incorporation of an isotope coded affinity tag (PhIAT) in phosphopeptides. Model phosphopeptides were differentially derivatized with light or heavy PhIAT reagent, respectively, and then identified and quantified by mass spectrometry [74].

Incorporation of ^{15}N amino acids, ^{2}H-leucine or ^{13}C-leucine into all proteins is a viable method for comprehensive isotope labeling of proteins. This method is applicable to prokaryote or eukaryote cell culture systems, which can be grown for days in a medium that contains the isotopically labeled precursor molecules [75–77]. The ^{15}N labeling method has been used for relative quantitation of yeast protein phosphorylation under various growth conditions [76]. Another strategy for differential labeling is to convert all peptides (obtained after digestion of protein populations) into chemically derivatized species. N- or C-terminal reagents have been developed for this purpose. Again, the use of heavy and light reagents enables relative quantitation of two populations of proteins by mass spectrometry [78]. However, the N- or C-terminal reagents also react with amino acid side chain amino groups and carboxyl groups, respectively, unless special precautions are taken.

In conclusion, several strategies that are potentially useful for quantitative phosphoproteome analysis have been reported. It is clear, however, that robust and efficient methods for quantitative phosphoproteome analysis by mass spectrometry remain to be established. It is also clear that such techniques are likely to be based on stable isotope labeling of proteins combined with a mass spectrometry based read-out of relative amounts of modified species.

5. Bioinformatics tools for phosphoprotein sequence analysis and mass spectrometry data interpretation

Detailed information of the status of modification of a majority of proteins is not available in the current annotated databases simply because it is currently impossible to predict and the experimental data is lacking. Several phosphorylation site prediction algorithms are available that will use prior knowledge of well-characterized phosphoproteins to predict probable phosphorylation sites in a query protein (Table 1). The algorithms are useful for rational design of experimental approaches aimed at determination of the actual utilized phosphorylation sites. Prediction methods may work very well for some phosphorylation motifs/domains in proteins, but they often tend to generate many candidate phosphorylation sites. Thus, putative phosphorylation sites should always be confirmed by experimentation, using direct protein analysis or by indirect methods, such as genetic engineering where, for example, a putative phosphate acceptor residue would be changed to an alanine residue [79].

Experimental data obtained from mass spectrometry analysis of phosphoproteins, including assignment of phosphorylation sites, is definitive proof of the existence of modified forms of a protein. As described in the previous sections, phosphopeptide identification and sequencing is conveniently pursued by MALDI-TOF-MS (peptide mapping) and MALDI or ESI tandem mass

Table 1

Publicly available Internet-based mass spectrometry tools and protein sequence analysis services

Peptide mass mapping (MS)	
ProFound	http://prowl.rockefeller.edu/
Mascot	http://www.matrixscience.com
PeptIdent	http://www.expasy.org
FindMod	http://www.expasy.org
ProteinProspector	http://prospector.ucsf.edu/
MS/MS data interpretation (peptide sequencing)	
Mascot	http://www.matrixscience.com
Sonar	http://65.219.84.5/service/prowl/sonar.html
ProteinProspector	http://prospector.ucsf.edu/
Phosphorylation site prediction	
ScanSite	http://scansite.mit.edu/
NetPhos	http://www.cbs.dtu.dk
Prosite	http://www.expasy.org
Prediction of protein domains and function	
SMART	http://smart.embl-heidelberg.de/
InterPro Scan	http://www.ebi.ac.uk/interpro/scan.html
ProtFun	http://www.cbs.dtu.dk

spectrometry, including capillary LC-MS/MS, for peptide sequencing. The mass spectrometry data is used, in turn, as an input to a database search engine that queries a protein or DNA sequence database. The output from the database query is the assignment of peptide masses and sequences to the cognant protein or gene sequence. The commonly used search engines, such as Mascot, Sequest, Profound and Sonar, are easily configured to take into account partial covalent modifications of individual amino acids, including phosphorylation of serine, threonine and tyrosine residues. The candidate-modified peptides will then be listed in the output report. Candidate phosphopeptides are often identified this way. It is important to perform further experiments when the initial phosphopeptide assignments are based on mass determination alone. This is achieved by alkaline phosphatase treatment and MS analysis or by phosphopeptide sequencing by MS/MS. In general, the database queries should only be performed using high resolution and high mass accuracy MS and MS/MS data, and with stringent parameter settings to avoid a high level of false positive identifications/assignments.

6. Summary

Phosphoproteome analysis by mass spectrometry is in its infancy, but this research field is developing very quickly. Integration of various technologies for phosphoprotein enrichment, phosphopeptide detection and quantitation and phosphopeptide sequencing is clearly needed. These methods have to be coupled to novel bioinformatics tools for analysis and visualization of the protein data that is obtained. Methods are continuously being refined at all these levels, most notably in sample preparation methods, tandem mass spectrometry techniques and bioinformatics software. Very soon, these developments will enable comprehensive and quantitative analysis of the dynamic phosphorylation events in cells.

Acknowledgements

I thank past and present members of the Protein Research Group and the Center for Experimental Bioinformatics at the University of Southern Denmark for their contributions to the development of protein mass spectrometry.

References

1. Cohen, P., The regulation of protein function by multisite phosphorylation — a 25 year update. *Trends Biochem. Sci.*, **25(12)**, 596–601 (2000).
2. Blume-Jensen, P. and Hunter, T., Oncogenic kinase signalling. *Nature*, **411(6835)**, 355–365 (2001).
3. Yan, J.X., Packer, N.H., Gooley, A.A. and Williams, K.L., Protein phosphorylation: technologies for the identification of phosphoamino acids. *J. Chromatogr. A*, **808(1/2)**, 23–41 (1998).
4. Sickmann, A. and Meyer, H.E., Phosphoamino acid analysis. *Proteomics*, **1(2)**, 200–206 (2001).
5. Quadroni, M. and James, P., Phosphopeptide analysis. *Exs*, **88**, 199–213 (2000).
6. McLachlin, D.T. and Chait, B.T., Analysis of phosphorylated proteins and peptides by mass spectrometry. *Curr. Opin. Chem. Biol.*, **5(5)**, 591–602 (2001).
7. Mann, M., Ong, S.E., Gronborg, M., Steen, H., Jensen, O.N. and Pandey, A., Analysis of protein phosphorylation using mass spectrometry: deciphering the phosphoproteome. *Trends Biotechnol.*, **20(6)**, 261–268 (2002).
8. Griffin, T.J., Goodlett, D.R. and Aebersold, R., Advances in proteome analysis by mass spectrometry. *Curr. Opin. Biotechnol.*, **12(6)**, 607–612 (2001).
9. Godovac-Zimmermann, J. and Brown, L.R., Perspectives for mass spectrometry and functional proteomics. *Mass Spectrom. Rev.*, **20(1)**, 1–57 (2001).
10. Jensen, O.N., Modification-specific proteomics: strategies for systematic studies of post-translationally modified proteins, In: Blackstock, W. and Mann, M. (Eds.), *Proteomics: A Trends Guide*. Elsevier, London, 2000, pp. 36–42.
11. Mann, M. and Jensen, O.N., Proteomic analysis of post-translational modifications. *Nat. Biotechnol.*, **21(3)**, 255–261 (2003).
12. Soskic, V., Gorlach, M., Poznanovic, S., Boehmer, F.D. and Godovac-Zimmermann, J., Functional proteomics analysis of signal transduction pathways of the platelet-derived growth factor beta receptor. *Biochemistry*, **38(6)**, 1757–1764 (1999).

13. Pandey, A., Podtelejnikov, A.V., Blagoev, B., Bustelo, X.R., Mann, M. and Lodish, H.F., Analysis of receptor signaling pathways by mass spectrometry: identification of vav-2 as a substrate of the epidermal and platelet-derived growth factor receptors. *Proc. Natl Acad. Sci. USA*, **97(1)**, 179–184 (2000).
14. Yamagata, A., Kristensen, D.B., Takeda, Y., Miyamoto, Y., Okada, K., Inamatsu, M. and Yoshizato, K., Mapping of phosphorylated proteins on two-dimensional polyacrylamide gels using protein phosphatase. *Proteomics*, **2(9)**, 1267–1276 (2002).
15. Fey, S.J. and Larsen, P.M., 2D or not 2D. Two-dimensional gel electrophoresis. *Curr. Opin. Chem. Biol.*, **5(1)**, 26–33 (2001).
16. Gronborg, M., Kristiansen, T.Z., Stensballe, A., Andersen, J.S., Ohara, O., Mann, M., Jensen, O.N. and Pandey, A., A mass spectrometry-based proteomic approach for identification of serine/threonine-phosphorylated proteins by enrichment with phospho-specific antibodies: identification of a novel protein, Frigg, as a protein kinase A substrate. *Mol. Cell. Proteom.*, **1(7)**, 517–527 (2002).
17. Marcus, K., Immler, D., Sternberger, J. and Meyer, H.E., Identification of platelet proteins separated by two-dimensional gel electrophoresis and analyzed by matrix assisted laser desorption/ionization-time of flight-mass spectrometry and detection of tyrosine-phosphorylated proteins. *Electrophoresis*, **21(13)**, 2622–2636 (2000).
18. Kalo, M.S. and Pasquale, E.B., Multiple in vivo tyrosine phosphorylation sites in EphB receptors. *Biochemistry*, **38(43)**, 14396–14408 (1999).
19. Andersson, L. and Porath, J., Isolation of phosphoproteins by immobilized metal (Fe3 +) affinity chromatography. *Anal. Biochem.*, **154(1)**, 250–254 (1986).
20. Neville, D.C., Rozanas, C.R., Price, E.M., Gruis, D.B., Verkman, A.S. and Townsend, R.R., Evidence for phosphorylation of serine 753 in CFTR using a novel metal-ion affinity resin and matrix-assisted laser desorption mass spectrometry. *Protein Sci.*, **6(11)**, 2436–2445 (1997).
21. Cao, P. and Stults, J.T., Phosphopeptide analysis by on-line immobilized metal-ion affinity chromatography–capillary electrophoresis–electrospray ionization mass spectrometry. *J. Chromatogr. A*, **853(1–2)**, 225–235 (1999).
22. Cao, P. and Stults, J.T., Mapping the phosphorylation sites of proteins using on-line immobilized metal affinity chromatography/capillary electrophoresis/electrospray ionization multiple stage tandem mass spectrometry. *Rapid Commun. Mass Spectrom.*, **14(17)**, 1600–1606 (2000).
23. Posewitz, M.C. and Tempst, P., Immobilized gallium(III) affinity chromatography of phosphopeptides. *Anal. Chem.*, **71(14)**, 2883–2892 (1999).
24. Stensballe, A., Andersen, S. and Jensen, O.N., Characterization of phosphoproteins from electrophoretic gels by nanoscale Fe(III) affinity chromatography with offline mass spectrometry analysis. *Proteomics*, **1(2)**, 207–222 (2001).
25. Xhou, W., Merrick, B.A., Khaledi, M.G. and Tomer, K.B., Detection and sequencing of phosphopeptides affinity bound to immobilized metal ion beads by matrix-assisted laser desorption/ionization mass spectrometry. *J. Am. Soc. Mass Spectrom.*, **11(4)**, 273–282 (2000).
26. Ficarro, S.B., McCleland, M.L., Stukenberg, P.T., Burke, D.J., Ross, M.M., Shabanowitz, J., Hunt, D.F. and White, F.M., Phosphoproteome analysis by mass spectrometry and its application to *Saccharomyces cerevisiae*. *Nat. Biotechnol.*, **20(3)**, 301–305 (2002).
27. Vener, A.V., Harms, A., Sussman, M.R. and Vierstra, R.D., Mass spectrometric resolution of reversible protein phosphorylation in photosynthetic membranes of *Arabidopsis thaliana*. *J. Biol. Chem.*, **276(10)**, 6959–6966 (2001).
28. Resing, K.A., Johnson, R.S. and Walsh, K.A., Mass spectrometric analysis of 21 phosphorylation sites in the internal repeat of rat profilaggrin, precursor of an intermediate filament associated protein. *Biochemistry*, **34(29)**, 9477–9487 (1995).

29. Jaffe, H., Veeranna and Pant, H.C., Characterization of serine and threonine phosphorylation sites in beta-elimination/ethanethiol addition-modified proteins by electrospray tandem mass spectrometry and database searching. *Biochemistry*, **37(46)**, 16211–16224 (1998).
30. Weckwerth, W., Willmitzer, L. and Fiehn, O., Comparative quantification and identification of phosphoproteins using stable isotope labeling and liquid chromatography/mass spectrometry. *Rapid Commun. Mass Spectrom.*, **14(18)**, 1677–1681 (2000).
31. Molloy, M.P. and Andrews, P.C., Phosphopeptide derivatization signatures to identify serine and threonine phosphorylated peptides by mass spectrometry. *Anal. Chem.*, **73(22)**, 5387–5394 (2001).
32. Adamczyk, M., Gebler, J.C. and Wu, J., Selective analysis of phosphopeptides within a protein mixture by chemical modification, reversible biotinylation and mass spectrometry. *Rapid Commun. Mass Spectrom.*, **15(16)**, 1481–1488 (2001).
33. Oda, Y., Nagasu, T. and Chait, B.T., Enrichment analysis of phosphorylated proteins as a tool for probing the phosphoproteome. *Nat. Biotechnol.*, **19(4)**, 379–382 (2001).
34. Zhou, H., Watts, J.D. and Aebersold, R., A systematic approach to the analysis of protein phosphorylation. *Nat. Biotechnol.*, **19(4)**, 375–378 (2001).
35. Vihinen, H. and Saarinen, J., Phosphorylation site analysis of Semliki forest virus nonstructural protein 3. *J. Biol. Chem.*, **275(36)**, 27775–27783 (2000).
36. Loughrey Chen, S., Huddleston, M.J., Shou, W., Deshaies, R.J., Annan, R.S. and Carr, S.A., Mass spectrometry-based methods for phosphorylation site mapping of hyperphosphorylated proteins applied to Net1, a regulator of exit from mitosis. *Mol. Cell. Proteom.*, **1(3)**, 186–196 (2002).
37. Bykova, N.V., Stensballe, A., Egsgaard, H., Jensen, O.N. and Moller, I.M., Phosphorylation of formate dehydrogenase in potato tuber mitochondria. *J. Biol. Chem.*, (2003).
38. Shevchenko, A., Wilm, M., Vorm, O. and Mann, M., Mass spectrometric sequencing of proteins silver-stained polyacrylamide gels. *Anal. Chem.*, **68(5)**, 850–858 (1996).
39. Ma, Y., Lu, Y., Zeng, H., Ron, D., Mo, W. and Neubert, T.A., Characterization of phosphopeptides from protein digests using matrix-assisted laser desorption/ionization time-of-flight mass spectrometry and nanoelectrospray quadrupole time-of-flight mass spectrometry. *Rapid Commun. Mass Spectrom.*, **15(18)**, 1693–1700 (2001).
40. Janek, K., Wenschuh, H., Bienert, M. and Krause, E., Phosphopeptide analysis by positive and negative ion matrix-assisted laser desorption/ionization mass spectrometry. *Rapid Commun. Mass Spectrom.*, **15(17)**, 1593–1599 (2001).
41. Asara, J.M. and Allison, J., Enhanced detection of phosphopeptides in matrix-assisted laser desorption/ionization mass spectrometry using ammonium salts. *J. Am. Soc. Mass Spectrom.*, **10(1)**, 35–44 (1999).
42. Erdjument-Bromage, H., Lui, M., Lacomis, L., Grewal, A., Annan, R.S., McNulty, D.E., Carr, S.A. and Tempst, P., Examination of micro-tip reversed-phase liquid chromatographic extraction of peptide pools for mass spectrometric analysis. *J. Chromatogr. A*, **826(2)**, 167–181 (1998).
43. Gobom, J., Nordhoff, E., Mirgorodskaya, E., Ekman, R. and Roepstorff, P., Sample purification and preparation technique based on nano-scale reversed-phase columns for the sensitive analysis of complex peptide mixtures by matrix-assisted laser desorption/ionization mass spectrometry. *J. Mass Spectrom.*, **34(2)**, 105–116 (1999).
44. Annan, R.S. and Carr, S.A., Phosphopeptide analysis by matrix-assisted laser desorption time-of-flight mass spectrometry. *Anal. Chem.*, **68(19)**, 3413–3421 (1996).
45. Metzger, S. and Hoffmann, R., Studies on the dephosphorylation of phosphotyrosine-containing peptides during post-source decay in matrix-assisted laser desorption/ionization. *J. Mass Spectrom.*, **35(10)**, 1165–1177 (2000).

46. Hoffmann, R., Metzger, S., Spengler, B. and Otvos, L. Jr., Sequencing of peptides phosphorylated on serines and threonines by post-source decay in matrix-assisted laser desorption/ionization time-of-flight mass spectrometry. *J. Mass Spectrom.*, **34(11)**, 1195–1204 (1999).
47. Hoffmann, R., Wehofsky, M. and Metzger, S., Neutral loss analysis in MALDI-MS using an ion-gate scanning mode. *Anal. Chem.*, **73(20)**, 4845–4851 (2001).
48. Liao, P.C., Leykam, J., Andrews, P.C., Gage, D.A. and Allison, J., An approach to locate phosphorylation sites in a phosphoprotein: mass mapping by combining specific enzymatic degradation with matrix-assisted laser desorption/ionization mass spectrometry. *Anal. Biochem.*, **219(1)**, 9–20 (1994).
49. Larsen, M.R., Sorensen, G.L., Fey, S.J., Larsen, P.M. and Roepstorff, P., Phospho-proteomics: evaluation of the use of enzymatic de-phosphorylation and differential mass spectrometric peptide mass mapping for site specific phosphorylation assignment in proteins separated by gel electrophoresis. *Proteomics*, **1(2)**, 223–238 (2001).
50. Zhang, X., Herring, C.J., Romano, P.R., Szczepanowska, J., Brzeska, H., Hinnebusch, A.G. and Qin, J., Identification of phosphorylation sites in proteins separated by polyacrylamide gel electrophoresis. *Anal. Chem.*, **70(10)**, 2050–2059 (1998).
51. Qin, J. and Chait, B.T., Identification and characterization of posttranslational modifications of proteins by MALDI ion trap mass spectrometry. *Anal. Chem.*, **69(19)**, 4002–4009 (1997).
52. Bennett, K.L., Stensballe, A., Podtelejnikov, A.V., Moniatte, M. and Jensen, O.N., Phosphopeptide detection and sequencing by matrix-assisted laser desorption/ionization quadrupole time-of-flight tandem mass spectrometry. *J. Mass Spectrom.*, **37(2)**, 179–190 (2002).
53. Lee, C.H., McComb, M.E., Bromirski, M., Jilkine, A., Ens, W., Standing, K.G. and Perreault, H., On-membrane digestion of beta-casein for determination of phosphorylation sites by matrix-assisted laser desorption/ionization quadrupole/time-of-flight mass spectrometry. *Rapid Commun. Mass Spectrom.*, **15(3)**, 191–202 (2001).
54. Baldwin, M.A., Medzihradszky, K.F., Lock, C.M., Fisher, B., Settineri, T.A. and Burlingame, A.L., Matrix-assisted laser desorption/ionization coupled with quadrupole/orthogonal acceleration time-of-flight mass spectrometry for protein discovery, identification, and structural analysis. *Anal. Chem.*, **73(8)**, 1707–1720 (2001).
55. Annan, R.S., Huddleston, M.J., Verma, R., Deshaies, R.J. and Carr, S.A., A multidimensional electrospray MS-based approach to phosphopeptide mapping. *Anal. Chem.*, **73(3)**, 393–404 (2001).
56. Carr, S.A., Huddleston, M.J. and Annan, R.S., Selective detection and sequencing of phosphopeptides at the femtomole level by mass spectrometry. *Anal. Biochem.*, **239(2)**, 180–192 (1996).
57. Wilm, M., Neubauer, G. and Mann, M., Parent ion scans of unseparated peptide mixtures. *Anal. Chem.*, **68(3)**, 527–533 (1996).
58. Neubauer, G. and Mann, M., Mapping of phosphorylation sites of gel-isolated proteins by nanoelectrospray tandem mass spectrometry: potentials and limitations. *Anal. Chem.*, **71(1)**, 235–242 (1999).
59. Steen, H., Kuster, B., Fernandez, M., Pandey, A. and Mann, M., Detection of tyrosine phosphorylated peptides by precursor ion scanning quadrupole TOF mass spectrometry in positive ion mode. *Anal. Chem.*, **73(7)**, 1440–1448 (2001).
60. Steen, H., Kuster, B., Fernandez, M., Pandey, A. and Mann, M., Tyrosine phosphorylation mapping of the epidermal growth factor receptor signaling pathway. *J. Biol. Chem.*, **277(2)**, 1031–1039 (2002).

61. Bateman, R.H., Carruthers, R., Hoyes, J.B., Jones, C., Langridge, J.I., Millar, A. and Vissers, J.P., A novel precursor ion discovery method on a hybrid quadrupole orthogonal acceleration time-of-flight (Q-TOF) mass spectrometer for studying protein phosphorylation. *J. Am. Soc. Mass Spectrom.*, **13(7)**, 792–803 (2002).
62. Zubarev, R.A., Horn, D.M., Fridriksson, E.K., Kelleher, N.L., Kruger, N.A., Lewis, M.A., Carpenter, B.K. and McLafferty, F.W., Electron capture dissociation for structural characterization of multiply charged protein cations. *Anal. Chem.*, **72(3)**, 563–573 (2000).
63. Stensballe, A., Jensen, O.N., Olsen, J.V., Haselmann, K.F. and Zubarev, R.A., Electron capture dissociation of singly and multiply phosphorylated peptides. *Rapid Commun. Mass Spectrom.*, **14(19)**, 1793–1800 (2000).
64. Shi, S.D., Hemling, M.E., Carr, S.A., Horn, D.M., Lindh, I. and McLafferty, F.W., Phosphopeptide/phosphoprotein mapping by electron capture dissociation mass spectrometry. *Anal. Chem.*, **73(1)**, 19–22 (2001).
65. Zappacosta, F., Huddleston, M.J., Karcher, R.L., Gelfand, V.I., Carr, S.A. and Annan, R.S., Improved sensitivity for phosphopeptide mapping using capillary column HPLC and microionspray mass spectrometry: comparative phosphorylation site mapping from gel-derived proteins. *Anal. Chem.*, **74(13)**, 3221–3231 (2002).
66. Schlosser, A., Bodem, J., Bossemeyer, D., Grummt, I. and Lehmann, W.D., Identification of protein phosphorylation sites by combination of elastase digestion, immobilized metal affinity chromatography, and quadrupole-time of flight tandem mass spectrometry. *Proteomics*, **2(7)**, 911–918 (2002).
67. Schlosser, A., Pipkorn, R., Bossemeyer, D. and Lehmann, W.D., Analysis of protein phosphorylation by a combination of elastase digestion and neutral loss tandem mass spectrometry. *Anal. Chem.*, **73(2)**, 170–176 (2001).
68. MacCoss, M.J., McDonald, W.H., Saraf, A., Sadygov, R., Clark, J.M., Tasto, J.J., Gould, K.L., Wolters, D., Washburn, M., Weiss, A., Clark, J.I. and Yates, J.R. III, Shotgun identification of protein modifications from protein complexes and lens tissue. *Proc. Natl Acad. Sci. USA*, **99(12)**, 7900–7905 (2002).
69. Campbell, D.G. and Morrice, N., Identification of protein phosphorylation sites by a combination of mass spectrometry and solid phase Edman sequencing. *J. Biomol. Tech.*, **13**, 119–130 (2002).
70. Haydon, C.E., Watt, P.W., Morrice, N., Knebel, A., Gaestel, M. and Cohen, P., Identification of a phosphorylation site on skeletal muscle myosin light chain kinase that becomes phosphorylated during muscle contraction. *Arch. Biochem. Biophys.*, **397(2)**, 224–231 (2002).
71. Knebel, A., Morrice, N. and Cohen, P., A novel method to identify protein kinase substrates: eEF2 kinase is phosphorylated and inhibited by SAPK4/p38delta. *EMBO J.*, **20(16)**, 4360–4369 (2001).
72. MacDonald, J.A., Mackey, A.J., Pearson, W.R. and Haystead, T.A., A strategy for the rapid identification of phosphorylation sites in the phosphoproteome. *Mol. Cell. Proteom.*, **1(4)**, 314–322 (2002).
73. Gygi, S.P., Rist, B., Gerber, S.A., Turecek, F., Gelb, M.H. and Aebersold, R., Quantitative analysis of complex protein mixtures using isotope-coded affinity tags. *Nat. Biotechnol.*, **17(10)**, 994–999 (1999).
74. Goshe, M.B., Veenstra, T.D., Panisko, E.A., Conrads, T.P., Angell, N.H. and Smith, R.D., Phosphoprotein isotope-coded affinity tags: application to the enrichment and identification of low-abundance phosphoproteins. *Anal. Chem.*, **74(3)**, 607–616 (2002).
75. Smith, R.D., Pasa-Tolic, L., Lipton, M.S., Jensen, P.K., Anderson, G.A., Shen, Y., Conrads, T.P., Udseth, H.R., Harkewicz, R., Belov, M.E., Masselon, C. and Veenstra, T.D., Rapid

quantitative measurements of proteomes by Fourier transform ion cyclotron resonance mass spectrometry. *Electrophoresis*, **22(9)**, 1652–1668 (2001).

76. Oda, Y., Huang, K., Cross, F.R., Cowburn, D. and Chait, B.T., Accurate quantitation of protein expression and site-specific phosphorylation. *Proc. Natl Acad. Sci. USA*, **96(12)**, 6591–6596 (1999).
77. Ong, S.E., Blagoev, B., Kratchmarova, I., Kristensen, D.B., Steen, H., Pandey, A. and Mann, M., Stable isotope labeling by amino acids in cell culture, SILAC, as a simple and accurate approach to expression proteomics. *Mol. Cell. Proteom.*, **1(5)**, 376–386 (2002).
78. Munchbach, M., Quadroni, M., Miotto, G. and James, P., Quantitation and facilitated de novo sequencing of proteins by isotopic N-terminal labeling of peptides with a fragmentation-directing moiety. *Anal. Chem.*, **72(17)**, 4047–4057 (2000).
79. Madec, E., Stensballe, A., Kjellström, S., Cladiere, L., Obuchowski, M., Jensen, O.N. and Seror, S.J., Mass spectrometry and site-directed mutagenesis identify several autophosphorylated residues required for activity of PrkC, a Ser/Thr kinase from *Bacillus subtilis*. *J. Mol. Biol.* (2003), in press.

Proteome Analysis. Interpreting the Genome.
D.W. Speicher (editor)

Chapter 7

Protein identification by in-gel digestion and mass spectrometry

KATHERYN A. RESING[a,*] and NATALIE G. AHN[a,b]

[a]*Department of Chemistry and Biochemistry, University of Colorado, Boulder, CO 80309, USA*
[b]*Howard Hughes Medical Institute, University of Colorado, Boulder, CO 80309, USA*

* Corresponding author. Tel.: +1-303-492-5519; fax: +1-303-492-2439. E-mail: resing@stripe.colorado.edu (K.A. Resing).

1. Introduction

The success of genome sequencing projects has offered new tools for protein identification; of particular interest to protein chemists are database search engines that allow identification of proteins from peptide mass fingerprinting or sequence fragmentation information. In combination with high-sensitivity mass spectrometers and new ionization methods, these tools have driven a revolution in protein chemistry. One method now widely used is in-gel protease digestion of proteins, in which proteins are digested in situ from gel pieces excised after 1D or 2D-PAGE. Peptides extracted from the gel piece are analyzed by mass spectrometry, allowing identification of protein identity and modifications (see Chapters 5 and 6). The development of this method required the dovetailing of several analytical advances.

One of the early applications of in-gel digestion used endoproteinases or cyanogen bromide to partially digest proteins in 1D gel slices or lanes. This was followed by a second dimension SDS-gel to visualize the peptides [1]. After its publication in 1977, this method was widely used to confirm protein identities, as well as provide information on modifications. In-gel digestion was used in the eighties to produce peptide digests analyzed by gas phase sequencing [2]. In 1992, Rosenfeld et al. [3] reported a digestion protocol that is essentially identical to the methods currently used. A key innovation was the dehydration of the gel piece, followed by rehydration with protease, facilitating incorporation of protease into the gel.

Early studies involved separation of peptides by conventional reverse phase high pressure liquid chromatography (HPLC), followed by mass spectrometry or Edman sequencing of fractions. Subsequent methods directly coupled capillary HPLC columns to mass spectrometer interfaces, for online sample analysis [4–6]. Several manuscripts using this LC–MS approach for peptide sequencing were published soon afterwards [7–12]. At the same time, methods were developed for analysis of unseparated peptides by matrix assisted laser desorption ionization (MALDI) mass spectrometry [13,14]. By 1995, computer search algorithms for identifying proteins from peptide mass spectral data began to be widely available [8,15–23], and the pace of publication in this area markedly increased. In 1996, Wilm and Mann reported a nanospray interface that allowed direct analysis of digests by mass spectrometry without prior separation of the peptides [24,25].

These advances, together with contributions from many other laboratories provided methods that greatly accelerated the identification of proteins separated by 1D and 2D gel electrophoresis. This was particularly true for cataloguing proteins separated by 2D-PAGE, where marked improvements in resolution and reproducibility had been achieved by development of immobilized pH gradients for isoelectric focusing [26,27]. Characterization of proteins by electroblotting and N-terminal Edman sequencing in early studies resulted in greater success with bacterial proteins, which are not as compromised as eukaryotic proteins by N-terminal blocking. The introduction of in-gel digestion greatly facilitated identifications of eukaryotic proteins. Pioneering studies in this area reported methods and computer programs for peptide mass fingerprinting and sequence search analysis and these enhanced methods made establishing databases of proteins expressed in human myocardial cells, melanoma cells, and yeast feasible [15,16,28,29]. The term 'proteomics', used to describe the sum of proteins expressed in a given cell type, was coined by Williams, Humphery-Smith and co-workers [30,31], and ensuing publications have popularized the use of this word to describe this exciting new field. A flurry of manuscripts since the introduction of proteomics as a new field have demonstrated the utility of this approach in a wide array of applications, opening a door into the realm of biomolecular mass spectrometry for biomedical and biological scientists.

2. In-gel digestion

2.1. Gel electrophoresis and staining

Two essential requirements for in-gel digestion prior to mass spectrometry analysis are efficient peptide recovery and minimal contamination. Because peptides are extracted from gel pieces by diffusion, it is best to cast gels with the minimal cross-linker and acrylamide concentrations needed to resolve the proteins of interest. Gel preparation, electrophoresis, staining, storage, and digestion should be carried out under conditions that minimize contamination from lab dust or the user. For example, sheep keratins are frequently identified in samples handled by a person wearing a fuzzy wool sweater through airborne contamination! Although gels can be run on the bench top with suitable precautions, in-gel digestions should be carried out in a dust-free area, such as a laminar flow hood. In this laboratory, a small clean hood (Baker IV-22) is routinely used as a digestion workstation.

Various staining methods can be used to visualize proteins. Excellent recoveries are usually obtained with Coomassie Blue (e.g. 0.5% w/v in 10% acetic acid, 15% isopropanol) or colloidal Coomassie, followed by destaining with 10% acetic acid and rinsing into water. More sensitive staining methods that show good recovery for in-gel digestion include reverse-staining with Zn^{2+}–SDS–imidazole [32–34], and SyproRuby fluorescence staining; the latter is highly sensitive when visualized with fluorescence image analyzers [35]. Silver staining recipes which use aldehyde fixation offer the highest sensitivity [36,37], but generally result in lowest recovery from in-gel digestion compared with the stains described earlier. Therefore, many silver staining protocols [38] including commercial 'MS-compatible' staining recipes (e.g. SilverQuest, Invitrogen) use methanol or ethanol fixation, which significantly improve recovery at a cost of ~5-fold decreased staining intensity, compared to aldehyde-based recipes. After staining and protein visualization, gels are rinsed into water (e.g. 3 × 30 min) to remove destaining reagents. Typically, comparable sensitivities are obtained using alcohol-based low-fixation silver staining, SyproRuby fluorescence staining, and Zn^{2+}–SDS–imidazole protocols.

2.2. Gel excision

An X-ACTO knife, razor blade, or metal punch (wash with ethanol to remove residual oil) can be used to manually excise proteins within gel pieces. In addition, plastic pipet tips are cheap and easy to use as gel punches. P1000 pipet tips are cut with an X-ACTO knife to the desired diameter, and then shaved around the inside to create a sharp beveled edge. The tip is inserted onto a P1000 pipetman or a plastic transfer pipet (e.g. Sarstedt 3.5 ml). With the gel immersed in water, expel some of the air from the pipet, punch out the gel spot, and immediately release the pipet plunger to draw the gel piece into the tip after excision. Then transfer the gel piece to a clean 1.5 ml microcentrifuge tube. Minimizing excess gel material increases peptide recovery; therefore, sacrificing a small amount of protein at the edges is preferred over excess gel volume. A blank gel piece should be analyzed in parallel as a control for contamination. Alternatively, an increasing number of vendors sell robotic equipment for automated excision of protein spots from gels. These devices use circular excision types. Hence, 1D gel bands must be excised as a series of overlapping small circles.

In general, proteins should be recovered and digested immediately after destaining gels, particularly when using silver staining recipes. If gels must be stored for any length of time, they should be washed several times in deionized water and stored at 4°C in airtight containers or zip-lock bags. If digestion cannot be carried out immediately after excision, then gel pieces should be stored in airtight microfuge tubes at −80°C. Some investigators destain, reduce, and alkylate the gel pieces before freezing (see later).

2.3. Digestion

Gloves should be worn and changed often, and HPLC grade water and reagent grade chemicals should be used to prepare solutions. A tube rack mounted on a vortex mixer ($60\ min^{-1}$) is useful for agitating the samples during each incubation. Several vendors (e.g. Applied Biosystems, BioRad, etc.) sell robotic instrumentation for automated in-gel digestion.

Proteins stained with Coomassie can be excised and processed directly. Gels reverse-stained with Zn^{2+}–SDS–imidazole are destained with citric acid or EDTA prior to processing. Gel pieces silver stained with either alcohol or aldehyde fixatives should be decolorized with Farmer's reagent [39] to remove precipitated silver, which improves peptide recovery by about 3-fold. In treating gels with this reagent, solutions of 100 mM sodium thiosulfate (0.62 g $NaS_2O_3{\cdot}H_2O$/25 ml H_2O) and 30 mM potassium ferricyanide (0.25 g $K_3Fe(CN)_6$/25 ml H_2O) are freshly made and equal volumes are mixed. The resulting solution (200 μl) is added to the gel piece in a 1.5 ml tube, followed by incubation for 5–10 min until the brown stain disappears due to oxidation of Ag to Ag^+. The solution is removed and the gel rinsed 3 × 5 min with 1.5 ml water. An alternative method uses hydrogen peroxide to remove precipitated silver [40].

After any required destaining steps, proteins in excised gel pieces are reduced and alkylated to minimize disulfide cross-linking and cysteine oxidation. Reduction can be performed by adding 0.2 ml 10 mM dithiothreitol in 100 mM NH_4HCO_3, pH 8 (stock solutions made fresh) to the gel piece, followed by incubation at 60°C with agitation for 30 min. The reducing solution is removed and replaced with 0.2 ml of 55 mM iodoacetamide in 100 mM NH_4HCO_3 pH 8, and incubated at room temperature in the dark for 30 min. The alkylating solution is removed and the gel piece rinsed for 20 min in 0.5 ml 100 mM NH_4HCO_3.

The second step involves dehydrating the gel piece with increasing amounts of organic solvent. The NH_4HCO_3 rinse solution is removed and replaced with 0.5 ml 50% acetonitrile, 100 mM NH_4HCO_3 pH 8, and the gel piece is incubated for 20 min at room temperature. After removing the 50% acetonitrile buffer solution, gel pieces are transferred to a clean 0.65 ml tube. If the gel pieces are long and skinny, they are cut into 2–4 pieces using a clean razor blade, so that they fit tightly into the bottom of the digestion tube. Dehydration is carried out by adding 50 μl of 100% acetonitrile to the gel piece followed by incubation for 15 min at room temperature. The gel piece should shrink noticeably and become opaque. Excess acetonitrile is removed and the sample then dried completely in a Speedvac centrifuge (~10 min). Use of a Speedvac with an oil-free pump and an easily cleaned trap (e.g. Heto), minimizes potential contamination at this step.

The transfer tubes should be low peptide binding, particularly if low amounts of protein are analyzed. Many brands of polypropylene tubes work well (e.g. Titertubes (BioRad), nonsilanized PCR tubes (Midwest Scientific, Sarstedt), and Optimum Bestubes (Life Science Products)). Alternatively, 96-well plates that are low peptide binding (Nunc) can be used for high throughput protocols. Peptide binding can be assessed by incubating solution digests of protein standards (e.g. myoglobin digested with trypsin) in the tube for 24 h, followed by analysis of recovery.

The third step incorporates protease into the gel by reswelling. Trypsin is usually the enzyme of choice, because it is small and relatively easy to diffuse into the gel, and it produces positively charged peptides favorable for mass spectrometry. Because proteins in gel pieces are often present at concentrations near or below the substrate K_m of the protease, longer incubation times and minimized gel volumes are essential for efficient proteolysis.

Normally, trypsin autolyzes rapidly, so that little active protease is present after 2–3 h. Therefore, most investigators use trypsin preparations where surface lysines have been covalently modified, to delay autolysis and allow longer digestion times. Stock solutions (1 mg/ml) of modified trypsin in 50 mM acetic acid (pH 3) can be made from lyophilized enzyme or frozen liquid (vendors include Promega, Roche, Princeton Separations, etc.). Lyophilized trypsin should be rehydrated carefully, so that the small flake of protease is not lost when opening the vial. Do not vortex the trypsin solution — it will resuspend easily by drawing the solution up and down gently with a pipet. It is not necessary to preincubate the trypsin before using it, and the stock can be stored for short periods of time on ice until needed. Leftover trypsin stock solution can be stored frozen in aliquots for future use.

The stock solution of modified trypsin is diluted to 0.02 mg/ml with 25 mM NH_4HCO_3, pH 8. Reswell the sample into a volume of trypsin that is slightly greater than the volume of the original gel piece. A preferred method is to add the trypsin in three or four aliquots to a final volume that is 2–3 μl greater than the volume of the original gel piece, waiting 5 min between each aliquot in order to allow the gel pieces to reswell. After the final addition of trypsin, the gel reswells for 10 min, then excess trypsin solution is removed and replaced with a volume of 25 mM NH_4HCO_3 that just covers the gel piece (approximately twice the gel volume). The gel pieces must stay hydrated during the digestion, and the extra volume of buffer allows peptides to diffuse out of the gels. Samples are incubated 16–24 h at 37°C, when using modified trypsin.

2.4. Peptide extraction

Peptide recovery is an aspect of the procedure that is least reproducible, and several variations on extraction methods have been reported. The optimal protocol depends on the ionization technique to be used, which can include LC–ESI, MALDI, or nanospray.

Concentrated formic acid or trifluoroacetic acid is typically added to a final concentration of 5% (v/v) to acidify the digestion mixture prior to extraction. Sonicate the supernatant + gel for 30 min in a bath sonicator, then transfer the supernatant to a new tube. For highly abundant proteins, the supernatant can be analyzed directly by MALDI- or LC–MS. Low abundance proteins usually require concentration by evaporation in a Speedvac; for optimal recovery, the sample should not be reduced to dryness. Some protocols suggest further extraction of the gel piece by adding 50–100 μl of 1% TFA/60% acetonitrile/40% water. The acetonitrile contracts the gel pieces, and the theory is that the peptides will be 'squeezed out'. Peptides in the pooled supernatants are then concentrated by Speedvac centrifugation under to ~10 μl, which also removes the organic solvent which is essential for subsequent binding of the peptides to the reversed phase resin for desalting. Use of additional extractions provides little improvement in peptide yield in this laboratory's experience, and recovery is reduced when samples are accidentally lyophilized to dryness. An alternative method for 'squeezing' the gel pieces is to place the pieces in buffer on top of a microspin column with a low protein binding frit, followed by centrifugation to extract the peptides.

2.5. Assessing peptide recovery

Recovery of peptides from in-gel digestions should be assessed by comparing yields with those from solution digestions. Low recovery from in-gel digestion reduces the sequence coverage obtained for peptide fingerprinting analysis. An analysis of myoglobin stained by various methods showed that many peptides were efficiently recovered from in-gel digestion, although larger, hydrophobic peptides (e.g. 1818 and 1881 Da) were poorly recovered (Fig. 1). The staining method significantly affects recovery, which was high with Coomassie R250 and Zn^{2+}–SDS–imidazole methods compared to solution digestion, but low with aldehyde- or methanol-fixed silver staining recipes. These differences are presumably due to protein cross-linking, and are less apparent when analyzing proteins at higher

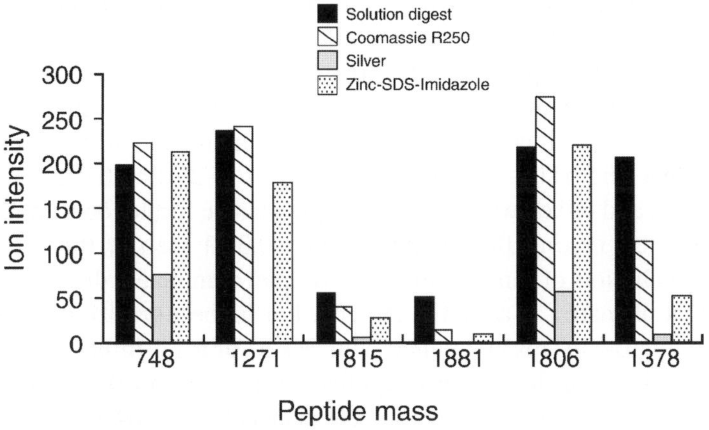

Fig. 1. Effects of protein staining on peptide recovery from in-gel digestion. Equal amounts of myoglobin (1 μg) were resolved by 1D-PAGE and stained with Coomassie R250, silver (formaldehyde fixation), or $ZnCl_2$–SDS–imidazole. Gel bands were excised and digested in-gel with trypsin, and the eluted peptides were analyzed by RP-HPLC (500 μm capillary POROS 20 R1) directly coupled to an electrospray tandem quadrupole mass spectrometer (Sciex API-III +). The sum of ion intensities for all charge states are shown for six representative peptides, and compared to the total intensity of peptides generated in a parallel solution digestion.

abundant levels [41]. Improved peptide recovery has been reported when using acid-labile surfactants, such as sodium 4-[(2-methyl-2-undecyl-1,3 dioxolan-4-yl)methoxyl]-1-propane sulfonate (ALS-1, Waters), in place of SDS during gel electrophoresis [42,43]. The acid-labile surfactant increases peptide ion signal-to-noise and protein sequence coverage due to higher efficiency extraction of peptides during in-gel digestion and subsequent ionization during MS.

3. Mass spectrometry

3.1. Sample preparation and data collection

Until recently, peptide analysis by MALDI-TOF required post-source decay for sequencing, which produces spectra that are not always easily interpretable. More recently, MALDI interfaces for ion trap, hybrid quadrupole-TOF, or TOF–TOF mass analyzers have made MALDI an accessible and easy method for analyzing in-gel digests [44–47]. Surprisingly, peptides fragment well from these MALDI sources, despite the fact that ions are singly charged. For highly abundant samples, the first supernatant from in-gel digests can be analyzed directly by MALDI-MS without further concentration. Peptide digests are mixed with matrix, and spotted onto gold-plated or stainless steel target plates. Typically α-cyano 4-hydroxy cinnamic acid (α-CHCA) or 2,5-dihydroxybenzoic acid (DHB) are efficient matrices for peptide analysis using MALDI-TOF or MALDI-quadrupole-TOF instruments [48,49]. Several protocols for co-crystallizing analyte and matrix are available online (e.g. http://prowl.rockefeller.edu/).

For low abundant samples, peptides are typically concentrated in this laboratory to 10–20 μl by Speedvac centrifugation, then desalted and further concentrated with pipet tips packed with reversed-phase resin (e.g. Millipore ZipTipC18; Harvard/Amika MicroTip). Samples are eluted in 1–2 μl with 50% acetonitrile, 1% formic acid and mixed with matrix. Samples can be further concentrated on the MALDI plate by mixing 0.3 μl sample with 0.3 μl matrix, then successively adding 0.3 μl aliquots of sample before the droplet completely dries; an additional 0.3 μl aliquot of matrix is added at the end. For highest sensitivity, peptides are adsorbed to micro-ZipTips and eluted with ~1 μl of a 1:1 mixture of matrix and 50% acetonitrile, then the eluate is applied to the plate in 0.3 μl aliquots with drying between applications [50]. Sample concentration when using α-CHCA matrix can also be achieved using hydrophobic masked plates purchased commercially [51], or made by stretching Teflon sealing tape (Fisher) onto the plate, securing it with tape, then drying aliquots of sample + matrix onto the Teflon surface [52].

Alternative methods for analyzing in-gel digests include nanospray-MS or microionspray LC–MS. In nanospray-MS [24,25,53], samples are desalted and concentrated on reversed-phase resin microtip columns, and the eluates are transferred into gold-coated glass needles which are then placed near the orifice of an ESI interface. If a closed-tip needle is used, the tip is touched against a metal surface to open it; alternatively, more expensive open-tipped needles can be used. Upon applying a high voltage to the needle, peptides are emitted over a period of more than 20 min, allowing time for analysis of multiple peptides by tandem mass spectrometry (MS/MS). Nanospray spectra can be subject to chemical noise or complicated by orifice fragmentation. High sensitivity microionspray LC–MS also can be carried out using capillary reversed-phase columns [54,55]. Columns of 150 μm inner diameter can be run at 1–2 μl/min, which allows nebulization-assisted ionization of peptides resolved by reversed-phase chromatography. Higher sensitivity can be achieved using columns as small as 75 μm, although these require nanospray nebulization methods.

3.2. Peptide mass fingerprinting

Two MS approaches are commonly used to identify proteins by in-gel digestion — peptide mass fingerprinting or tandem mass spectrometry. Peptide mass fingerprinting compares observed peptide masses from MALDI-TOF or ESI–MS spectra, and compares them against calculated masses based on known protein sequences and specificity of proteolytic cleavages ([15–19] with a good example in ref. [56]). Even as few as 3–5 peptide masses may be sufficient to identify the protein, provided that high mass accuracy measurements (<30 ppm) are attainable [57].

Analysis of peptide mass fingerprinting spectra should first identify and exclude peaks with masses corresponding to common contaminants such as autolyzed trypsin and human keratin, as well as contaminants in blank gel pieces. The remaining masses are then used to search databases, as described later. It is critical to avoid using too many masses in a database search because 'noise' peaks will increase the probability of randomly making an incorrect protein match. It is seldom useful to include a number of masses greater than the expected number of peptides. Usually, 10–15 mass values (minimally at least five mass values) are input for proteins of ~50 kDa. If a statistically significant 'hit' is not found in the first search, removing lower intensity ions from the input list may enhance success in a second search.

Once candidate proteins are identified, the spectra are re-examined for peaks corresponding to predicted peptides that were initially unidentified. Statistical scoring of the top candidate protein, and the difference in the score between the top and next best candidate provides an important indicator of accuracy in protein identification. Confidence increases when the same protein is identified with a high score in replicate digestions [19]. Peptides from trypsin are useful as indicators of peptide recovery, and also serve as standards for internal MS calibration, to enable

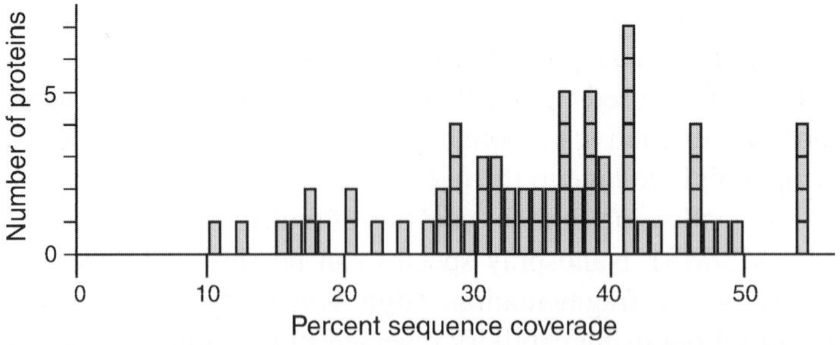

Fig. 2. Sequence coverage of proteins analyzed by peptide mass fingerprinting. Sixty-nine proteins resolved on 2D-gels and visualized by methanol-based silver staining [37] were excised, decolorized with Farmer's reagent and processed by in-gel digestion. Digests were concentrated on ZipTip-C18 columns and analyzed for peptide mass using a PE Voyager DE-STR MALDI-TOF mass spectrometer. Proteins were identified by peptide mass fingerprinting and later confirmed by LC/MS/MS sequencing. Shown are percent sequence coverages represented by peptides observed in MALDI-TOF spectra.

higher mass accuracy of unknown peptides. Masses from porcine trypsin that can be used for calibration are listed on the Protein Prospector website (http://prospector.ucsf.edu).

Fig. 2 shows representative sequence coverages from MALDI-TOF peptide fingerprinting of 69 protein spots harvested from 2D gels stained with silver using methanol fixation. For low abundance proteins, gel pieces were harvested from replicate gels to yield ~0.5 pmol sample prior to in-gel digestion. The sequence coverage for the identified proteins ranged 15%–55%, with average recovery of 35%. The peptide fingerprinting method correctly identified ~70% of the proteins, as confirmed by subsequent ESI-MS/MS sequencing.

Parallel analysis of protein digests by ESI–MS vs. MALDI-TOF generally show higher sequence coverage with ESI–MS than MALDI-TOF (Table 1). Use of C4 or perfusion resins during LC–MS will give higher recovery of larger peptides, while C18 resins allow recovery of smaller peptides with higher chromatographic resolution, which is important in high-throughput analyses. Thus, ESI–MS provides greater coverage than MALDI, and may be advantageous for mass fingerprinting, when performed on instruments with high mass accuracy (e.g. quadrupole TOF).

3.3. Protein identification using MS/MS spectra

The above mass fingerprinting approach is particularly successful with organisms containing simple genomes and few splice variants. With more complex organisms, such as mammals, the probability of identifying a protein from the peptide

Table 1

Recovery of peptides from an in-gel digest of Uroplakin II (47 kDa)

Peptide mass (Da)	MALDI-TOF	ESI–LC/MS (Poros 20 R1)	ESI–LC/MS (C18)	Peptide sequence
175				R
294			Present	FK
375			Present	LQD
391				TDR
476			Present	DVSR
565			Present	TGPYK
644			Present	GPSLDR
767			Present	AEVYASK
949			Present	NASVQDSTK
1263	Present	Present	Present	ASEILNAYLIR
1493	Present	Present	Present	LTLYSAIDTWPGR
1498	Present	Present	Present	SLGTSEPSYTSVNR
1508	Present	Present	Present	TPLSSTFQQTSGGR
1865	Present	Present	Present	TDRLTLYSAIDTWPGR
1936		Present	Present	AAAFDLTPCSDSPSLDAVR
2027		Present	Present	GDADGATSHDSQITQEAVPK
2521		Present	Present	YVLVNMSSGLVQDQTWSDPIR
2951				Glycosylated┐ VGTNGTCLLDPNFQGLC NPPLSAATEYR
3377				SGGMIVITSILGSLPFFLLIGFAGAIVL SLVDR
6006				VNLQPQLASVTFATNNPTLTTVALE KPDCMFDSSAALHGTYEVYLYVLV DSASFR

Compared are peptide coverages of an in-gel tryptic digest analyzed by MALDI-TOF vs. ESI–LC/MS. LC/MS was performed using perfusion (POROS 20 R1) or C18 (Phenomenex Columbus) reversed phase resins packed in 500 μm fused silica capillary columns. Data were collected on PE Voyager DE-STR MALDI-TOF and Sciex API-III + tandem quadrupole mass spectrometers.

mass fingerprint is lower and requires confirmation using MS/MS data or other confirmatory methods. MS/MS data is obtained by analyzing daughter ions generated by peptide fragmentation using triple quadrupole, q-TOF, TOF–TOF, or ion trap instruments, with MALDI, capillary LC/MS, or nanospray interfaces.

Knowledge of peptide chemistry is helpful for interpreting MS/MS spectra and in selected cases assigning peptide sequences by de novo sequencing. Trypsin cleaves specifically and nearly quantitatively at the C-terminus of Lys or Arg, but rarely between Lys-Pro or Arg-Pro. Often, cleavage efficiency is reduced at Lys or

Arg residues adjacent to acidic residues, such as Asp, Glu, phosphoSer, phosphoThr, or phosphoTyr. Because the amino terminal residue of each peptide (except for the C-terminal peptide) is basic, tryptic peptides are usually doubly or triply charged in electrospray MS. Multiply charged ions have a higher probability of fragmenting efficiently, generally yielding both b and y product ions, which are produced by cleavage of amide bonds and extend from the cleavage sites to the N- and C-termini, respectively. These may produce multiply charged fragment ions as well, which can be identified on instruments with enough resolution to distinguish isotopically related ions. In general, MS/MS database search algorithms are more successful with smaller peptides (<3000 Da) and with singly or doubly charged forms. An example of an MS/MS spectrum confirming peptide sequence from b and y fragment ions is shown in Fig. 3A.

Care must be exercised in the event that peptides are covalently modified. Commonly observed mass changes are +16 or +32, reflecting methionine oxidation to sulfoxides or sulfones, respectively. With less frequency, oxidation is also observed at tryptophan, cysteine, and histidine. Some labile peptides undergo dehydration (−18 Da), loss of ammonia (−17 Da), or deamidation of Gln and Asn, respectively, to Glu and Asp (+1 Da) in the acidic MALDI matrices or during ESI; this is exacerbated by prolonged storage of the sample. N-terminal Gln residues can also cyclize with loss of ammonia (−17 Da). Free cysteine can react with various alkylating agents, including acrylamide (+71 Da), oxidized acrylamide (+86 Da), or sulfhydryl reductants used for sample preparation (e.g. +76 Da for 2-mercaptoethanol). For this reason, reduction and alkylation of protein samples before SDS-PAGE is advised. Alkylation with iodoacetamide at low pH (e.g. pH 6.8) may produce a −48 Da artifact, due to loss of CH_3SH from methionine [58].

Post-translational modifications can be identified from peptide mass and MS/MS data, and several database searching algorithms allow MS and MS/MS searching against covalently modified peptide sequences (see Chapters 5 and 6). A comprehensive listing of protein covalent modifications can be viewed at the Delta Mass web site, developed by Ken Mitchelhill (http://www.abrf.org/ABRF/ResearchCommittees/deltamass.html). However, because sequence coverage is low for many proteins, in-gel digestion is not the method of choice for systematically mapping covalent modifications. In addition, in-gel digestion methods often preclude effective analysis of proteins with covalent modifications of higher mass, such as glycosylation (see Chapter 5), which interfere with peptide diffusion out of the gel pieces. Solution digestion of purified proteins is a more efficient approach for mapping covalent modifications. Often, proteins can be affinity purified to enrich modified forms, resolved by SDS-PAGE, then analyzed by MS. This approach can also be applied to the peptides themselves. For example, methods have been described to enrich phosphorylated proteins and peptides (see Chapter 6) by biotinylation followed by avidin binding, or by affinity purification using immobilized metal affinity chromatography [59–62].

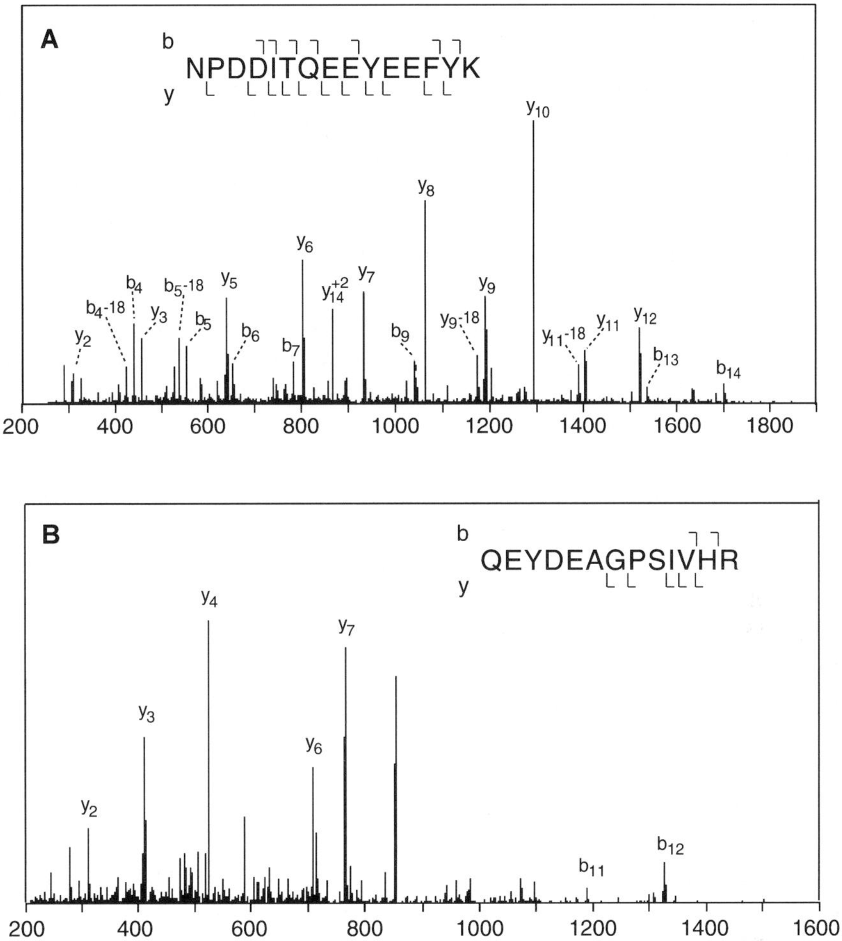

Fig. 3. Examples of MS/MS spectra and results from database searching using the Mascot algorithm. (A) Spectrum of a peptide assigned by Mascot to the sequence NPDDITQEEYEEFYK, which shows an example of a good match. High coverage of b and y ions is observed, summarized by the indicated cleavage sites that generate these ions. Most of the fragment ions are singly charged; however, note the appearance of the doubly charged ion, y_{14}^{+2}, which reports preferential cleavage N-terminal to a proline residue. Dehydration products (− 18 Da) are typically found in spectra of peptides containing carboxylic or alcohol side groups. Finally, most of the high-intensity fragment ions are accounted for by the candidate sequence, thus the assignment is most likely correct. (B) Spectrum of peptide assigned by Mascot to the sequence QEYDEAGPSIVHR, which shows an example of a poor match. Low coverage of fragment ions is observed and several high intensity fragment ions are unaccounted for. In addition, y_6 is less intense than y_7; thus the expected preferential cleavage at the proline N terminus is not apparent, and expected cleavage C-terminal to Asp is not observed. This sequence is most likely misassigned.

3.4. Database searching algorithms

The availability of publicly accessible, web-based analysis tools has been critical for successful analysis of MS and MS/MS data derived from in-gel digests. Online email networks, such as the one supported by The Association of Biomolecular Resource Facilities (ABRF) has also helped greatly in disseminating information and expertise on the use of these methods. Many companies and individual researchers have protocols available on websites, and several labs have developed web-based computer algorithms for searching mass spectrometric data against sequence databases. Examples of widely used search engines that enable analysis by peptide mass fingerprinting, peptide sequencing, and account for post-translational modifications are Protein Prospector (http://prospector.ucsf.edu), Profound (http://prowl.rockefeller.edu/), and PeptideSearch (http://www.narrador.embl-heidelberg.de) [8,17,23,63], which are user friendly and accessible with program documentation on public websites. Programs for commercial licensing are Sequest (Thermo Finnigan), Sonar (Genomic Solutions), and Mascot (Matrix Sciences) [22,64,65]. Mascot is also available on a public website for analyzing limited numbers (<300) of datafiles. Vendors of mass spectrometry instrumentation include database search engines as part of MS software packages. Much information is available online, and many links are available at BasePeak (http://www.spectroscopynow.com/; choose Proteomics then Link).

4. Troubleshooting

4.1. Using standards

Determining the source of problems with in-gel digestion and mass spectrometry can be difficult due to numerous steps in the protocol. Use of standard proteins is vital for troubleshooting, e.g. working out the protocols with myoglobin and BSA. Both proteins have peptides that easily diffuse out of gels, allowing high recovery. Furthermore, myoglobin has no cysteine sulfhydryls and is easily digested by trypsin, whereas BSA has many cysteine sulfhydryls, and provides a good control for peptide recovery after reduction and alkylation.

4.2. What if there are no peptides in the MS spectra?

4.2.1. Is the protease active?

If sample peptides are absent in the MS spectrum, the first thing to observe is whether peaks corresponding to trypsin autolysis products are present or absent. If the in-gel digestion is performed under conditions where trypsin is incorporated into the gel piece and removed from the surrounding solution, the autolysis products report on the protease activity within the gel piece. Absence of autolysis products means that the protease may have been inactivated or incorrectly diluted. Another cause of reduced proteolysis is inadequate neutralization of acid by NH_4HCO_3, e.g. due to inadequate removal of destaining solution. A positive control for proteolysis should be included by digesting a protein standard in solution, in parallel with in-gel digestions.

4.2.2. Is there sufficient peptide recovery?

Extraction depends on efficiency of diffusion from the gel. Too little buffer outside the gel piece (e.g. due to condensation of water at the top of the tube) will inhibit diffusion. Most of the diffusion occurs during the digestion, so simply spinning the liquid down does not allow time for diffusion. The sensitivity of the MS approach used to collect data may not be adequate for the level of protein analyzed, particularly when working with silver stained gels. Parallel solution digestion of a standard protein provides a good test for instrument sensitivity. Peptides can also be lost during desalting and concentration. We have seen lot variations in commercial ZipTips that led to irreversible adsorption of peptides, and therefore recommend testing each new lot of ZipTips with a standard myoglobin digest.

4.3. Common contaminants

Contaminants can be so prominent that they obscure weaker ions. Although peptides from autolyzed proteases are inevitable, human K1 and K10 keratins derived from lab dust are common contaminants in mass spectra. Keratin contamination can be minimized by careful sample handling. For contaminants that are difficult to remove, LC/MS analysis using a long gradient may allow separation and detection of low-abundance peptides of interest.

Compounds such as detergents or glycerol, which are often carried through during gel processing, will inhibit crystallization for MALDI and will also interfere with LC/MS analysis. LC/MS and nanospray analyses are also severely compromised by polymers (usually multiples of 44 Da), often derived from plasticizers and mold release compounds contaminating tubes, pipet tips, and even glassware. Sample containers can be tested for contamination by mock extraction in the absence of protein and, analysis of extracted polymers by mass spectrometry. Formic acid is a useful solvent for removing almost any polymeric contamination. Containers should be washed with 88% formic acid, followed by multiple rinses with water. Buffers can also become contaminated with diatoms, bacteria, and mold, which secrete polymers. This is particularly a problem with SDS and Tris. In some cases, acrylamide can also contribute contaminants, particularly if many gel pieces are pooled for in-gel digestions. Longer prewashing of samples will usually eliminate contaminants derived from acrylamide, and buffer blanks and gel pieces cut from blank areas are useful controls for buffer contaminants.

4.4. Evaluating database 'hits'

Problems are commonly encountered in evaluating the significance of protein identification by database searching. Several search engine web sites have good discussions of statistical scoring. Troubleshooting methods can be performed using peptide masses corresponding to known proteins as positive controls. More severe are false positive candidates from analysis of MS or MS/MS spectra, which preclude accurate protein identification.

Heterogeneity is a concern when more than one protein is present in a stained spot or gel band. This complicates peptide mass fingerprinting results, and results in many observed peptides that cannot be identified. The best approach is to analyze these samples by MS/MS, although some peptide mass finger-printing database searching algorithms allow facile searching of unidentified peptides.

To verify the results of database searches using MS/MS data, the MS/MS spectra should be examined to determine whether observed ions are consistent with the amino acid composition and chemistry of the candidate sequence. In general, $>80\%$ of total intensity in MS/MS spectra should be accounted for by product ions, internal product ions, or ions produced by covalent modification during mass spectrometry (e.g. dehydration, deamination). A good assignment should account for all major ions (those with signal to noise > 15). Additional criteria for reliability of sequence assignments include: (i) expected b or y fragment ions showing successive cleavages over several residues; (ii) efficient cleavages at Ile/Leu, Glu/Asp, and Pro residues; and (iii) immonium ions corresponding to residues in the sequence. Absence of preferential cleavage N-terminal to Pro and C-terminal to Asp is a common diagnostic for incorrect assignments. Examples of good and poor sequence matches to MS/MS spectra are shown in Fig. 3A and B.

5. Summary

Analysis of proteins by in-gel digestions provides a convenient, rapid, and easy analytical technique that can be applied using any modern mass spectrometer. Furthermore, improved methods for recovery of peptides from gels should become available in the near future. As highly sensitive mass spectrometers are becoming widespread and easier to use, more laboratories can exploit this technique to tackle biology and clinical problems.

Acknowledgements

We are indebted to former and current colleagues who have made these protocols work in our labs, particularly Jennifer Yeh, Timothy Lewis, John Hunt, Lauren Aveline, Karen Jonscher, David Friedman, Karine Bernard, Yukihito Kabuyama, Rebecca Schweppe, and Claire Haydon.

References

1. Cleveland, D.W., Fischer, S.G., Kirschner, M.W. and Laemmli, U.K., Peptide mapping by limited proteolysis in sodium dodecyl sulfate and analysis by gel electrophoresis. *J. Biol. Chem.*, **252**, 1102–1106 (1977).
2. Eckerskorn, C. and Lottspeich, F., Internal amino acid sequence analysis of proteins separated by gel electrophoresis after tryptic digestion in polyacrylamide matrix. *Chromatographia*, **28**, 92–94 (1989).
3. Rosenfeld, J., Capdevielle, J., Guillemot, J.C. and Ferrara, P., In-gel digestion of proteins for internal sequence analysis after one- or two-dimensional gel electrophoresis. *Anal. Biochem.*, **15**, 173–179 (1992).
4. Kassel, D.B., Musselman, B.D. and Smith, J.A., Primary structure determination of peptides and enzymatically digested proteins using capillary liquid chromatography/mass spectrometry and rapid linked-scan techniques. *Anal. Chem.*, **63**, 1091–1097 (1991).
5. Davis, M.T. and Lee, T.D., Analysis of peptide mixtures by capillary high performance liquid chromatography: a practical guide to small-scale separations. *Protein Sci.*, **1**, 935–944 (1992).
6. Hess, D., Covey, T.C., Winz, R., Brownsey, R.W. and Aebersold, R., Analytical and micro-preparative peptide mapping by high performance liquid chromatography/electrospray mass spectrometry of proteins purified by gel electrophoresis. *Protein Sci.*, **2**, 1342–1351 (1993).
7. Resing, K.A., Johnson, R.S. and Walsh, K.A., Characterization of protease processing sites during conversion of rat profilaggrin to filaggrin. *Biochemistry*, **32**, 10036–10045 (1993).
8. Mann, M. and Wilm, M., Error-tolerant identification of peptides in sequence databases by peptide sequence tags. *Anal. Chem.*, **66**, 4390–4399 (1994).
9. Matsumoto, H., Kurien, B.T., Takagi, Y., Kahn, E.S., Kinumi, T., Komori, N., Yamada, T., Hayashi, F., Isono, K. and Pak, W.L., Phosrestin I undergoes the earliest light-induced phosphorylation by a calcium/calmodulin-dependent protein kinase in *Drosophila* photoreceptors. *Neuron*, **12**, 997–1010 (1994).
10. Hellman, U., Wenstedt, C., Gonez, J. and Heldin, C.H., Improvement of an "in-gel" digestion procedure for the micropreparation of internal protein fragments for amino acid sequencing. *Anal. Biochem.*, **224**, 451–455 (1995).
11. Norbeck, J. and Blomberg, A., Gene linkage of two-dimensional polyacrylamide gel electrophoresis resolved proteins from isogene families in *Saccharomyces cerevisiae* by microsequencing of in-gel trypsin generated peptides. *Electrophoresis*, **16**, 149–156 (1995).
12. Jeno, P., Mini, T., Moes, S., Hintermann, E. and Horst, M., Internal sequences from proteins digested in polyacrylamide gels. *Anal. Biochem.*, **224**, 75–82 (1995).
13. Billeci, T.M. and Stults, J.T., Tryptic mapping of recombinant proteins by matrix-assisted laser desorption/ionization mass spectrometry. *Anal. Chem.*, **65**, 1709–1716 (1993).
14. Zhang, W., Czernik, A.J., Yungwirth, T., Aebersold, R. and Chait, B.T., Matrix-assisted laser desorption mass spectrometric peptide mapping of proteins separated by two-dimensional gel electrophoresis: determination of phosphorylation in synapsin I. *Protein Sci.*, **3**, 677–686 (1994).

15. Henzel, W.J., Billeci, T.M., Stults, J.T., Wong, S.C., Grimley, C. and Watanabe, C., Identifying proteins from two-dimensional gels by molecular mass searching of peptide fragments in protein sequence databases. *Proc. Natl Acad. Sci. USA*, **90**, 5011–5015 (1993).
16. Clauser, K.R., Hall, S.C., Smith, D.M., Webb, J.W., Andrews, L.E., Tran, H.M., Epstein, L.B. and Burlingame, A.L., Rapid mass spectrometric peptide sequencing and mass matching for characterization of human melanoma proteins isolated by two-dimensional PAGE. *Proc. Natl Acad. Sci. USA*, **92**, 5072–5076 (1995).
17. Zhang, W. and Chait, B.T., ProFound: an expert system for protein identification using mass spectrometric peptide mapping information. *Anal. Chem.*, **72**, 2482–2489 (2000).
18. Pappin, D.J.C., Hojrup, P. and Bleasby, A.J., Rapid identification of proteins by peptide mass fingerprinting. *Curr. Biol.*, **3**, 327–332 (1993).
19. James, P., Quadroni, M., Carafoli, E. and Gonnet, G., Protein identification in DNA databases by peptide mass fingerprinting. *Protein Sci.*, **3**, 1347–1350 (1994).
20. Taylor, J.A., Walsh, K.A. and Johnson, R.S., Sherpa: a Macintosh-based expert system for the interpretation of electrospray ionization LC/MS and MS/MS data from protein digests. *Rapid Commun. Mass Spectrom.*, **10**, 679–687 (1996).
21. Mortz, E., O'Connor, P.B., Roepstorff, P., Kelleher, N.L., Wood, T.D., McLafferty, F.W. and Mann, M., Sequence tag identification of intact proteins by matching tandem mass spectral data against sequence data bases. *Proc. Natl Acad. Sci. USA*, **93**, 8264–8267 (1996).
22. Eng, J.K., McCormack, A.L. and Yates, J.R., An approach to correlate tandem mass spectral data of peptides with amino acid sequences in a protein database. *J. Am. Soc. Mass Spectrom.*, **5**, 976–989 (1994).
23. Clauser, K.R., Baker, P.R. and Burlingame, A.L., Role of accurate mass measurement (+/ − 10 ppm) in protein identification strategies employing MS or MS/MS and database searching. *Anal. Chem.*, **71**, 2871–2882 (1999).
24. Wilm, M. and Mann, M., Analytical properties of the nanoelectrospray ion source. *Anal. Chem.*, **68**, 1–8 (1996).
25. Wilm, M.A., Shevchenko, A., Houthaeve, T., Breit, S., Schweigerer, L., Fotsis, T. and Mann, M., Femtomole sequencing of proteins from polyacrylamide gels by nano-electrospray mass spectrometry. *Nature*, **379**, 466–469 (1996).
26. Westermeier, R., Postel, W., Weser, J. and Gorg, A., High-resolution two-dimensional electrophoresis with isoelectric focusing in immobilized pH gradients. *J. Biochem. Biophys. Methods*, **8**, 321–330 (1983).
27. Gorg, A., Obermaier, C., Boguth, G., Harder, A., Scheibe, B., Wildgruber, R. and Weiss, W., The current state of two-dimensional electrophoresis with immobilized pH gradients. *Electrophoresis*, **21**, 1037–1053 (2000).
28. Muller, E.C., Thiede, B., Zimny-Arndt, U., Scheler, C., Prehm, J., Muller-Werdan, U., Wittmann-Liebold, B., Otto, A. and Jungblut, P., High-performance human myocardial two-dimensional electrophoresis database: edition 1996. *Electrophoresis*, **17**, 1700–1712 (1996).
29. Shevchenko, A., Jensen, O.N., Podtelejnikov, A.V., Sagliocco, F., Wilm, M., Vorm, O., Mortensen, P., Shevchenko, A., Boucherie, H. and Mann, M., Linking genome and proteome by mass spectrometry: large-scale identification of yeast proteins from two dimensional gels. *Proc. Natl Acad. Sci. USA*, **93**, 14440–14445 (1997).
30. Wasinger, V.C., Cordwell, S.J., Cerpa-Poljak, A., Yan, J.X., Gooley, A.A., Wilkins, M.R., Duncan, M.W., Harris, R., Williams, K.L. and Humphery-Smith, I., Progress with gene-product mapping of the mollicutes: *Mycoplasma genitalium*. *Electrophoresis*, **16**, 1090–1094 (1995).
31. Wilkins, M.R., Sanchez, J.C., Gooley, A.A., Appel, R.D., Humphery-Smith, I., Hochstrasser, D.F. and Williams, K.L., Progress with proteome projects: why all proteins expressed by a genome should be identified and how to do it. *Biotechnol. Genet. Engng Rev.*, **13**, 19–50 (1996).

32. Ortiz, M.L., Calero, M., Fernandez-Patron, C., Patron, C.F., Castellanos, L. and Mendez, E., Imidazole–SDS–Zn reverse staining of proteins in gels containing or not SDS and microsequence of individual unmodified electroblotted proteins. *FEBS Lett.*, **296**, 300–304 (1992).
33. Ferreras, M., Gavilanes, J.G. and Garcia-Segura, J.M., A permanent Zn^{+2} reverse staining method for the detection and quantification of proteins in polyacrylamide gels. *Anal. Biochem.*, **213**, 206–212 (1993).
34. Castellanos-Serra, L., Vallin, A., Proenza, W., Le Caer, J.P. and Rossier, J., An optimized procedure for detection of proteins on carrier ampholyte isoelectric focusing and immobilized pH gradient gels with imidazole and zinc salts: its application to the identification of isoelectric focusing separated isoforms by in-gel proteolysis and mass spectrometry analysis. *Electrophoresis*, **22**, 1677–1685 (2001).
35. Lopez, M.F., Berggren, K., Chernokalskaya, E., Lazarev, A., Robinson, M. and Patton, W.F., A comparison of silver stain and SYPRO Ruby protein gel stain with respect to protein detection in two-dimensional gels and identification by peptide mass profiling. *Electrophoresis*, **21**, 3673–3683 (2000).
36. Blum, H., Beier, H. and Gross, H.J., Improved silver staining of plant proteins, RNA and DNA in polyacrylamide gels. *Electrophoresis*, **8**, 93–99 (1987).
37. Rabilloud, T., A comparison between low background silver diammine and silver nitrate protein stains. *Electrophoresis*, **13**, 429–439 (1992).
38. Shevchenko, A., Wilm, M., Vorm, O. and Mann, M., Mass spectrometric sequencing of proteins from silver-stained polyacrylamide gels. *Anal. Chem.*, **68**, 850–858 (1996).
39. Gharahdaghi, F., Weinberg, C.R., Meagher, D.A., Imai, B.S. and Mische, S.M., Mass spectrometric identification of proteins from silver-stained polyacrylamide gels: a method for the removal of silver ions to enhance sensitivity. *Electrophoresis*, **20**, 601–605 (1999).
40. Sumner, L.S., Wolf-Sumner, B., White, S.P. and Asirvatham, V.S., Silver stain removal using H_2O_2 for enhanced peptide mass mapping by MALDI-TOF mass spectrometry. *Rapid Commun. Mass Spectrom.*, **16**, 160–168 (2002).
41. Shevchenko, A. and Shevchenko, A., Evaluation of the efficiency of in-gel digestion of proteins by peptide isotopic labeling and MALDI mass spectrometry. *Anal. Biochem.*, **296**, 279–283 (2001).
42. Ross, A.R., Lee, P.J., Smith, D.L., Langridge, J.I., Whetton, A.D. and Gaskell, S.J., Identification of proteins from two-dimensional polyacrylamide gels using a novel acid-labile surfactant. *Proteomics*, **2**, 928–936 (2002).
43. Konig, S., Schmidt, O., Rose, K., Thanos, S., Besselmann, M. and Zeller, M., Sodium dodecyl sulfate versus acid-labile surfactant gel electrophoresis: comparative proteomic studies on rat retina and mouse brain. *Electrophoresis*, **24**, 751–756 (2003).
44. Wattenberg, A., Organ, A.J., Schneider, K., Tyldesley, R., Bordoli, R. and Bateman, R.H., Sequence dependent fragmentation of peptides generated by MALDI quadrupole time-of-flight (MALDI Q-TOF) mass spectrometry and its implications for protein identification. *J. Am. Soc. Mass Spectrom.*, **13**, 772–783 (2002).
45. Krutchinsky, A.N., Kalkum, M. and Chait, B.T., Automatic identification of proteins with a MALDI-quadrupole ion trap mass spectrometer. *Anal. Chem.*, **73**, 5066–5077 (2001).
46. Baldwin, M.A., Medzihradszky, K.F., Lock, C.M., Fisher, B., Settineri, T.A. and Burlingame, A.L., Matrix-assisted laser desorption/ionization coupled with quadrupole/orthogonal acceleration time-of-flight mass spectrometry for protein discovery, identification, and structural analysis. *Anal. Chem.*, **73**, 1707–1720 (2001).
47. Yergey, A.L., Coorssen, J.R., Backlund, P.S., Blank, P.S., Humphrey, G.A., Zimmerberg, J., Campbell, J.M. and Vestal, M.L., De novo sequencing of peptides using MALDI/TOF–TOF. *J. Am. Soc. Mass Spectrom.*, **13**, 784–791 (2002).

48. Cohen, S.L. and Chait, B.T., Influence of matrix solution conditions on the MALDI-MS analysis of peptides and proteins. *Anal. Chem.*, **68**, 31–37 (1996).
49. Katayama, H., Nagasu, T. and Oda, Y., Improvement of in-gel digestion protocol for peptide mass fingerprinting by matrix-assisted laser desorption/ionization time-of-flight mass spectrometry. *Rapid Commun. Mass Spectrom.*, **15**, 1416–1421 (2001).
50. Gobom, J., Nordhoff, E., Mirgorodskaya, E., Ekman, R. and Roepstorff, P., Sample purification and preparation technique based on nano-scale reversed-phase columns for the sensitive analysis of complex peptide mixtures by matrix-assisted laser desorption/ionization mass spectrometry. *J. Mass Spectrom.*, **34**, 105–116 (1999).
51. Schuerenberg, M., Luebbert, C., Eickhoff, H., Kalkum, M., Lehrach, H. and Nordhoff, E., Prestructured MALDI-MS sample supports. *Anal. Chem.*, **72**, 3436–3442 (2000).
52. Yuan, X. and Desiderio, D.M., Protein identification with teflon as matrix-assisted laser desorption/ionization sample support. *J. Mass Spectrom.*, **37**, 512–524 (2002).
53. Körner, R., Wilm, M., Morand, K., Schubert, M. and Mann, M., Nano electrospray combined with a quadrupole ion trap for the analysis of peptides and protein digests. *J. Am. Soc. Mass Spectrom.*, **7**, 150–156 (1996).
54. Emmett, M.R. and Caprioli, R.M., Micro-electrospray mass spectrometry — ultra high-sensitivity analysis of peptides and proteins. *J. Am. Soc. Mass Spectrom.*, **5**, 605–613 (1994).
55. Gale, D.C. and Smith, R.D., Small-volume and low flow-rate electrospray ionization mass spectrometry of aqueous samples. *Rapid Commun. Mass Spectrom.*, **7**, 1017–1021 (1993).
56. Wigge, P.A., Jensen, O.N., Holmes, S., Soues, S., Mann, M. and Kilmartin, J.V., Analysis of the Saccharomyces spindle pole by matrix-assisted laser desorption/ionization (MALDI) mass spectrometry. *J. Cell Biol.*, **141**, 967–977 (1998).
57. Jensen, O.N., Podtelejnikov, A.V. and Mann, M., Identification of the components of simple protein mixtures by high-accuracy peptide mass mapping and database searching. *Anal. Chem.*, **69**, 4741–4750 (1997).
58. Lapko, Y.N., Smith, D.L. and Smith, J.B., Identification of an artifact in the mass spectrometry of proteins derivatized with iodoacetamide. *J. Mass Spectrom.*, **35**, 572–575 (2000).
59. Oda, Y., Nagasu, T. and Chait, B.T., Enrichment analysis of phosphorylated proteins as a tool for probing the phosphoproteome. *Nat. Biotechnol.*, **19**, 379–382 (2001).
60. Zhou, H., Ranish, J.A., Watts, J.D. and Aebersold, R., Quantitative proteome analysis by solid-phase isotope tagging and mass spectrometry. *Nat. Biotechnol.*, **20**, 512–515 (2002).
61. Andersson, L. and Porath, J., Isolation of phosphoproteins by immobilized metal (Fe^{3+}) affinity chromatography. *Anal. Biochem.*, **154**, 250–254 (1986).
62. Ficarro, S.B., McCleland, M.L., Stukenberg, P.T., Burke, D.J., Ross, M.M., Shabanowitz, J., Hunt, D.F. and White, F.M., Phosphoproteome analysis by mass spectrometry and its application to *Saccharomyces cerevisiae*. *Nat. Biotechnol.*, **20**, 301–305 (2002).
63. Fenyö, D., Identifying the proteome: software tools. *Curr. Opin. Biotechnol.*, **11**, 391–395 (2000).
64. Yates, J.R., Morgan, S.F., Gatlin, C.L., Griffin, P.R. and Eng, J.K., Method to compare collision-induced dissociation spectra of peptides: potential for library searching and subtractive analysis. *Anal. Chem.*, **70**, 3557–3565 (1998).
65. Perkins, D.N., Pappin, D.J., Creasy, D.M. and Cottrell, J.S., Probability-based protein identification by searching sequence databases using mass spectrometry data. *Electrophoresis*, **20**, 3551–3567 (1999).

Proteome Analysis. Interpreting the Genome.
D.W. Speicher (editor)

Chapter 8

The use of accurate mass and time tags based upon high-throughput Fourier transform ion cyclotron resonance mass spectrometry for global proteomic characterization

DAVID G. CAMP II* and RICHARD D. SMITH

Macromolecular Structure and Dynamics, Environmental Molecular Sciences Laboratory, Battelle, Pacific Northwest National Laboratory, P.O. Box 999/MSIN K8-98, Richland, WA 99352, USA

* Corresponding author. Tel.: +1-509-376-7535; fax: +1-509-376-7722. E-mail: dave.camp@pnl.gov (D.G. Camp).

1. Introduction

The ability to study how the biologically active molecular constituents of a cell or organism change and interact following a perturbation provides a foundation for understanding the function(s) of their component parts, and ultimately how the system operates. Developing this ability requires a global perspective in both modeling and experimentation to study changes in cellular systems under different conditions, e.g. to identify single gene or sets of gene products most sensitive to perturbations or to establish a set of nodes most sensitive to multiple small perturbations, needed to produce a specific systems-level response. Methods that simultaneously assess the abundances of thousands of expressed genes at the mRNA level are now broadly applied [1,2] with more or less success. However, post-transcriptional processes play a major role in determining protein abundances and modification states, and protein abundances can show poor correlation with mRNA levels [3–5]. Thus, considerable attention is now focused on the proteome, the complement of proteins expressed by a particular cell, organism, or tissue at a given time or under a specific set of environmental conditions.

Current proteome analysis capabilities are predominantly based on protein separations using two-dimensional polyacrylamide gel electrophoresis (2D PAGE), a method that can resolve up to thousands of putative protein 'spots'. Proteome coverage in 2D PAGE is problematic for proteins that have very high or low isoelectric points ($\lesssim 3.5$ and $\gtrsim 9.5$) or extremes in molecular weight, and also for membrane proteins (due to solubility issues during sample processing). In composite, these kinds of proteins typically account for more than half of all proteins. The number of spots can be poorly correlated with the number of various proteins detected and the detected proteins are predominantly in high abundance based on their codon bias [5]. Furthermore, a single gene can give rise to multiple spots [5] due to co- and post-translational modifications, degradation intermediates, and alternative expression (e.g. alternative splicing of mRNAs, translational frame shifts). Conventionally, protein identification after 2D PAGE involves the separate extraction, digestion, and analysis of each spot using mass spectrometry (MS) [6–8], but this approach remains lacking in proteome coverage, sensitivity, dynamic range, throughput, and the precision needed to discern small, but often biologically important, changes in protein abundances. The sensitivity of 2D PAGE is generally limited to femtomole levels [8,9] by the need to visualize the protein spot on the gel and to subsequently process and analyze the spot. The largest 2D PAGE/MS study reported to date identified 502 proteins from *Haemophilus influenzae* [10]. Similarly, the most comprehensive yeast proteome 2D PAGE/MS studies published to date (the broadest of which identified 279 proteins [11], and a combined total of only ~500 [7,11–14]) provide a skewed codon bias distribution [15], indicating that only the more abundant proteins were detected. Many important regulatory proteins are expressed at such low levels (e.g. <1000 copies per cell) that their detection is precluded unless 2D PAGE is preceded by extensive fractionation of large quantities of protein and/or processing of a large numbers of gels. Finally, the precision of protein abundance determinations using 2D PAGE is based on comparison of protein spot intensities, limiting the capability for discerning subtle differences in protein abundances for large numbers of proteome-wide measurements (e.g. from time course studies).

Many alternative proteomics technologies presently being evaluated employ a separation methodology combined with some form of MS, most typically applied after protein digestion using specific proteases (e.g. trypsin). Analysis of polypeptides sufficiently large enough for protein identification, typically in the $\gtrsim$ 5- to 10-mer size range depending on sequence uniqueness for a specific organism, is now effectively achieved by MS [16–19]. A widely used approach involves MS selection of a polypeptide which is then dissociated to form fragments whose mass-to-charge (m/z) ratios are measured. This 'tandem' MS/MS analysis provides primary sequence-related information that allows the peptide, and most often its parent protein, to be identified provided the peptide sequence is contained in an appropriate database [20]. MS/MS analysis of only one polypeptide is often sufficient for protein identification [20–24].

A current, strong trend in proteomic characterization involves approaches that utilize liquid phase separations in direct combination with MS. Recently, Washburn et al. demonstrated this approach by identifying 1484 yeast proteins by a 2D capillary liquid chromatography (LC)–MS/MS strategy, in which peptides were separated in the first dimension by cation exchange LC into 15 fractions that were subsequently separated by reversed phase LC [25]. While this work demonstrates the potential for broad proteome coverage based on the analysis of highly complex polypeptide mixtures, the approach still leaves much to be desired in terms of speed, sensitivity, comprehensiveness and confidence of protein identifications, and the quantitative utility of the measurement method. The major bottleneck associated with such approaches is the need to conduct tandem MS measurements on extremely large numbers of putative peptides in order to achieve confident identifications.

In this review the technological basis and progress towards a new global proteomics strategy are described that uses peptide accurate mass measurements augmented by information from separations (e.g. LC retention times) to provide large improvements in sensitivity, dynamic range, comprehensiveness, and throughput. The use of accurate mass and time (AMT) tags serves to eliminate the need for routine MS/MS measurements [26]. As a case study, the research efforts in this laboratory illustrate the role of AMT tags within the broader context of a state-of-the-art proteomics effort. This strategy exploits high-resolution capillary LC separations combined with Fourier transform ion cyclotron resonance mass spectrometry (FTICR). AMT tags represent peptide biomarkers and can be used to confidently identify proteins based on the high mass measurement accuracy

(MMA) provided by FTICR combined with LC elution times. Once identified using MS/MS, these biomarkers provide the foundation for subsequent high throughput studies using only AMT tags to identify and quantify the proteins expressed within a cell system. Key attractions of this approach include the feasibility of completely automated high-confidence protein identifications, extensive proteome coverage, and the capability for exploiting stable-isotope labeling methods for high precision abundance measurements [27]. Additional developments described in this review include methods for more effective coverage of membrane proteins [28], for dynamic range expansion of proteome measurements [29], and for multi-stage separations that promise to enable more focused analyses, to further extend the quality of measurements, and also extend measurements to more complex proteomes.

2. Overall experimental and data processing approach

The current approach to generating AMT tags involves a two-stage process (Fig. 1). First, the proteome sample is digested (e.g. with trypsin) and analyzed by high-efficiency capillary LC–MS/MS using either a conventional (e.g. LCQ ion trap or Q-TOF) mass spectrometer operating in a data-dependent mode or using FTICR to obtain tentative peptide identifications or potential mass tags (PMTs). The PMTs are subsequently validated as AMT tags provided the predicted peptide's accurate mass is observed using FTICR in a corresponding sample and at an equivalent elution time [27].

In earlier studies in this laboratory, MS/MS methods generated PMTs initially based on identification 'scores' produced by the SEQUEST search program, in which identifications are based on similarity (cross-correlation) of the spectrum with a set of peaks predicted from the most common known peptide fragmentation processes. Due to the nature of the analysis, the results invariably range from low scores where identifications are highly doubtful to high scores where identifications are quite reliable with no clear line of demarcation. Using only the highest scores for identification, fewer proteins are identified; however, uncritical use of lower scores results in many false identifications. Conventionally, many MS/MS spectra need to be manually examined to establish acceptable confidence in identifications. This process generally results in discarding a substantial fraction of the peptides identified with lower scores, and serves to increase confidence to an extent that is difficult to quantify. In the approach described here, using highly accurate mass measurements provides an additional, high-quality 'test' for tentative peptide identifications that can be applied to the data analysis. An advantage of automated validation of AMT tags from PMTs is the increased confidence in peptide identifications. Once a protein has been identified using AMT tags, its subsequent identification and quantitation in other

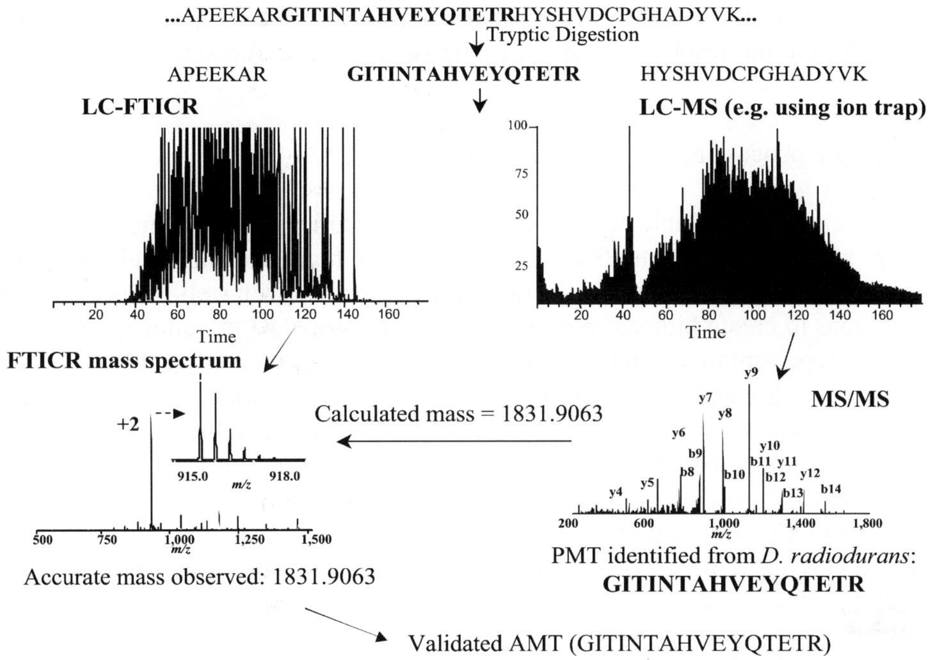

Fig. 1. The AMT tag strategy. Experimental steps involved in establishing an AMT tag are illustrated by the identification of an AMT tag for elongation factor Tu (EF-Tu). Peptides are automatically selected for collisional induced dissociation (CID) and tentatively identified as a potential mass tag (PMT) using an automated search program (SEQUEST). In this example a tryptic peptide from EF-Tu (in bold) was identified by MS/MS using an ion trap mass spectrometer. The accurate mass of this PMT was calculated based on its sequence (i.e. 1831.9063 Da) and its elution time recorded. In the second stage, the same proteome sample is analyzed under the same LC–MS conditions using a high-field FTICR mass spectrometer. An AMT tag is established when a peptide eluting at the same time and corresponding to the calculated mass (within 1 ppm) of the PMT identified in the first stage is observed. This peptide then functions as a biomarker to identify this particular protein in all subsequent experiments analyzing a proteome sample from a specific organism. In the LC–FTICR analysis of the same sample, a doubly charged peptide was observed at this same elution time, having a mass within 1 ppm (i.e. 1831.9063 Da) of the calculated mass of this peptide. This peptide is then considered an AMT tag for EF-Tu within the *D. radiodurans* proteome.

studies is based on peptide accurate mass FTICR measurements (in correlation with the elution time of the measured peptides), with much greater sensitivity than with conventional MS instrumentation.

Additionally, once an AMT tag has been established, it can be used to confidently identify a specific protein in subsequent proteome studies. By eliminating the need to re-establish the identity of a peptide using MS/MS analyses, multiple high-throughput studies focused on measuring changes in relative protein abundances between two (or more) different proteomes are facilitated. In such comparative studies, stable isotope labeling methods can be used as a means to measure protein relative abundances, a process that also benefits from the resolution and sensitivity of the FTICR measurements.

3. Sample processing

Tightly controlled sample preparations are key to quality proteomic measurements using high-sensitivity capillary LC coupled to high-throughput electrospray ionization (ESI) FTICR MS. The sample process begins with well-defined culture conditions to create biomass representative of a given state of cellular material.

In a representative study, *Saccharomyces cerevisiae* (yeast), *Deinococcus radiodurans* R1 and murine B16 melanoma cells were cultured in appropriate media, harvested, the proteins were extracted from the cells, and tryptically digested as previously described [30,31]. MS studies indicated that digestion was complete for most detected proteins based on trypsin cleavage specificity (i.e. C-terminal of Lys and Arg residues), although some proteins displayed up to four missed cleavages, which is consistent with the greater stability of some proteins. Such incomplete digestion products were explicitly included in the data analysis, and did not significantly affect results.

3.1. Fractionation of complex peptide mixtures

Greater sensitivity for initial PMT peptide identifications was obtained by employing an initial peptide fractionation step. This was performed by gradient ion exchange chromatography as previously described [27]. Fifteen-second fractions were collected and pooled for capillary LC–MS/MS analyses for PMT identifications.

3.2. Preparation of membrane proteins

The preparation of *D. radiodurans* membrane proteins employed a combination of carbonate extraction and methanol-assisted solubilization to enrich for integral membrane proteins. A bacterial pellet was diluted in 50 mM Tris (pH 7.3) using a 1:3 (mg/ml) cells-to-buffer ratio. The cell suspension was lysed with three presses in a French Pressure Cell (Amnico, Rochester, NY) at 16,000 psi. Unbroken cells and large debris were removed by centrifugation at 3200*g* for 10 min at 4°C. The supernatant containing soluble membrane vesicles was collected.

Membrane proteins were isolated from vesicles in the retained supernatant using a slightly modified carbonate fractionation procedure [32,33]. The supernatant was diluted (1:10 (v/v), supernatant-to-buffer ratio) with ice-cold 100 mM sodium carbonate (pH 11). At 4°C the solution was agitated using a shaker/rotisserie (Labquake, Conroe, TX) for 1.5 h and then centrifuged at

115,000*g* for 1 h. The resulting supernatant was discarded, the membrane pellet rinsed with deionized water, washed in 50 mM Tris (pH 7.3), and pelleted at 115,000*g* for 20 min. The pellet was resuspended in 50 mM ammonium bicarbonate (pH 8.0) and the proteins were thermally denatured using a previously described in-solution technique [34].

The resuspended and thermally denatured proteins were solubilized by dilution to a final concentration of 60% methanol (v/v) combined with intermittent vortexing and sonication. Tryptic digestion was performed using a mixed organic–aqueous solvent proteolysis technique previously described [35]. The proteolysis was performed for 5 h at 37°C (1:20 (w/w), trypsin-to-protein ratio) and the reaction stopped by snap freezing with liquid nitrogen. The sample was stored at − 80°C until LC–MS/MS or LC–FTICR analysis.

3.3. Stable-isotope labeling methods

The strategy of metabolic labeling allows the entire proteome of an organism to be investigated by determining relative protein abundances of cells cultured under two distinct growth conditions. This was accomplished in this study by measuring the ratios of isotopically labeled peptides produced from equal numbers of cells from each culture state.

For stress studies with hydrogen peroxide or radiation, *D. radiodurans* cells were cultured on TGY media until mid-log phase, H_2O_2 was added to a final concentration of 60 μM, and the cells were further incubated for 2 h and then promptly harvested. Cells cultured in ^{15}N-labeled media (Bioexpress; Cambridge Isotopes) [31] were harvested at mid-log phase. ^{15}N cultured and H_2O_2 (^{14}N) stressed *D. radiodurans* were mixed, processed, and analyzed as described earlier. The two versions of each peptide AMT tag, differing in mass by the number of nitrogen atoms, allowed the pair to be identified in an automated fashion with high confidence. It was also possible to distinguish the two versions due to the distinctive isotopic distribution of ^{15}N-labeled peptides.

^{14}N/^{15}N-metabolic labeling was combined with a commercially available Cys-affinity tag, iodoacetyl–polyethylene oxide(PEO)–biotin, to derivatize and isolate Cys-polypeptides [31], similar to the ICAT™ approach described by Gygi et al. [36]. The protein mixture was denatured by the addition of guanidine hydrochloride to a final concentration of 6 M combined with boiling. The denatured proteins were subsequently reduced by incubating in a final concentration of 5 mM tributylphosphine for 1 h at 37°C. Iodoacetyl–PEO–biotin was then added to an estimated 5-fold excess of the number of Cys residues and incubated while stirring for 90 min in the dark. The iodoacetyl–PEO–biotin derivatized sample was desalted into 100 mM NH_4HCO_3 (pH 8.4) with 5 mM EDTA, and digested with trypsin (1:50 (w/w), enzyme-to-protein ratio) overnight at 37°C. After boiling for 5 min, the samples were loaded onto an avidin column and washed with five bed volumes of 50 mM NH_4HCO_3 (pH 8.4). The bound Cys-polypeptides were eluted using 30% acetonitrile with 0.4% trifluoroacetic acid.

Phosphoprotein isotope-coded affinity tag (PhIAT) labeling [37,38] was employed for determining the relative abundances of phosphorylated proteins under different cellular states. The PhIAT approach is a post-extraction, stable isotope, affinity-labeling technique that combined the methodology described earlier for the iodoacetyl–PEO–biotin derivatization of free sulfhydryls immediately following the selective de-phosphorylation of phosphoserine and/or phosphothreonine residues via hydroxide ion-mediated, beta-elimination of the phosphate group and the subsequent Michael addition of 1,2-ethanedithiol (EDT) to the newly formed alpha, beta-unsaturated residues (Fig. 2). The addition of EDT, synthesized with either four aliphatic hydrogen atoms (D0-EDT) or four aliphatic deuterium atoms (D4-EDT), provided the stable isotopic label. The free sulfhydryls present on the EDT-labeled proteins were reacted using iodoacetyl–PEO–biotin to produce PhIAT-labeled proteins. The light isotopically labeled (PhIAT-D0) and heavy isotopically labeled (PhIAT-D4) peptide pairs differ in

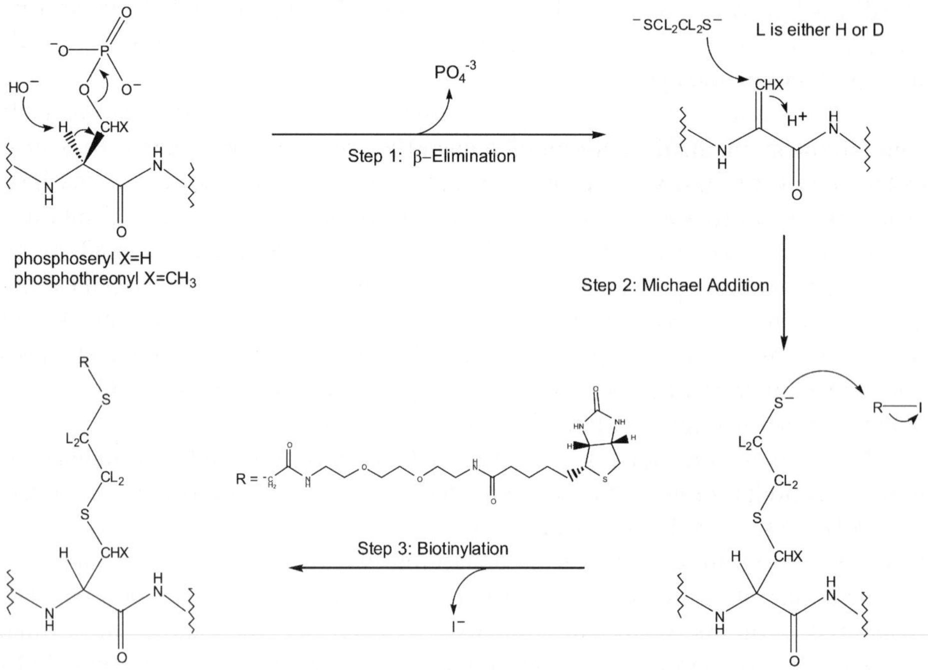

Fig. 2. Phosphoprotein isotope-coded affinity tag (PhIAT) labeling approach. Proteins containing phosphoseryl ($X = H$) or phosphothreonyl ($X = CH_3$) residues are isotopically labeled and biotinylated. After proteolytic digestion, biotinylated peptides are isolated from non-phosphorylated peptides via immobilized avidin affinity chromatography. The ability to quantitate the extent of phosphorylation between two identical peptides extracted from different sources is based on the use of a light ($L = H$, EDT-D0) and heavy ($L = D$, EDT-D4) isotopic version of 1,2-ethanedithiol.

mass by 4 Da. Prior to the implementation of the PhIAT labeling method, important steps were taken to block cysteine sulfhydryl groups and to remove O-linked carbohydrate moieties when investigating phosphorylated species. Depending on the organism being studied, quantitation and identification of both phosphorylation and O-linked glycosylation sites can potentially be evaluated using the PhIAT approach. Tryptic digestion of PhIAT-labeled proteins, followed by affinity chromatographic isolation of the biotinylated peptides by immobilized avidin, permitted enrichment of the PhIAT-labeled peptides while simultaneously reducing the complexity of the sample for mass spectrometric measurements.

For each of the stable isotope labeling methods described earlier the relative abundance of the two labeled versions of each peptide was determined by integrating the areas under each peak from the appropriate set of mass spectra. This process was automated based on the detection of the same accurate mass species in a set of contiguous spectra. The two versions of each peptide were sometimes observed in different proximal spectra due to differences in intensity and small differences in relative elution times.

4. High-resolution separations

Optimized sample loading for nanoLC/nanoESI–MS is often desirable, provided sufficient sample is available for MS detection, because both mass and peak concentrations are increased if separation quality is retained. For a complex sample with a wide range of relative abundances, low-abundance components are more readily detected by increased sample loading. The maximum sample loading in LC/ESI–MS is limited by the column sample capacity (beyond which the separation quality degrades) and by possible effects due to the ESI process. Previously reported work has shown that an 85 cm × 150 μm i.d. packed capillary (containing 3 μm C18-bonded porous particles) is capable of loading up to hundreds of micrograms of polypeptides, and that the number of detected peptide species increased from ~9000 to ~60,000 by increasing sample loading from 5 to 50 μg using FTICR (Fig. 3) [39].

For the study described herein, packed capillary columns of various inner diameters were made 'in-house' as described [40] and LC experiments were carried out at constant pressure on a home-built capillary LC system. The mobile phases were stably delivered at the maximal pressure of 10,000 psi using ISCO Model 100DM pumps (ISCO, Lincoln, NB), and mixed in a steel mixing compartment (~1.5 ml) equipped with a magnetic stirring bar, before entering the separation capillary.

Under identical experimental conditions and using a quadrupole time-of-flight (QTOF) MS, loading 25–1500 ng (by dilution to specific concentrations) of

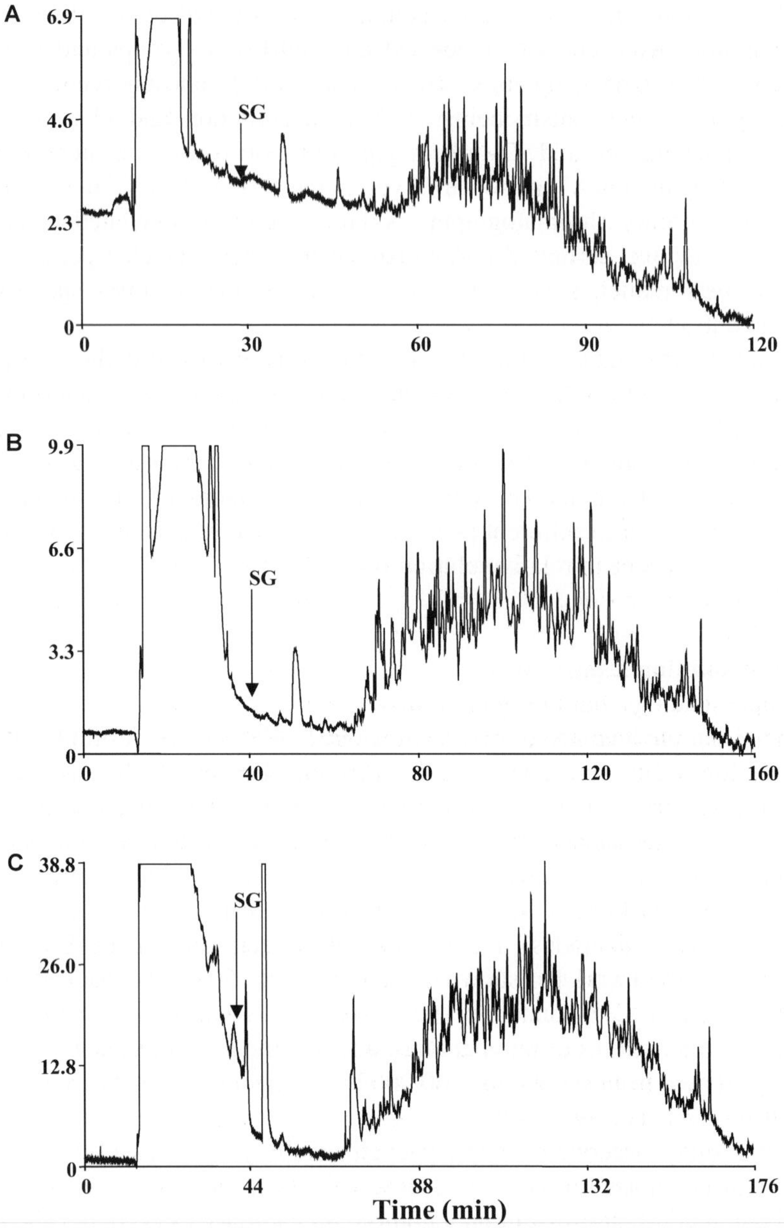

Fig. 3. High-pressure capillary liquid chromatography. Packed capillary reversed phase LC sample capacity evaluation for *D. radiodurans* tryptic digests. Conditions: 1 m × 150 μm i.d. fused-silica capillary packed with 5 μm C18-bonded particles (300 Å pores, Jupiter); constant pressure operation at 8000 psi; sample injection volume of 10 μl with concentrations of (A) 5; (B) 12.5; and (C) 50 μg/μl; after eluting the salt peak, the gradient was initiated with 100% mobile phase A (H_2O 0.1% (v/v) TFA) and ran to 75% B (H_2O/CH_3CN, 10:90 0.1% (v/v) TFA) in 180 min. Vertical axis is relative UV adsorption magnitude (215 nm). SG marks the start of the gradient.

a sample to a 29.7 μm i.d. packed capillary demonstrated that many additional species become detectable with increased sample loading [40]. Increasing sample loading from 25 to 1000 ng raised the number of detected species from 144 to 2037, or approximately 14-fold. Using a 29.7 μm i.d. packed capillary with a flow rate of ~75 nl/min, the QTOF MS detected only a limited number of peptides for <25 ng of this very complex polypeptide mixture. Studies that employ sensitive FTICR mass spectrometers [41–43] should significantly decrease sample size requirements. However, because all mass spectrometers have a sensitivity limit, the use of even smaller inner diameter packed capillaries may further improve the sensitivity for such complex mixtures.

The sample capacity or maximum sample loading was examined using a 29.7 μm i.d. packed capillary column. Sample loadings of 1500 ng resulted in a peak broadening by ~30%. Approximately 1000 ng was determined to be the maximum sample-loading amount (i.e. sample capacity) of this column. The separation quality for loadings from 25 to 1000 ng was evaluated and equivalent separation efficiencies (peak capacities of $\sim 10^3$) were observed. As a result the intensity of the MS peaks was used to characterize the relative ESI–MS response. The MS peak intensities of individual components in the complex mixture increased linearly with the total sample loading (and simultaneously, the total sample concentration in the column). Such *quantitative* behavior is observed for low sample loadings, but *non-quantitative* (i.e. non-linear intermediate response, between *mass sensitive* and *concentration sensitive* behavior) is obtained for the higher loadings. This can be attributed to the ESI performance for higher peptide concentrations that effectively 'compete' for the limited charge available on electrospray droplet surfaces and thus leading to increased 'ion suppression' effect for some species.

The *non-quantitative* behavior of ESI, most evident when using small diameter columns, has some significant implications for both quantitation and the dynamic range of measurements. Clearly, when dealing with complex mixtures having large differences in the relative concentrations of components, it is very possible to encompass both mass and concentration limited detection behaviors in a single separation. Thus, the best quantitative results are obtained for sample loadings that are sufficiently small so that the *quantitative* performance is obtained for all components (and where discrimination effects in ESI will also be minimal). However, such operation places the greatest demands upon ESI–MS sensitivity. On the other hand, the most reproducible performance (e.g. the best S/N) is achieved with higher sample loadings, but quantitation can be problematic for very complex samples where many components co-elute and suppression effects will generally occur. These considerations indicate an aspect of this behavior that can be advantageous. For complex samples, concentrations that span the range from *quantitative* and *non-quantitative* ESI performance at different elution times result in an effectively compressed dynamic range; i.e. ionization efficiency is suppressed for elution of more abundant components, but is maximized for low abundance species. This can be of significant benefit in quantitative applications such as those involving MS/MS to identify peptides in complex mixtures. In such applications, the dynamic range of measurements tends to extend beyond what would be expected based simply upon the relative abundances of sample components.

5. Chromatographic separations coupled to FTICR

In this study, both 5000 and 10,000 psi reversed phase packed capillary (150 μm i.d. × 360 μm o.d. fused silica; Polymicro Technologies, Phoenix, AZ, 5 μm C18, 300 Å pores, Phenomenex, Terrance, CA) LC separations were used to obtain peak capacities of up to 1000, significantly increasing the effective dynamic range and sensitivity of MS measurements (Fig. 3) [39]. The 11.4 T FTICR assembled at the authors' laboratory used an ESI interface comprised of a heated metal capillary inlet, an electrodynamic ion funnel [44,45], and three radio frequency quadrupoles for collisional ion focusing and highly efficient ion accumulation and transport to the cylindrical ICR cell for analysis [43,46]. FTICR simultaneously provides high sensitivity, resolution, dynamic range, and MMA for large numbers of peptides [39,41,47]. Mass spectra were acquired with $\sim 10^5$ resolution, with each spectrum requiring about 5 s (approximately equivalent to the time required for one MS/MS measurement). Analysis of the large data sets arising from capillary LC–FTICR experiments was performed in an automated fashion using ICR-2LS. Time domain signals were apodized (Hanning) and zero-filled twice before fast Fourier transform to produce mass spectra. Isotopic distributions and charge states in the mass spectra were deconvoluted to provide neutral peptide masses. The accurate *m/z* values of the parent ions and of the fragments were used for database searching. To obtain the desired 1 ppm MMA, a calibration program was developed that first uses the multiple charge states (e.g. 2+, 3+) for any protonated peptides produced by ESI to provide a correction for shifts due to large variations in trapped ion population [48]. In initial work, this involved the use of 'lock masses' (i.e. confidently known species that serve as effective FTICR internal calibrants) for each spectrum derived from commonly occurring peptides that were identified with high confidence from one capillary LC separation using FTICR MS/MS with accurate mass measurements as described elsewhere [49], and automated as part of the 'ICR-2LS' software package developed in this laboratory. More recently, improved methods of calibrant ion introduction to each spectrum increased the overall MMA [50].

In addition to the MS/MS measurements using FTICR, large numbers of polypeptides were tentatively identified from *D. radiodurans* using multiple capillary LC separations with a conventional ion trap MS (LCQ, ThermoFinnigan Corp.)

using data-dependent MS/MS analyses and the SEQUEST identification program [51] searching against a genome sequence derived data base. A multi-run MS/MS strategy [52] was used to segment *m/z* ranges and increase the number of peptide PMTs using the search/identification program SEQUEST and a minimum cross-correlation score of 2. These analyses identified large numbers of polypeptide PMTs, of which ~70% were then validated as AMT tags based on the detection of a species having the predicted mass for the PMT to < 1 ppm at the corresponding elution time in the FTICR analysis. This automated process increases the confidence for polypeptide identifications and allows the validated AMT tags to be used in subsequent experiments without the need for MS/MS.

6. Generation of accurate mass and time tags and their utilization

Several strategies were applied in initial work with *D. radiodurans* to increase the number of AMT tags so as to subsequently and routinely allow lower-level proteins to be analyzed. Proteins were enzymatically digested into peptide fragments (e.g. using trypsin) to produce tens to hundreds of potentially detectable peptides (and modified peptides) from each protein, and perhaps 10^5 to $> 10^6$ total peptides, depending upon proteome complexity, the dynamic range of the measurements, etc. This complex peptide mixture was then analyzed by combined high-resolution capillary separations with LC–MS/MS and LC–FTICR. Samples were analyzed several times using the same capillary LC–MS/MS strategy, but with different *m/z* ranges and with the 'exclusion' of parent ions that were previously selected for MS/MS, resulting in the selection of different peptides and generation of many additional PMTs. Beyond variations in instrument approaches, proteome samples extracted from cells harvested at different growth phases (i.e. mid-log, stationary phase, etc.) or cultured under a variety of different conditions (i.e. nutrients, perturbations) were also analyzed. By varying growth conditions and harvesting stages, the potential pool of PMTs increases significantly because the absolute number of proteins collectively present in the different samples is significantly greater than the number expressed by the organism under a single growth condition. Finally, because any additional sample fractionation will increase the overall dynamic range achievable, peptide fractions first separated off-line by ion exchange chromatography were also analyzed and again resulting in the generation of large numbers of additional PMTs for peptides that would otherwise have too low abundance for conventional MS/MS analyses. Note that any number of alternative sample fractionation and analysis strategies can be performed to increase the number of PMTs and AMT tags generated, and that the extra efforts at this stage are more than off-set by the resulting ability to make subsequent comprehensive proteome measurements with much greater sensitivity and speed. PMT generation efforts for *D. radiodurans* continued for over 200 different ion trap MS/MS runs, until the rate of generation of novel PMTs decreased significantly. Since the analysis procedure can be totally automated, this corresponds to a one-time effort requiring approximately 3 weeks using a single ion trap instrument, and additional experience might be expected to further reduce the number of runs required for PMT generation.

Thus, while a substantial number of samples are analyzed to generate the MS/MS spectra used for generating the set of AMT tags for a specific organism, this initial investment of effort obviates the need for routine use of MS/MS in future analyses. The dividends for such an investment are most importantly realized in proteome studies designed to quantify changes in the relative abundance of proteins as a function of time or environment. While fractionation of samples prior to LC–MS/MS analysis is necessary to generate a large number of useful spectra using conventional instruments, these lower-level peptides can subsequently be routinely detected using the FTICR and LC elution times without the need for sample fractionation. Without the need to re-identify the expressed proteins through time-consuming MS/MS analyses and extensive sample fractionation, studies designed to quantify the relative abundances of proteins between two distinct proteome samples can be completed in a high-throughput manner with low attomole level sensitivity and also provide broad proteome coverage [27].

7. The dynamic range of proteome coverage

The extent of proteome coverage for any approach depends substantially on the achievable dynamic range of the MS measurements, which in turn depends significantly upon the resolution (or peak capacity) of the separation step(s) preceding MS analyses and any overall constraints due to sensitivity. The example in Fig. 4 shows that the dynamic range obtainable in a single FTICR mass spectrum obtained during an LC separation exceeds 10^3. If the dynamic range measurement is performed using an on-line LC–MS instrument, and given an elution time peak width for a highly abundant species of 30–40 s and an instrument duty cycle of 3–5 s, this effectively spreads the acquisition of the abundant species over approximately 10 spectra. In order for a lower abundant species to be measured, in theory, only one spectrum containing this species needs to be acquired. Thus, a rough estimate of an increase in dynamic range of 10 from online LC is made and an overall dynamic range of 10^4 is obtained from the combination of LC and MS for the detection of peptides. Furthermore, if the aim is protein identification, then a significant (perhaps 10-fold) increase in effective dynamic range will result due to the variable ESI or detection efficiency for different peptide sequences, and is estimated that the dynamic range of approximately 10^4–10^5 can be achieved [30].

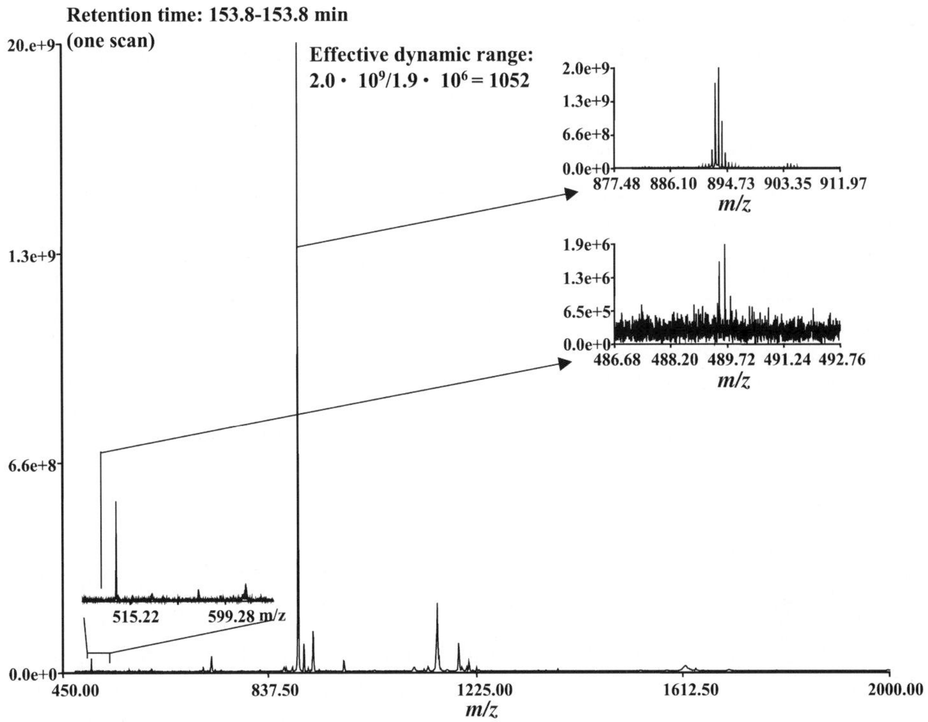

Fig. 4. FTICR dynamic range. Effective dynamic range of a single FTICR mass spectrum from a capillary LC–FTICR analysis of a global yeast soluble protein tryptic digest obtained under capillary LC data acquisition conditions.

The power of MS for protein identification derives from the specificity of mass measurements for either the intact peptides or their fragments after dissociation in MS/MS measurements. It is implicitly based upon the relatively small number of possible peptide sequences for a specific organism compared to the total number of possible sequences (Table 1). The distinctiveness of peptide sequences increases with size, but in practice the utility of increased size for identification is mitigated by the increased likelihood that a peptide will be unpredictably modified. Though much smaller than the number of possible sequences, the number of potentially distinguishable peptide *masses*, given sufficient resolution and accuracy, also dwarfs the number of predicted peptides from any organism. As shown in Table 2, an ideal tryptic digestion of all yeast proteins would produce 194,239 peptides having masses between 500 and 4000 Da, the range typically studied by MS. Of these, 34% are unique at ± 0.5 ppm MMA. (A larger fraction is unique if constrained by additional information resulting from any prior sample fractionation steps or the use of LC elution times.) These distinctive peptide masses would cover 98% and 96.6% of all predicted *S. cerevisiae* and *Caenorhabditis elegans* proteins, respectively.

Table 1

Number of possible peptides and number predicted for three organisms from tryptic digestion

Length	Possible		Predicted number of peptides[a]		
	Sequences[b]	Masses[c]	*D. radiodurans*	*S. cerevisiae*	*C. elegans*
10-mers	10^{13}	2×10^{7}	3471	11,275	30,623
20-mers	10^{26}	7×10^{10}	1292	3463	9475
30-mers	10^{39}	2×10^{13}	494	1278	3602
40-mers	10^{52}	1×10^{15}	195	405	1295

[a] Predicted from the identified open reading frames and applying the cleavage specificity of trypsin.
[b] Assumes 20 possible distinguishable amino acid residues.
[c] The number of peptides of length r potentially distinguishable by mass based upon the number of possible combinations of n different amino acids, $(n+r-1)!/r!(n-1)!$. The actual number of possible masses is somewhat smaller due to some mass degeneracy. The number of distinguishable peptides in actual measurements depends upon the MS resolution.

Thus, given sufficient MMA, a peptide mass measurement augmented by the use of LC separation times can often be confidently attributed to a single protein within the constraints provided by a single genome sequence and its predicted proteome. The AMT tag strategy obviates the routine requirement of MS/MS for peptide identification, and subsequently reduces the quantity of sample needed for such measurements. Because the masses of many peptides are generally obtained in each mass spectrum and require equivalent or less time than one MS/MS measurement, the increase in throughput is at the least equal to the average number of peptides in each spectrum. In practice, the increase in either throughput or proteome coverage is even greater since the lower abundance peptides are often not analyzed by conventional MS/MS approaches, or require the need for additional time for extended ion accumulation or spectrum averaging to yield spectra of sufficient

Table 2

Predicted number of peptides for ideal global tryptic digestions

Organism	All peptides	Unique[a] (%)	ORF coverage[b] (%)	Cys-peptides	Unique[a] (%)	ORF coverage[b] (%)
D. radiodurans	60,068	51.4	99.4	4,906	87.2	66
E. coli	84,162	48.6	99.1	11,487	83.6	80
S. cerevisiae	194,239	33.9	98	27,483	72.7	84
C. elegans	527,863	20.9	96.6	108,848	52.5	92

All peptides or Cys-containing peptides in mass range of 500–4000 Da, assuming ideal trypsin cleavage specificity.
[a] Percent unique to ± 0.5 ppm (by mass not using elution time).
[b] Percent of predicted ORFs (predicted proteins) covered by unique peptides.

quality for confident peptide identification. Thus, the AMT tag approach provides increased sensitivity, coverage and throughput, and facilitates quantitative studies involving many analyses of different perturbations or time points.

7.1. Dynamic range expansion by DREAMS FTICR MS

The large variation among protein relative abundances that have potential biological significance in mammalian systems (>6 orders of magnitude) presents a major challenge for proteomics. While FTICR has a demonstrated capability for ultra-sensitive characterization of biopolymers (e.g. achieving sub-attomole detection limits) [43,53], the maximum dynamic range for a single mass spectrum (i.e. without the use of spectrum averaging or summation) is typically constrained to about 10^3. An important factor conventionally limiting achievable FTICR sensitivity and dynamic range is the maximum charge capacity of either the external ion accumulation device or the FTICR mass analyzer itself. Prolonged ion accumulation is helpful during the LC elution of low-abundance components (i.e. during the 'valleys' in chromatograms) and potentially allows measurable signals to be obtained for otherwise undetectable species, thereby increasing the effective overall dynamic range of proteome measurements. Unfortunately, 'overfilling' of external multipole ion traps by high-abundance species often results in a biased accumulation process in which parts of the *m/z* range are selectively retained or lost [54] and/or extensive ion activation and dissociation occurs well before sufficient populations of low-level species can be accumulated [55,56]. Approaches for minimizing or eliminating such overfilling of the 2D quadrupole accumulation trap have recently been developed in the authors' laboratory [57]. However, even when artifacts associated with trap overfilling are avoided, allowing use of greatly prolonged accumulation times, the ion capacity of an FTICR trap beyond which unacceptable space charge effects occur (e.g. peak coalescence) still limits the dynamic range achievable in a single spectrum to $\sim 10^3$ for complex peptide mixtures [30]. Thus, the species present at low relative abundances are effectively 'masked' by the presence of co-eluting highly abundant species. Under such conditions the overall dynamic range achievable depends on factors that include: (i) the quality of the preceding sample processing and LC separations steps and (ii) the extent to which the maximum capacity of the FTICR trap can be effectively utilized in generating each spectrum. When combined with high-resolution LC separations, the overall effective dynamic range of FTICR-based proteome measurements is approximately 10^4–10^5 [30]. While the overall dynamic range can certainly be increased further by the use of additional 'up-front' separation dimensions, in principle to any desired extent, doing so necessarily results in selective sample losses, increased sample size requirements, and greatly decreased overall throughput since many more samples will have to be analyzed to achieve broad proteome coverage.

Application of the Dynamic Range Enhancement Applied to Mass Spectrometry (DREAMS) FTICR approach [29,58] allows the use of longer accumulation times for species present at low abundance eliminating the most highly abundant species during the ion accumulation process. This provides an overall improvement in sensitivity for any lower abundance species not separated from high abundance species (e.g. by chromatography), and thus extends the overall dynamic range of proteome measurements. Methods for FTICR dynamic range expansion based on ion manipulations in the FTICR cell using selective ejection of the most abundant species by excitation of their magnetron motion in the FTICR cell have recently been demonstrated [58]. Unfortunately, this approach is not readily compatible with online separations because it (i) requires transient introduction of a relatively high gas pressure in the FTICR cell that must be removed before analysis and (ii) takes longer to apply as magnetic field increases due to the longer times for ion ejection by magnetron excitation [59]. More recently results showed that the FTICR performance is unaffected when the ion ejection step is performed in an external 2D quadrupole [58]. While DREAMS reports [29,58] have illustrated the basic principles of the technique using either a yeast global tryptic digest or a murine B16 cell tryptic digest, recent work described improved peptide abundance measurements by employing $^{14}N/^{15}N$ peptide pairs for quantitative measurements, and illustrated significant increases in the number of peptides for which precise abundance ratios can be determined [60].

Capillary LC–FTICR results showing the total ion chromatogram (TIC) and portions of the mass spectra acquired during the normal spectrum acquisition process and the alternating DREAMS spectrum acquisition process obtained using a 50 cm long column and 120 min reversed phase gradient are displayed in Fig. 5. The total amount of sample injected was approximately 5 μg. The reconstructed TIC (top of Fig. 5) from the FTICR spectra was acquired during an RPLC separation of the mixture of identical aliquots of a natural isotopic abundance and ^{15}N-labeled version of a mouse B16 cells and representative spectrum obtained using broadband mode acquisition and a 100 ms accumulation time. The corresponding TIC (bottom of Fig. 5) and representative spectrum were obtained using RF-only selective acquisition and 300 ms accumulation time. Resonant frequencies for RF-only dipolar excitation were identified during a broadband ion acquisition (top of Fig. 5) and up to five species having relative abundance $>10\%$ were then data-dependently ejected during a selective acquisition that immediately followed (bottom of Fig. 5). The use of nominally 'unbiased' ion accumulation conditions in the 2D quadrupole resulted in a much greater gain in sensitivity for low-level components in the DREAMS spectrum than would be expected from the difference in ion accumulation times alone.

The overall speed of the DREAMS process was limited by both software for data processing, data transfer times, the generation of the DREAMS rf waveforms and the ion cooling times used to improve FTICR spectrum quality. The total time

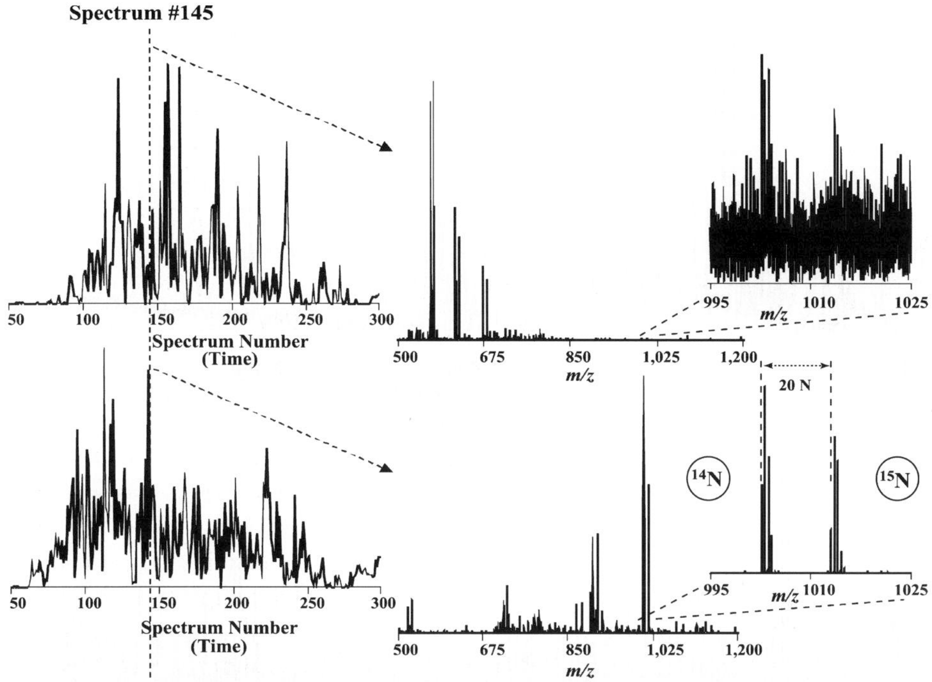

Fig. 5. The DREAMS approach. Two mass spectra are compared for one point in a capillary LC separation obtained using a 50 cm long column and 120 min reversed phase gradient to illustrate dynamic range enhancement capability using DREAMS. The data corresponding to the normal and DREAMS spectra (left top and bottom) allow two chromatograms to be reconstructed. Comparison of the full mass spectra shows the normal spectrum (middle top) is dominated by a number of major peptide ions, most prominently three pairs of $^{14}N/^{15}N$-labeled peptides in the $500 < m/z < 700$ range. The information from this spectrum was used 'on the fly' to apply dipolar RF excitation in the 2D quadrupole at the secular frequencies corresponding to the *m/z* of these major ions during ion accumulation for the next spectrum. The species in this *m/z* region were effectively ejected prior to accumulation and the FTICR spectrum is then dominated by a much different set of species (middle bottom). As a single example selected from many similar cases, the inset right shows that the S/N for a peptide pair at $m/z \sim 1000$ is greatly improved from a level at which no effective identification could be obtained, to a level at which a very precise relative abundance ratio (AR) can be determined for the peptide pair, with the gain in the S/N in this case of ~50.

required for acquisition of both the normal and DREAMS spectra pair was ~20 s. In order to provide a basis for a more quantitative evaluation of the additional information acquired with the DREAMS analysis a longer, high performance, reversed phase gradient separation, where the minimum peak width is about 1 min, was used. Fig. 6 shows reconstructed chromatograms from the FTICR spectra acquired during a single high-pressure capillary LC separation of a 1:1 mixture of ^{14}N- and ^{15}N-labeled mouse B16 tryptic peptides using a 120 cm long column and a nearly 15 h gradient, with the data being acquired for ~10 h

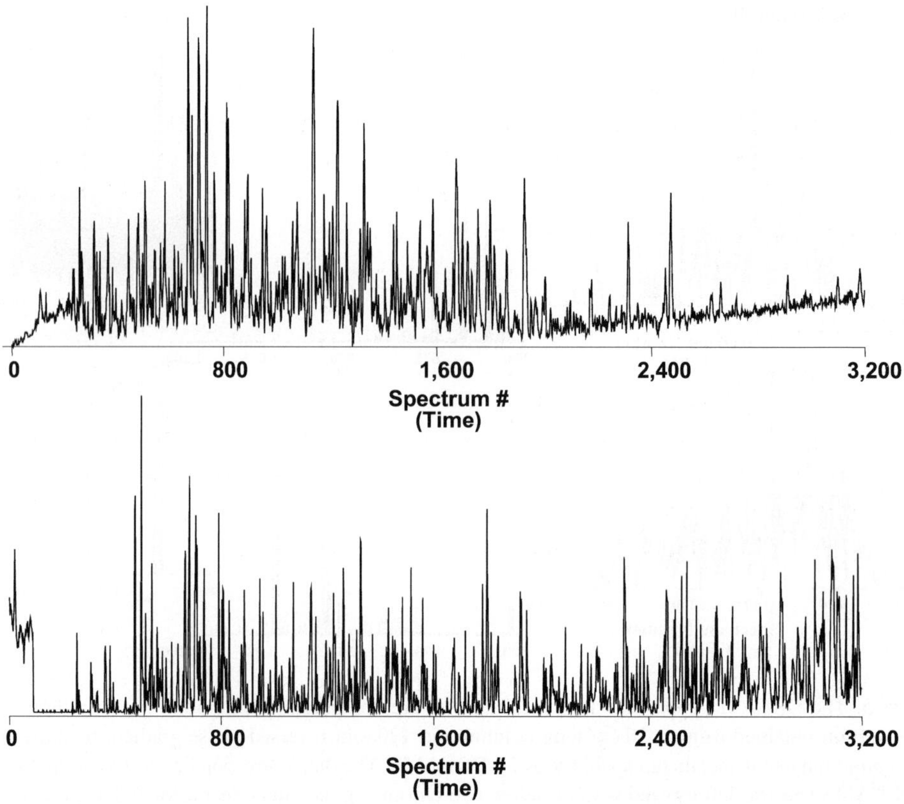

Fig. 6. Comparison of DREAMS versus normal FTICR acquisition. Total ion chromatograms (TICs) reconstructed from FTICR mass spectra acquired using broadband mode acquisition with 100 ms accumulation time (top) and with RF-only selective (i.e. DREAMS) acquisition and 1 s accumulation time (bottom) during a capillary LC separation of soluble proteins that were digested with trypsin from a 1:1 mixture of ^{14}N- and ^{15}N-labeled mouse B16 cells.

(started before the first peaks were expected to elute). The average LC peak widths spanned from 3 to more than 10 spectra for both the normal and DREAMS chromatograms. It is evident from the quite different chromatographic profiles that many additional species were detected with much greater signal intensities in the DREAMS spectra, particularly at longer retention times. However, the chromatograms only provide a qualitative view of the added information from the DREAMS approach, and one that is obviously limited to the more abundant species.

In a further evaluation of the utility of the DREAMS approach, the number of $^{14}N/^{15}N$-labeled peptide pairs that could be confidently assigned to both the normal and DREAMS spectra sets for the separation shown in Fig. 6 were used as a figure of merit utilizing methods as described by Pasa-Tolic et al. [60].

Fig. 7 shows a 2D display that includes spots for only the peptide pairs measured from the capillary LC–FTICR analysis of Fig. 6. Fig. 7 resulted from substantial post-processing of the capillary LC–FTICR analysis to convert the mass spectral information into molecular masses and to identify the subset of $^{14}N/^{15}N$-labeled peptide pairs. Analysis of the normal spectra resulted in detection of a total of 9896 $^{14}N/^{15}N$-labeled peptide pairs (Fig. 7, left). The average AR for the peptide pairs was 1.05, only a slight deviation for the nominal value of 1.0 expected. The standard deviation for the peptide pairs was 0.28. The second set of DREAMS mass spectra revealed 8856 $^{14}N/^{15}N$-labeled peptide pairs (Fig. 7, right), of which 7917 were 'new' peptide pairs not detected in the normal spectra, and 939 that were lower level peptide pairs also observed in the normal spectra. The average AR for the peptides detected in the DREAMS portion of the analysis was 1.015, and the standard deviation was 0.31. Note that the assigned peptide pairs accounted for less than 25% of the total number of species detected in the analysis that presumably contained contaminants or other mixture components and specifically excluded pairs of peptides as described [60]. Finally, the combined

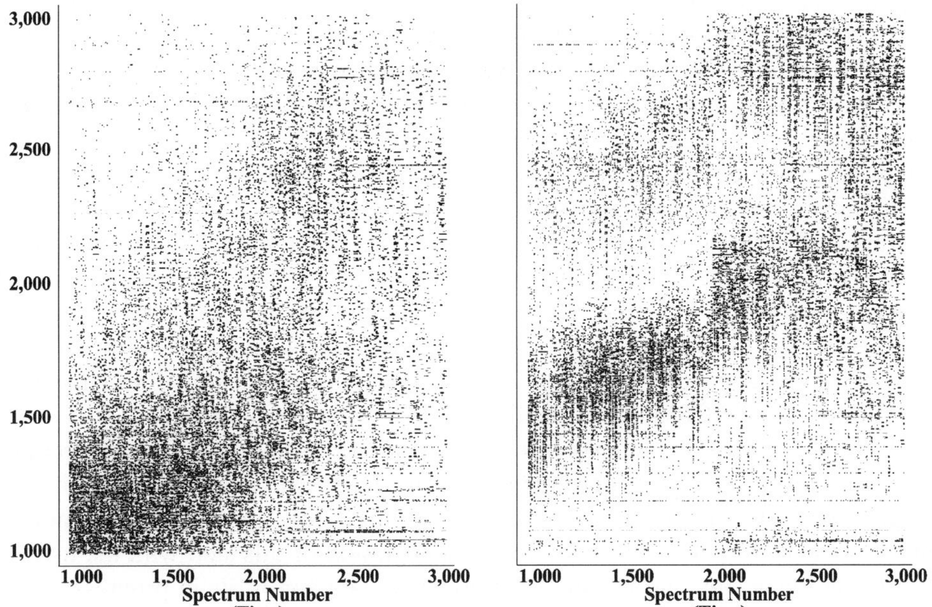

Fig. 7. Peptide detection with DREAMS. A 2D display showing the molecular masses and separation number for peptide pairs detected from a 1:1 mixture of ^{14}N- and ^{15}N-labeled peptides from mouse B16 cells (see Fig. 6). The left panel shows peptide pairs detected in the normal spectrum set and the right panel shows peptide pairs detected in the DREAMS spectrum set. A total of 17,813 unique peptide pairs were detected using conservative criteria [60], which represented an 80% increase over the number observed from the normal spectrum set alone.

total of 17,813 unambiguous and unique peptide pairs gave an average AR of 1.035 with a standard deviation of 0.29.

While different data analysis approaches (e.g. using different peptide pair selection criteria) will lead to slightly different results, the key conclusion is that the number of detected peptides is greatly increased with the DREAMS methodology. Indeed, the number of peptides for which quantitative information could be obtained from the relative abundances of peptide pairs was increased by 80% using the criteria described earlier.

The analysis of a mixture of identical aliquots of a natural isotopic abundance and ^{15}N-labeled version of *D. radiodurans* cells addressed whether the increase in the number of observed peptides translated into markedly better proteome coverage. DREAMS LC–FTICR analysis of the mixture resulted in detection of 9559 peptide pairs in the 'normal' set of spectra and 7926 peptide pairs in the DREAMS set of spectra. The *D. radiodurans* AMT tag database, assembled in the initial demonstration of an AMT tag approach for high-throughput quantitative proteomics [27,61], uniquely identified 1672 peptides, corresponding to 862 open reading frames (ORFs), in the normal set of spectra and 1748 peptides, corresponding to 941 ORFs, in the DREAMS set of spectra, for a total of 2507 peptides and 1446 ORFs. 835 (48%) unique peptides, corresponding to 548 (62%) ORFs, identified in the DREAMS dataset were 'new' (i.e. not identified in the normal dataset). Thus, both the number of identified peptides and the number of proteins for which quantitative information could be obtained from the relative abundances of peptide pairs was increased by ~50%. The overall proteome coverage can be compared with recent work by Washburn et al. [62] using their multidimensional protein identification procedure (MudPIT) [25], who reported an average of 869 yeast proteins (13.8% of the predicted larger set of yeast proteins) observed in the first dimension in analyses of 1:1, 5:1, and 10:1 mixtures of *S. cerevisiae* grown in ^{14}N- and ^{15}N-enriched minimal media. In the present work, 1446 *D. radiodurans* proteins (46.4% of the predicted *D. radiodurans* proteome) were observed as peptide pairs and also obtained good precision ARs in a single DREAMS LC–FTICR analysis using the AMT tag-based approach.

Additional efforts are underway for improving the resolution of the ion ejection step (currently 30–50) and for implementing the DREAMS analysis approach as a routine part of the accurate mass tag strategy for quantitative proteome measurements. In particular, the effective integration of this step with the use of a variable ion accumulation time for automated gain control is in progress. This step is important for enabling use of absolute peak intensities, and also for eliminating small errors associated with AR measurements due to possible variations in sensitivity during peak elution and the small elution time offsets associated with the different stable isotope-labeled version of each peptide. Because this capability would more effectively exploit the optimum trap capacity for each spectrum, it is anticipated that this combination of capabilities would further increase the dynamic range in proteome measurements. Presently, a 5 s ion cooling step is utilized to optimize the quality of the resulting data. However, when this is done in conjunction with DREAMS analysis, the result is either the need for extended analysis time, or effectively decreased chromatographic resolution and significantly decreased proteome coverage. The potential of decreasing the ion cooling time by up to an order of magnitude has been recently demonstrated using an adiabatic cooling scheme that involves the ramping of the FTICR ion trapping well voltages [63]. The combination of this step with decreased data transfer and processing times promises to result in substantially reduced overall analysis time.

8. Demonstration of global proteomic characterization

The Gram positive, non-motile, red-pigmented, non-pathogenic bacterium, *D. radiodurans*, was selected as the model organism to demonstrate the global proteomic approach described in this review. The *D. radiodurans* strain R1 genome contains two chromosomes, one megaplasmid and one plasmid [64] with 3116 predicted protein-encoding ORFs (ftp://ftp.tigr.org/pub/data/dradiodurans/GDR.pep) [65] (71 ORFs predicted to contain frame shifts are not included in this analysis). Proteomic measurements provide physical validation that individual ORFs actually encode a protein. A 2D visualization of a portion of one analysis (Fig. 8) illustrates the ability to identify proteins from the peptides detected for *D. radiodurans* (grown in a defined minimal medium [66] and harvested at mid-log phase).

The composite analysis of over 200 LC–MS/MS analyses of peptides from collective culture conditions yielded PMT identifications for 9159 peptides having a SEQUEST score above 2. Measurements verified 6997 AMT tags with high confidence corresponding to 1910 ORFs from *D. radiodurans*, representing 61% of the predicted ORFs (Table 3), the broadest proteome coverage for an organism achieved to date. In a functional category breakdown (Table 4), coverages for categories associated with housekeeping functions were significantly higher than hypothetical proteins. In a single FTICR analysis, the masses for ~1500 AMT tags are typically detected corresponding to ~700 ORFs (depending on the culture condition), and 15%–20% of the *D. radiodurans* proteome. Since the method uses two distinct MS measurements, matching accurate mass with MS/MS measurements, the use of the FTICR in these analyses lends higher confidence to those peptides identified while reducing the total number of peptides detected [27].

A major challenge for proteomics is the large variation in the relative abundances of proteins having potential biological significance (>5 orders of magnitude). The disadvantage of conventional proteomic studies which employ 2D PAGE separations coupled with protein identification by MS analysis is that

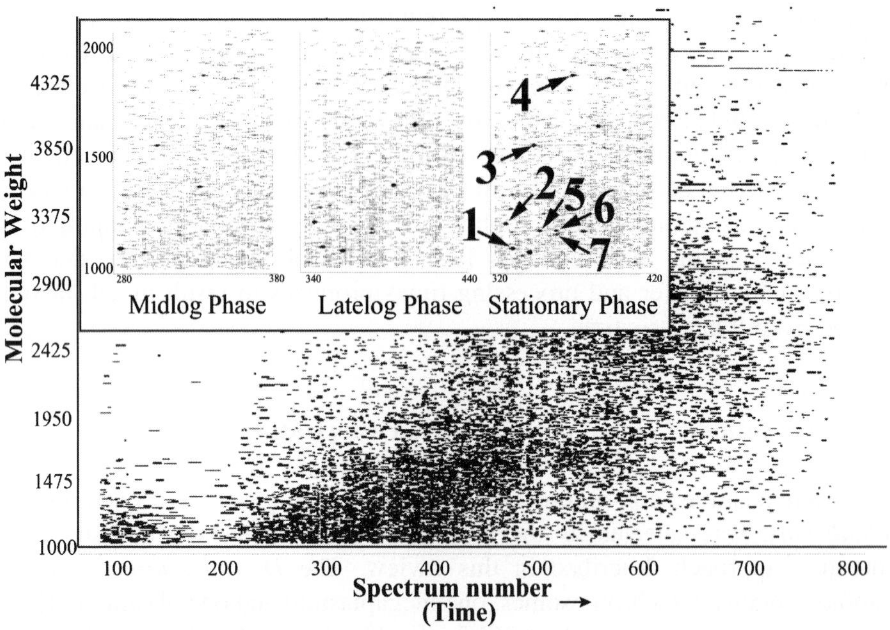

Fig. 8. Selected protein identifications by LC–FTICR peptide analysis from *D. radiodurans*. A 2D display of a capillary LC–FTICR analysis in which >50,000 putative polypeptides were detected from a tryptic digest of *Deinococcus radiodurans* proteins harvested in mid-log phase (OD_{600} 0.3–0.4; 30°C). The inset shows portions of displays for peptides from *D. radiodurans* harvested in mid-log (left), late-log (middle) and post-stationary (right) phases (spot size reflects relative abundance). Each individual spot corresponded to a peptide that can be identified along with the parent protein using AMT tags. Spot identifications: Spot 1: DR1314 (hypothetical protein); Spot 2: DR1790 (homolog of yellow/royal jelly protein of insects); Spot 3: DR1314 (hypothetical protein); Spot 4: DR0309 (elongation factor Tu); Spot 5: DR2577 (S-layer protein); Spot 6: DR0989 (membrane protein); and Spot 7: DR1124 (S-layer protein).

generally only higher abundance proteins are detected [8,9]. Work from this laboratory has previously shown that ~20 zmol (about 12,000 molecules) of protein can be detected during an FTICR analysis [67]. In this work, capillary LC–FTICR measurements provided an overall dynamic range of 10^4–10^5 [27].

Table 3

D. radiodurans proteome coverage

Category	Size	Predicted ORFs	Observed ORFs	Coverage (%)
Total	3.29 Mbp	3116	1910	61
Chromosome 1	2.65 Mbp	2633	1586	60
Chromosome 2	41.2 kbp	369	34	63
Mega plasmid	177 kbp	145	76	52
Small plasmid	46 kbp	40	14	35

Table 4

D. radiodurans proteome coverage by TIGR assigned functional category

Category	Total[a]	Seq. Cov.[b]	St. Dev.[c]	Observed[d]	Category (%)[e]
Amino acid biosynthesis	80	21.5	15.1	70	88
Cofactor biosynthesis, etc.	61	16.1	9.7	41	67
Cell envelope	77	17.7	19	62	81
Cellular processes	89	21.9	21.5	64	72
Central metabolism	154	14.7	10.5	111	72
DNA metabolism	81	13.2	10.1	55	68
Energy metabolism	199	24.7	20	152	76
Fatty acid metabolism	53	23.8	16.2	40	75
Conserved hypothetical	499	16.6	13.1	276	55
Phage related/transposon	47	11.7	3.7	9	19
Protein fate	86	22.2	20.6	65	76
Protein synthesis	114	38.8	24.2	100	88
Nucleic acid synthesis	53	19.2	14.7	42	79
Regulatory functions	126	15.5	10.8	76	60
Transcription	28	18.5	11.8	22	79
Transport proteins	191	16.5	15.5	138	72
Unknown function	176	15.4	12.4	107	61
Hypothetical	1002	18.7	14.2	479	48

[a] Number of ORFs assigned to each category by TIGR.
[b] Percentage of the protein sequence represented by AMT tags averaged over all the proteins observed in a single category.
[c] Standard deviation of the sequence coverage for the entire functional category.
[d] Number of ORFs detected in each category.
[e] Percentage of ORFs identified for each functional category.

Further expansion of the dynamic range of measurements using DREAMS technology [29] offers the potential to detect proteins at <1 copy per cell given practical cell populations (10^8 cells per analysis). For more abundant proteins, broad coverage of peptide fragments was achieved; e.g. ribosomal proteins were identified with an average of 9 AMT tags. Similarly, DNA polymerase I (Pol I), considered to be the most abundant polymerase in prokaryotic cells [68], is predicted to be present in about 400 copies per cell and was identified with 6 AMT tags providing 20% coverage of the protein.

Another method for estimating protein abundance in a cell population is the use of codon adaptation index (CAI) [69] or predicted highly expressed (PHX) [70] proteins, two separate algorithms devised to estimate protein abundance based on codon usage. While many of the proteins fall into a range corresponding to proteins expressed at high abundance, about 50% of the proteins detected have both CAI and PHX scores predicting low abundance. If codon usage were the only indicator of protein abundance, the increase in AMT tag coverage per ORF with higher CAI and PHX values (Table 5) would suggest a slight expected bias toward the detection of high-abundance proteins. However, the substantial number of ORFs identified with low PHX and CAI values also suggest that this approach effectively identifies lower abundance proteins as well.

The increased throughput of measurements provided by the use of AMT tags allows investigation into the expression pattern of detected proteins, under a range of environmental conditions (Fig. 9A). These results illustrate that many predicted proteins associated with 'housekeeping' functions are expressed under all conditions evaluated. Although the detection of an AMT tag unequivocally confirms the presence of the protein in the culture, the absence of AMT tags for a particular protein cannot exclude the protein's expression at a low level. Examination of the expression patterns can suggest potential functions for ORFs having no homology to any other sequenced protein (Fig. 9B). For example, many of these hypothetical proteins are detected under every culture condition, suggesting that they may fill 'housekeeping' function (e.g. DR1172, DR1245, DR1623 and DR1768; Fig. 9C). DR0253 (Fig. 9C) was detected only in post-stationary phase (regardless of the culture conditions), while DR0871 and DR1228 were detected only in minimal medium; DR0528, DR1591, and DR2450 were detected only when cells were exposed to various 'stress' conditions. While limited quantities of such information cannot unequivocally reveal specific protein function, it can help narrow the range of possible functions and identify ORFs for further study.

Table 5

Detected ORFs for different CAI and PHX values

CAI	Total[a]	Observed[b]	Coverage[c]	PHX	Total[a]	Observed[b]	Coverage[c]
0.1–0.2	24	9	10.3	0–0.6	542	370	14.8
0.2–0.3	94	33	14.6	0.6–0.7	729	436	17.5
0.3–0.4	292	120	16.3	0.7–0.8	657	387	19.1
0.4–0.5	786	404	14.6	0.8–0.9	451	256	19.1
0.5–0.6	1239	801	15.6	0.9–1	204	114	21.2
0.6–0.7	546	424	24.0	1–1.25	233	142	23.1
0.7–0.8	120	109	49.8	1.25–1.5	86	56	25.0
0.8–0.9	12	9	57.7	1.5–2	96	72	35.5
				2.0–3	21	19	43

[a] Total number of ORFs with Codon Adaptation Index (68) or Predicted Highly Expressed Proteins (69) for the bin ranges indicated.

[b] Number of observed ORFs with CAI or PHX values indicated.

[c] Percentage of observed proteins represented by AMT tags for all the proteins predicted in a single bin.

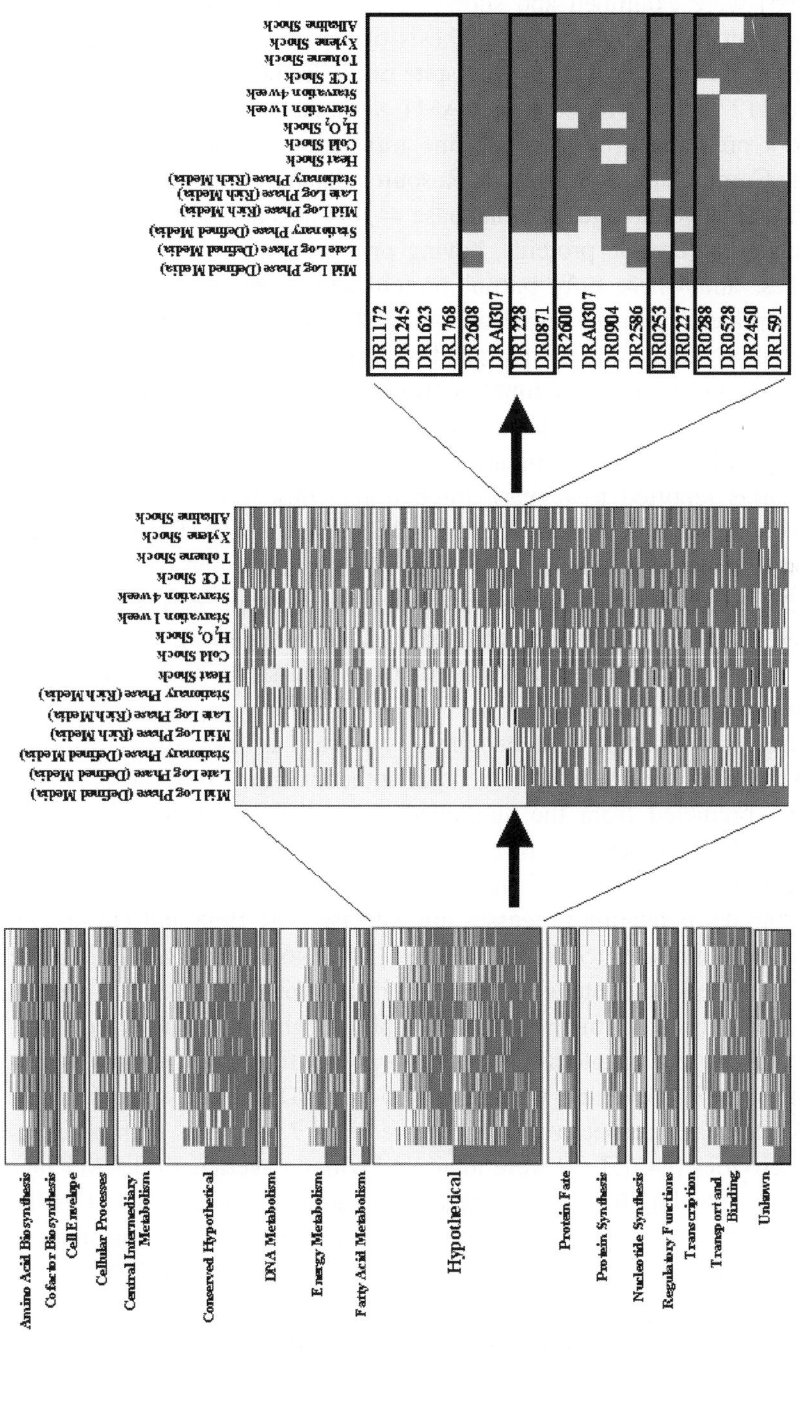

Fig. 9. Open reading frame expression in *D. radiodurans*. (A) The qualitative pattern of ORF expression detected using AMT tags for various conditions by TIGR assigned functional category [65] (light, AMT tags detected; dark, AMT tags not observed). Each culture condition was analyzed a minimum of two times. (B) An expansion of the pattern for predicted hypothetical proteins, illustrating similarities and differences in the patterns for protein expression under different culture conditions. (C) Expansion showing several representative hypothetical proteins.

Data for protein expression by genes with frame-shifts designated in the initial annotation [65] were examined and showed that *D. radiodurans* displays both direct and indirect pathways for the synthesis of asparaginyl-tRNA ($tRNA^{ASN}$) and glutaminyl-tRNA ($tRNA^{GLN}$) [71] in spite of the frameshift in the $tRNA^{ASN}$-synthetase (DR1270) [65] (based on an AMT tag analysis of peptides from all possible ORFs predicted from a six frame translation of the *D. radiodurans* genome). The Comprehensive Microbial Resource database (http://www.tigr.org/tigrscripts/CMR2/GenomePage3.spl?database = gdr) lists an additional 71 genes that are predicted to encode proteins, having probable frameshifts to unknown reading frames, and where only partial protein sequences were predicted. In addition to peptides for asparaginyl-tRNA synthetase, several other peptides derived from genes believed to be disrupted by frameshifts were observed. Some of these peptides are markers for functional proteins, suggesting that frameshift corrections may be occurring, perhaps at the level of transcription. For example, peptides predicted from DR0100 (putative single-stranded DNA-binding protein; SSB) [65] that is reported to contain three frameshifts were detected, which support the existence of at least some functional SSB protein. AMT tags were also identified for DR1624 (RNA helicase), DR2477 (3-hydroxyacyl-CoA dehydrogenase), and DRC0020 (putative modification methylase). AMT tags were detected in all three reading frames of ORF DR2477, suggesting that either *D. radiodurans* is able to mediate the required two frame-shifts or that there is a mistake in the genome sequence.

D. radiodurans is predicted to encode a spectrum of 148 stress response proteins of which four are predicted to contain authentic frame shifts [64]. A total of 74 proteins predicted from the annotated genome were identified with an average of 6 AMT tags per protein corresponding to 24% overall amino acid sequence coverage. Two classes of annotated *D. radiodurans* proteins shown to play roles in the detoxification processes are catalase (DR 1998 and DRA0259) and superoxide dismutase (SOD) (DR1279, DR1546 and DRA0202). The presence of both predicted catalases were confirmed with 48 (DR1998) and 37 (DRA0259) AMT tags covering ~80% and ~50% of the amino acid sequence, respectively. Of the SOD proteins, DR 1279 was identified with 14 AMT tags corresponding to a 87% coverage, whereas DR1546 and DRA0202 were identified with 2 and 3 AMT tags, respectively. Of the predicted 75 proteins with potential DNA repair activities [64,65], 39 were identified with an average of 3 AMT tags per protein, corresponding to 12% coverage of the amino acid sequence. Finally, RecA is central to homologous recombinational repair of irradiation-induced double strand breaks in *D. radiodurans* chromosomal and plasmid DNA [15, 72–74]. Five different RecA AMT tags (covering 34% of the amino acid sequence) were identified primarily in cells recovering from exposure to ionizing radiation.

9. Overcoming challenges to proteome-wide measurements

One approach to more comprehensive proteome measurements is to reduce overall sample complexity by fractionation and/or enrichment. Examples of methods that lead to a reduction in the complexity of the sample include the subcellular fractionation of organelles and cellular components [75], the selective enrichment of phosphopeptides by immobilized metal affinity chromatography (IMAC) [76], and the use of the isotope coded affinity tags (ICAT™) for labeling of cysteine-containing proteins [36]. The selective enrichment of proteins or peptides from a targeted portion of an entire proteome, usually designated as a subproteome, is expected to facilitate the identification of very low abundance proteins, and in the case of subcellular (organelle) fractionation, assist in deducing protein or organelle function under a given set of experimental conditions or well-defined environmental circumstances.

9.1. The membrane subproteome

The characterization of the membrane subproteome targets proteins inserted into the phospholipid bilayer membrane, defined as integral membrane proteins, which are important biological and pharmacological targets. Contemporary genomic analyses indicate that 20%–30% of all ORFs found in eukaryotic, eubacterial, and archaean organisms encode for integral membrane proteins [77]. Integral membrane proteins display two common membrane spanning tertiary structural motifs: (1) the α-helical transmembrane domain which is ubiquitous in the cytoplasmic membrane; and (2) the β-barrel, typically found in the outer membranes of Gram-negative bacteria [78]. The α-helical transmembrane domains of integral cytoplasmic membrane proteins are composed of predominantly non-polar amino acids that are inserted into the highly, hydrophobic environment of the phospholipid bilayer while integral outer membrane proteins contain significantly fewer non-polar amino acid residues. Accounting for the properties of integral membrane proteins, the main challenge for large-scale MS analysis of the membrane subproteome are the requirements to achieve the dissolution of these hydrophobic proteins from the phospholipid bilayer and to maintain their solubility during the isolation and separation steps while simultaneously avoiding reagents which suppress the ESI process in the LC–MS/MS analysis [78–80].

Initial work with fractionation and enrichment of *D. radiodurans* membranes in combination with organic solvent-assisted solubilization of membrane proteins led to the identification of 268 unique integral membrane proteins using PMTs (conventional SEQUEST (ThermoFinnigan, San Jose, CA) scoring; tryptic peptide cleavages only, containing up to two missed cleavages; cross-correlation score ≥ 2.0; and delta-correlation score ≥ 0.1) from a single experiment [28]. This initial 24% coverage of the predicted integral membrane proteins represents a significant improvement in protein identification for conventional, low throughput LC–MS/MS analysis (~20 h for eight runs) of the bacterial membrane subproteome. These results demonstrate that hydrophobic peptides are readily detected using single-dimension, high-pressure, reversed phase capillary LC coupled with ESI–MS/MS analysis. Further analysis of hydrophobic peptides by high-throughput, high-MMA LC–FTICR (~2.5 h per single run) is necessary to complete the AMT tag process and convert the observed hydrophobic peptide PMTs into AMT tags.

In this initial study the identification of 37 (65%) out of 58 known integral cytoplasmic membrane proteins that contain four and more predicted transmembrane domains (by detecting at least one hydrophobic peptide spanning a transmembrane domain) demonstrated the efficacy of the organic solvent-assisted solubilization method. Out of 268 integral membrane proteins identified in this study, 22% (60 proteins) were annotated as hypothetical and having no significant similarity to any other proteins in the current protein databases accessible at SWISS-PROT, and 10% (28 proteins) were annotated as conserved hypothetical and had limited similarity to other proteins. Extension of this integral membrane protein preparation method to other microbes and organisms will facilitate an enhanced coverage of the membrane subproteome.

9.2. The phosphoproteome

Reversible phosphorylation in eukaryotic organisms at serine, threonine, and tyrosine residues, one of the best-characterized post-translational mechanisms for the regulation of signal transduction pathways, is a covalent modification estimated to be present on approximately 30% of all eukaryotic proteins. Examination of purified, intact proteins by Edman degradation or MS often simplifies the task of identifying the amount and position(s) of phosphorylation; however, large-scale identification of the phosphoproteome when using complex biological mixtures has proven elusive. A variety of experimental techniques to reduce overall sample complexity and enrich for phosphoproteins or phosphopeptides have been developed to investigate the phosphoproteome. These methods include anti-phosphotyrosine antibodies for enriching low abundance phosphotyrosine-containing proteins [81], IMAC enrichment of underivatized phosphopeptides [76] and derivatized phosphopeptides [82], chemical modification of phosphopeptides with solid-phase capture [83], and affinity tag labeling of phosphoserine and phosphothreonine [84].

The PhIAT strategy, developed in the authors' laboratory, specifically labels the phosphorylated amino acids serine and threonine of proteins in complex cellular lysates [37,38]. This method provides the direct determination of the exact sites of protein phosphorylation in complex mixtures that has previously only been accomplished using carefully purified proteins. By first attaching a stable isotopic label (D0-EDT or D4-EDT) to phosphorylated proteins, to which an affinity tag (PEO–biotin) is subsequently attached, the PhIAT-labeled peptides may be enriched by affinity chromatography following the tryptic digestion of the labeled proteins. The mass spectrometric analysis of the PhIAT-labeled peptides provides for the identification of the phosphorylated protein, the specific site of phosphorylation, and the subsequent quantitation of the phosphorylated protein in an experimental sample when compared to the abundance of the same phosphorylated protein in a control sample. This phosphoprotein labeling and peptide enrichment process is particularly suitable for the elucidation of low-abundance proteins present in the cell. Many of the key regulatory proteins that undergo reversible protein phosphorylation are low-abundance proteins that are responsible for a wide range of biological functions and activities.

Using two equal samples of beta-casein, it was demonstrated that the PhIAT method provides isotopic enrichment by affinity isolation and subsequent quantitation by MS (Fig. 10). The PhIAT modification does not interfere with the MS/MS fragmentation of phosphopeptides and the label remains bound to the peptide allowing subsequent site-specific identification [85]. The beta-casein test model also provided a performance demonstration of the ability of the PhIAT approach to efficiently modify, affinity extract, and identify low-level phosphopeptides in a mixture of high-abundance proteins from the same commercially

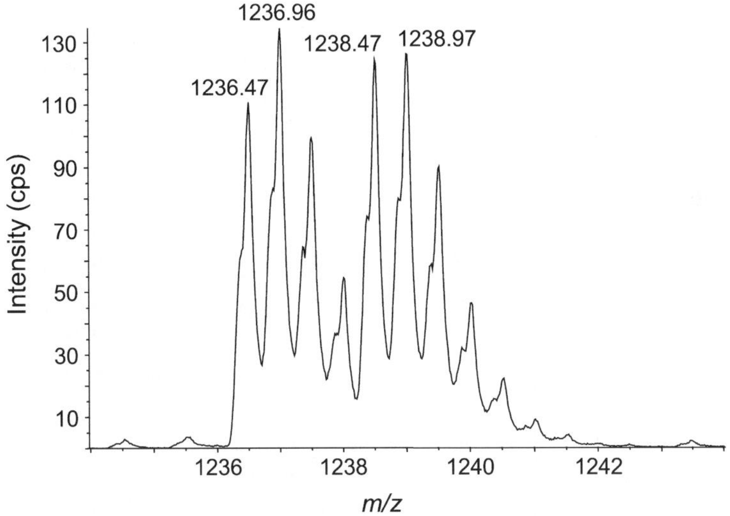

Fig. 10. Demonstration of PhIAT approach with a model protein. A mass spectrum of the PhIAT-D0/D4-labeled beta-casein phosphopeptide (FQpSEEQQQTEDELQDK; $[M + 2H]^{2+}$ ion pair) using reversed phase LC coupled directly on-line to a Perkin Elmer Sciex API QStar Pulsar hybrid quadrupole time-of-flight mass spectrometer where 10 μl of an approximately 0.5 μg/μl sample was injected. The PhIAT-labeled peptides were enriched using immobilized avidin affinity chromatography prior to LC–MS/MS analysis.

Table 6

Results of database searching the product ions using SEQUEST with the data acquired from LC–MS/MS analysis of tryptically digested PhIAT-D0/D4-labeled beta-casein

Peptide[a]	Protein	Calcd mass[b]	Obsd mass	Charge state	Xcorr[c]	Delta Cn[d]	Ions[e]	PhIAT label (*)[f]
K.IEKFQS * EEQQQTEDELQDK.I	Beta-casein	2843.2	2842.9	3	5.117	0.416	30/72	D0
K.IEKFQS * EEQQQTEDELQDK.I		2847.2	2848.4	3	4.778	0.342	28/72	D4
K.TVDMES * TEVFTK.K	Alpha$_{S2}$-casein	1877.2	1876.8	2	2.763	0.190	13/22	D0
K.TVDMES * TEVFTK.K		1881.2	1880.6	2	3.044	0.363	15/22	D4
K.VPQLEIVPNS * AEERLHSMK.E	Alpha$_{S1}$-casein	2668.2	2670.1	3	2.238	0.135	24/72	D0
K.YKVPQLEIVPNS * AEER.L		2366.8	2367.9	3	3.826	0.316	28/60	D4
K.[SC] * QAQPTTMAR.H	Kappa-casein	1684.1	1683.5	2	2.476	0.425	14/20	D0
K.[SC] * QAQPTTMAR.H		1684.1	1684.8	2	1.758	0.109	12/20	D0

[a] The amino acid residues appearing before and after the periods correspond to the residues proceeding and following the peptide in the protein sequence.
[b] Average peptide mass.
[c] Cross-correlation score of the peptide is based on the ‘fit’ of the MS/MS data to the theoretical distribution of ions produced for the peptide.
[d] Difference between the top two Xcorr scores for the given peptide.
[e] The total number of b and y ions identified/theoretical.
[f] PhIAT label, -S-CL_2CL_2-S-acetyl-PEO-biotin where L is H (D0) or D (D4).
[g] Based on the MS/MS data, the PhIAT-labeling site could only be narrowed down to the SC dipeptide segment.

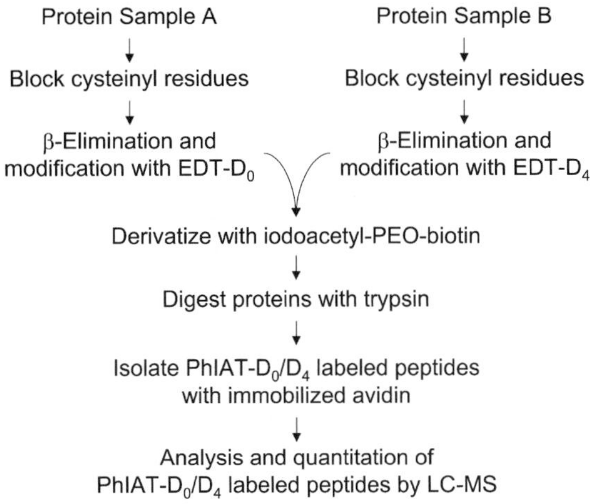

Fig. 11. Flow diagram for quantitation using PhIAT. The analytical scheme followed for the relative quantitation and isolation of phosphopeptides from two different samples using the PhIAT labeling strategy.

available beta-casein sample (Table 6). A preliminary study on the PhIAT labeling of whole cell lysates from yeast (*S. cerevisiae*), following the experimental scheme shown in Fig. 11, resulted in mass spectrometric detection of pairs of isotopic doublets. For the labeled pairs the isotopic distribution differed by 4 Da for each $[M + H]^+$ species and 1:1 stoichiometric labeling was observed for the equally mixed sample. Some representative PhIAT-D0 and PhIAT-D4 peptide pairs from the labeling of the yeast lysates are shown in Fig. 12. These results display the effectiveness of labeling phosphoproteins on a proteome-wide level in complex biological mixtures using the PhIAT method.

10. Summary

While strategies to identify and quantitate proteomes on both the intact protein and peptide level are being rapidly developed, at this point most proteomic studies are conducted on the peptide level. This is primarily due to the increased challenge of identifying intact proteins that have post-translational modifications in addition to other issues. Analysis of intact proteins, however, can provide important complementary information related to the origin of species observed in the peptide analysis of the same proteome. This additional information can only serve to bolster the confidence in peptide identification.

Regardless of which (or both) strategy is used for proteome characterization, the requirement for high-resolution separations cannot be underestimated. While conventional mass spectrometers are able to measure many ion species in a single

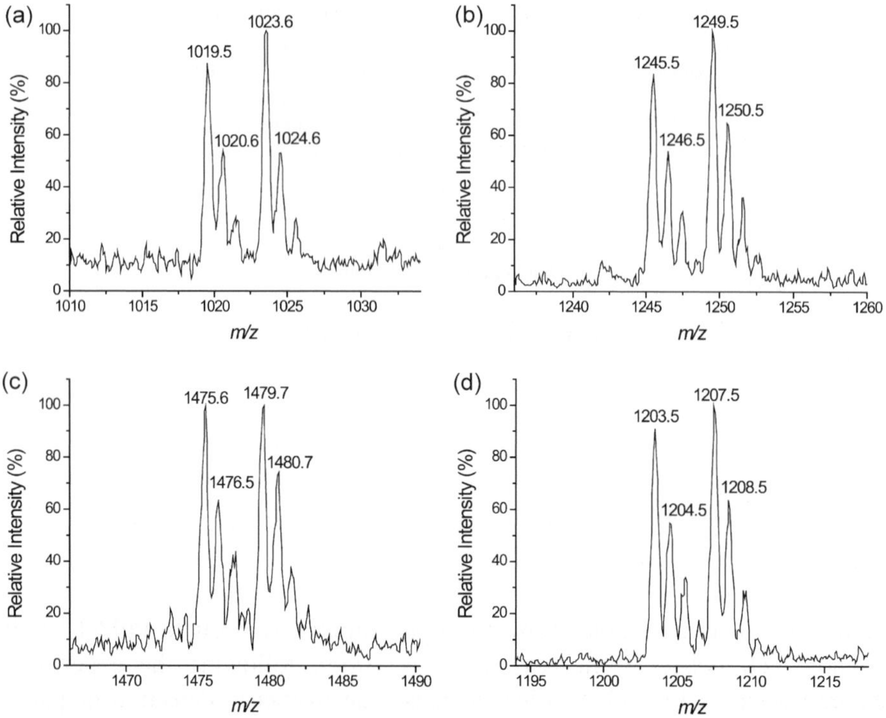

Fig. 12. PhIAT labeled yeast peptides. Representative mass spectra of pairs of PhIAT-D0/D4-labeled yeast (*S. cerevisiae*) peptides enriched by immobilized avidin affinity chromatography. Ten μl of a 1.0 μg/μl sample was analyzed by capillary reversed phase LC coupled directly online to a Finnigan MAT LCQ ion trap mass spectrometer. Each $[M + H]^+$ ion pair features the characteristics of the PhIAT labeling approach: the 4 Da mass difference between ion pairs and the 1:1 ratio indicative of the stoichiometric conversion of identical yeast phosphoproteins to PhIAT-D0- or PhIAT-D4-labeled peptides. Carboxymethylation was used to block the cysteinyl residues.

scan and FTICR can resolve as many as 10^3 species in a single spectrum during online separations, the best results are still obtained using optimal separations. This is primarily due to the increase in dynamic range that can be gained by separating low-abundance species from higher abundance species [86].

Identification of the numerous species is the most time-consuming step in proteome characterization. Strategies have been developed in this laboratory to increase the throughput of this step and to circumvent the need for extensive MS/MS identification of every species present in these complex mixtures. Work has been performed on the intact protein level using MS/MS as well as isotopic labeling to supplement the MW information with information concerning the number of a specific amino acid type in the protein. Most of the effort, however, has focused on the peptide level developing the concept of AMT tags and using Cys-constraints to aid in peptide identification.

Beyond identification, quantitative strategies to measure relative protein/peptide abundances employing isotopic labeling have been developed on both the cell growth level as well as the post-extraction level. While labeling at the cell culture level with isotopic-labeling media (i.e. ^{15}N-enriched, etc.) is attractive due to the ability to combine and process proteomic samples concurrently, post-extraction labeling techniques, such as the ICAT™ strategy, provide a more universal approach. While many prokaryotic systems can be labeled at the cell growth level using a variety of different labeling strategies, many of these options are not presently useful for studying mammalian cells in culture. At present, analysis of mammalian tissue samples definitely requires post-extraction labeling strategies.

Finally, the high MMA, resolution, dynamic range, and sensitivity achievable with ESI–FTICR provide an abundance of new opportunities in proteomics. These include the use of proteolytic fragments as AMT tags (i.e. as effective biomarkers) for protein identification. The use of high MMA to produce AMT tags will allow the identification of the protein constituents within a proteome in a complex mixture when combined with high-quality separations accomplished using such techniques as CIEF or capillary LC. For protein identification, conventionally done on a one protein at a time basis, the improved quality of mass measurements should enable greater speed and confidence. Given high MMA, it is easy to envision identifying complex mixtures of proteins based solely on the mass of a single peptide. Once proteins are identified, a major application of MS will become high throughput comparative studies of proteomes enabling changes from many different perturbations in proteome composition to be studied in a rapid fashion.

The possibilities for application of the new quantitative proteome measurement technology and methods described in this work are broad. While major challenges remain before the full potential of the technology is realized, particularly for obtaining useful quantitative measurements for low-abundance proteins and for extending the approach to modified proteins, it is clear that many useful applications of the technology to microbial systems are already tractable. The combination of sensitivity, dynamic range, and throughput should enable new types of studies to be contemplated. In particular, studies that would otherwise demand excessive quantities of protein or present too much complexity should now be tractable.

Most of the work to date has focused on relatively simple microbial systems. Mammalian proteomes pose a much greater challenge relative to microorganisms due to their greater complexity and range of relative protein abundances. Mouse B16 cells cultured in both ^{14}N- and ^{15}N-enriched media and affinity tagged with iodoacetyl–PEO–biotin have been used to isolate two Cys-containing versions of each peptide having masses that differ by the number of nitrogen atoms [31]. Both this and the ICAT strategy significantly reduce the complexity by isolating

Cys-peptides without significantly decreasing the proteome coverage. As shown in Table 2, approximately 90% of *C. elegans* (and presumably human) proteins are potentially identifiable using Cys-peptides. Initial results for mammalian proteomes have detected fewer peptides than observed for yeast, likely due to both the reduced complexity due to affinity selection of the Cys-peptides and the greater dynamic range for proteins, including the presence of a number of highly abundant proteins. The DREAMS capability presently under development for use with LC separations and the implementation of protein fractionation steps prior to LC–FTICR analysis should substantially increase the number of detectable species beyond the 100,000 detectable components recently experimentally demonstrated for soluble yeast proteins [39]. The human proteome can be estimated to yield a peptide mixture having a complexity of the order of several million components for an ideal tryptic digest. Although the actual complexity is certainly higher, only a fraction of the possible proteins will be expressed in any single cell type. A fractionation method that yields 10 distinct fractions combined with the capillary LC–FTICR approach would provide a combined theoretical peak capacity $>10^8$, potentially allowing mammalian proteomes to be studied without the need to apply Cys-peptide selection methods. The DREAMS FTICR technology is likely to provide the basis for a substantial further gain in the coverage of proteomic measurements. It is clear that continued technology development efforts along these lines can have a significant impact upon the practice of proteomics, particularly for applications where measurements of the highest quality and broadest scope are beneficial. Over the next couple of years the initial application of this new technology to the study of microbial systems should clarify its role.

Acknowledgements

The contributions of Ljiljana Pasa-Tolic, Mary Lipton, Gordon Anderson, Kim Hixson, Ron Moore, Rui Zhao, Mike Belov, Christophe Masselon, David Anderson, Nikola Tolic, Yufeng Shen, Eric Strittmatter, Nestor Rodriguez, Mike Goshe, and Josip Blonder to the work reviewed here are gratefully acknowledged. We thank the US Department of Energy Office of Biological and Environmental Research for long term support of the *D. radiodurans* research and the FTICR technology development, as well as the National Institutes of Health, through NCI (CA86340 and CA93306) and NCRR (RR12365) for support of portions of the reviewed research. Pacific Northwest National Laboratory is operated by Battelle Memorial Institute for the US Department of Energy under contract DE-AC06-76RLO 1830.

References

1. Adams, M.D., Serial analysis of gene expression: ESTs get smaller. *Bioessays*, **18**, 261–262 (1996).
2. Velculescu, V.E., Zhang, L., Zhou, W., Vogelstein, J., Basrai, M.A., Bassett, D.E.J., Hieter, P., Vogelstein, B. and Kinzler, K.W., Characterization of the yeast transcriptome. *Cell*, **88**, 243–251 (1997).
3. Anderson, L. and Seilhammer, J., A comparison of selected mRNA and protein abundances in human liver. *Electrophoresis*, **18**, 533–537 (1997).
4. Haynes, P.A., Gygi, S.P., Figeys, D. and Aebersold, R., Proteome analysis: biological assay or data archive? *Electrophoresis*, **19**, 1862–1871 (1998).
5. Gygi, S.P., Corthals, G.L., Zhang, Y., Rochon, Y. and Aebersold, R., Evaluation of two-dimensional gel electrophoresis-based proteome analysis technology. *Proc. Natl Acad. Sci. USA*, **97**, 9390–9395 (2000).
6. Yates, J.R., Speicher, S., Griffin, P.R. and Hunkapiller, T., Peptide mass maps: a highly informative approach to protein identification. *Anal. Biochem.*, **214**, 397–408 (1993).
7. Shevchenko, A., Jensen, O.N., Podtelejnikov, A.V., Sagliocco, F., Wilm, M., Vorm, O., Mortensen, P., Boucherie, H. and Mann, M., Linking genome and proteome by mass spectrometry: large-scale identification of yeast proteins from two dimensional gels. *Proc. Natl Acad. Sci. USA*, **93**, 14440–14445 (1996).
8. Wilm, M., Shevchenko, A., Houthaeve, T., Breit, S., Schweigerer, L., Fotsis, T. and Mann, M., Femtomole sequencing of proteins from polyacrylamide gels by nano-electrospray mass spectrometry. *Nature*, **379**, 466–469 (1996).
9. Shevchenko, A., Wilm, M., Vorm, O. and Mann, M., Mass spectrometric sequencing of proteins from silver stained polyacrylamide gels. *Anal. Chem.*, **68**, 850–858 (1996).
10. Langen, H. and Al, E., Two-dimensional map of the proteome of *Haemophilus influenzae*. *Electrophoresis*, **21**, 411–429 (2000).
11. Perrot, M., Two-dimensional gel protein database of *Saccharomyces cerevisiae*. *Electrophoresis*, **20**, 2280–2298 (1999).
12. Futcher, B., Latter, G.I., Monardo, P., Mclaughlin, C.S. and Garrels, J.I., A sampling of the yeast proteome. *Mol. Cell. Biol.*, **19**, 7357–7368 (1999).
13. Gygi, S.P., Rochon, Y., Franza, B.R. and Aebersold, R., Correlation between protein and mRNA abundance in yeast. *Mol. Cell. Biol.*, **19**, 1720–1730 (1999).
14. Garrels, J.I., Mclaughlin, C.S., Warner, J.R., Futcher, B., Latter, G.I., Kobayashi, R., Schwender, B., Volpe, T., Anderson, D.S., Mesquitafuentes, R. and Payne, W.E., Proteome studies of *Saccharomyces cerevisiae* — identification and characterization of abundant proteins. *Electrophoresis*, **18**, 1347–1360 (1997).
15. Kitayama, S. and Matsuyama, A., Mechanism for radiation lethality in *M. radiodurans*. *Int. J. Radiat. Biol. Relat. Stud. Phys. Chem. Med.*, **19**, 13–19 (1971).
16. Henzel, W.J., Billeci, T.M., Stults, J.T., Wong, S.C., Grimley, C. and Watanabe, C., Identifying proteins from two-dimensional gels by molecular mass searching of peptide fragments in protein sequence databases. *Proc. Natl Acad. Sci. USA*, **90**, 5011–5015 (1993).
17. Pappin, D.J., Hojrup, P. and Bleasby, A.J., Rapid identification of proteins by peptide-mass fingerprinting. *Curr. Biol.*, **3**, 327–332 (1993).
18. Mann, M., Hojrup, P. and Roepstorff, P., Use of mass spectrometric molecular weight information to identify proteins in sequence databases. *Biol. Mass Spectrom.*, **22**, 338–345 (1993).
19. James, P., Quadroni, M., Carafoli, E. and Gonnet, G., Protein identification by mass profile fingerprinting. *Biochem. Biophys. Res. Commun.*, **195**, 58–64 (1993).

20. Yates, J.R., McCormack, A.L. and Eng, J., Mining genomes with MS. *Anal. Chem.*, **68**, A534–A540 (1996).
21. McCormack, A.L., Schieltz, D.M., Goode, B., Yang, S., Barnes, G., Drubin, D. and Yates, J.R., Direct analysis and identification of proteins in mixtures by LC/MS/MS and database searching at the low-femtomole level. *Anal. Chem.*, **69**, 767–776 (1997).
22. Ducret, A., Vanoostveen, I., Eng, J.K., Yates, J.R. and Aebersold, R., High throughput protein characterization by automated reverse-phase chromatography electrospray tandem mass spectrometry. *Protein Sci.*, **7**, 706–719 (1998).
23. Link, A.J., Hays, L.G., Carmack, E.B. and Yates, J.R., Identifying the major proteome components of *Haemophilus influenzae* type-strain NCTC 8143. *Electrophoresis*, **18**, 1314–1334 (1997).
24. Yates, J.R., Mass spectrometry and the age of the proteome. *J. Mass Spectrom.*, **33**, 1–19 (1998).
25. Washburn, M.P., Wolters, D. and Yates, J.R., Large-scale analysis of the yeast proteome by multidimensional protein identification technology. *Nat. Biotechnol.*, **19**, 242–247 (2001).
26. Conrads, T.P., Anderson, G.A., Veenstra, T.D., Pasa-Tolic, L. and Smith, R.D., Utility of accurate mass tags for proteome-wide protein identification. *Anal. Chem.*, **72**, 3349–3354 (2000).
27. Smith, R.D., Anderson, G.A., Lipton, M.S., Pasa-Tolic, L., Shen, Y., Conrads, T.P., Veenstra, T.D. and Udseth, H.R., An accurate mass tag strategy for quantitative and high throughput proteome measurements. *Proteomics*, **2**, 513–523 (2002).
28. Blonder, J., Goshe, M.B., Moore, R.J., Pasa-Tolic, L., Masselon, C.D., Lipton, M.S. and Smith, R.D., Enrichment of integral membrane proteins for proteomics analysis using liquid chromatography–tandem mass spectrometry. *J. Proteome Res.*, **1** (2002) Web ASAP Article.
29. Belov, M.E., Anderson, G.A., Angell, N.H., Shen, Y., Tolic, N., Udseth, H.R. and Smith, R.D., Dynamic range expansion applied to mass spectrometry based on data-dependent selective ion ejection in capillary liquid chromatography Fourier transform ion cyclotron resonance for enhanced proteome characterization. *Anal. Chem.*, **73**, 5052–5060 (2001).
30. Shen, Y., Tolic, N., Zhao, R., Pasa-Tolic, L., Li, L., Berger, S.J., Harkewicz, R., Anderson, G.A., Belov, M.E. and Smith, R.D., High-throughput proteomics using high efficiency multiple-capillary liquid chromatography with on-line high performance ESI FTICR mass spectrometry. *Anal. Chem.*, **73**, 3011–3021 (2001).
31. Conrads, T.P., Alving, K., Veenstra, T.D., Belov, M.E., Anderson, G.A., Anderson, D.J., Lipton, M.S., Pasa-Tolic, L., Udseth, H.R., Chrisler, W.B., Thrall, B.D. and Smith, R.D., Quantitative analysis of bacterial and mammalian proteomes using a combination of cysteine affinity tags and 15N-metabolic labeling. *Anal. Chem.*, **73**, 2132–2139 (2001).
32. Molloy, M.P., Herbert, B.R., Slade, M.B., Rabilloud, T., Nouwens, A.S., Williams, K.L. and Gooley, A.A., Proteomic analysis of the *Escherichia coli* outer membrane. *Eur. J. Biochem.*, **267**, 2871–2881 (2000).
33. Fujiki, Y., Hubbard, A.L., Fowler, S. and Lazarow, P.B., Isolation of intracellular membranes by means of sodium carbonate treatment: application to endoplasmic reticulum. *J. Cell Biol.*, **93**, 97–102 (1982).
34. Park, Z.Y. and Russell, D.H., Identification of individual proteins in complex protein mixtures by high-resolution, high-mass-accuracy MALDI TOF-mass spectrometry analysis of in-solution thermal denaturation/enzymatic digestion. *Anal. Chem.*, **73**, 2558–2564 (2001).
35. Russell, W.K., Park, Z.Y. and Russell, D.H., Proteolysis in mixed organic-aqueous solvent systems: applications for peptide mass mapping using mass spectrometry. *Anal. Chem.*, **73**, 2682–2685 (2001).

36. Gygi, S.P., Rist, B., Gerber, S.A., Turecek, F., Gelb, M.H. and Aebersold, R., Quantitive analysis of complex protein mixtures using isotope-coded affinity tags. *Nat. Biotechnol.*, **17**, 994–999 (1999).
37. Goshe, M.B., Conrads, T.P., Panisko, E.A., Angell, N.H., Veenstra, T.D. and Smith, R.D., Phosphoprotein isotope-coded affinity tag approach for isolating and quantitating phosphopeptides in proteome-wide analyses. *Anal. Chem.*, **73**, 2578–2586 (2001).
38. Goshe, M.B., Veenstra, T.D., Panisko, E.A., Conrads, T.P., Angell, N.H. and Smith, R.D., Phosphoprotein isotope-coded affinity tags: application to the enrichment and identification of low abundance phosphoproteins. *Anal. Chem.*, **74**, 607–616 (2002).
39. Shen, Y., Zhao, R., Belov, M.E., Conrads, T.P., Anderson, G.A., Tang, K., Pasa-Tolic, L., Veenstra, T.D., Lipton, M.S., Udseth, H.R. and Smith, R.D., Packed capillary reversed-phase liquid chromatography with high-performance electrospray ionization Fourier transform ion cyclotron resonance mass spectrometry for proteomics. *Anal. Chem.*, **73**, 1766–1775 (2001).
40. Shen, Y., Zhao, R., Berger, S.J., Anderson, G.A., Rodriguez, N. and Smith, R.D., High-efficiency nanoscale liquid chromatography coupled on-line with mass spectrometry using nanoelectrospray ionization. *Anal. Chem.* (2002), in press.
41. Marshall, A.G., Hendrickson, C.L. and Jackson, G.S., Fourier transform ion cyclotron resonance mass spectrometry: a primer. *Mass Spectrom. Rev.*, **17**, 1–35 (1998).
42. Pasa-Tolic, L., Jensen, P.K., Anderson, G.A., Lipton, M.S., Peden, K.K., Martinovic, S., Tolic, N., Bruce, J.E. and Smith, R.D., High throughput proteome-wide precision measurements of protein expression using mass spectrometry. *J. Am. Chem. Soc.*, **121**, 7949–7950 (1999).
43. Belov, M.E., Gorshkov, M.V., Udseth, H.R., Anderson, G.A. and Smith, R.D., Zeptomole-sensitivity electrospray ionization–Fourier transform ion cyclotron resonance. *Anal. Chem.*, **72**, 2271–2279 (2000).
44. Kim, T., Tolmachev, V., Harkewicz, R., Prior, D.C., Anderson, G.A., Udseth, H.R., Smith, R.D., Bailey, T.H., Rakov, S. and Futrell, J.H., Design and implementation of a new electrodynamic ion funnel. *Anal. Chem.*, **72**, 2247–2255 (2000).
45. Belov, M.E., Gorshkov, M.V., Udseth, H.R., Anderson, G.A., Tolmachev, A.V., Prior, D.C., Harkewicz, R. and Smith, R.D., Initial implementation of an electrodynamic ion funnel with FTICR mass spectrometry. *J. Am. Soc. Mass Spectrom.*, **11**, 19–23 (2000).
46. Belov, M.E., Nikolaev, E.N., Anderson, G.A., Auberry, K.J., Harkewicz, R. and Smith, R.D., Electrospray ionization–Fourier transform ion cyclotron mass spectrometry using ion preselection and external accumulation for ultra-high sensitivity. *J. Am. Soc. Mass Spectrom.*, **12**, 38–48 (2001).
47. Jensen, P.K., Pasa-Tolic, L., Anderson, G.A., Horner, J.A., Lipton, M.S., Bruce, J.E. and Smith, R.D., Probing proteomes using capillary isoelectric focusing–electrospray ionization Fourier transform ion cyclotron resonance mass spectrometry. *Anal. Chem.*, **71**, 2076–2084 (1999).
48. Bruce, J.E., Anderson, G.A., Brands, M.D., Pasa-Tolic, L. and Smith, R.D., Obtaining more accurate FTICR mass measurements without internal standards using multiply charged ions. *J. Am. Soc. Mass Spectrom.*, **11**, 416–421 (2000).
49. Li, L., Masselon, C., Anderson, G.A., Pasa-Tolic, L., Lee, S.-W., Shen, Y., Zhao, R., Lipton, M.S., Conrads, T.P., Tolic, N. and Smith, R.D., High-throughput peptide identification from protein digests using data-dependent multiplexed tandem FTICR mass spectrometry coupled with capillary liquid chromatography. *Anal. Chem.*, **73**, 3312–3322 (2001).
50. Tang, K., Tolmachev, A.V., Nikolaev, E., Zhang, R., Belov, M.E., Udseth, H.R. and Smith, R.D., Independent control of ion transmission in a jet disrupter dual channel ion funnel ESI–MS interface. *Anal. Chem.* (2002), submitted for publication.

51. Yates, J.R., Eng, J.K., Mccormack, A.L. and Schieltz, D., Method to correlate tandem mass spectra of modified peptides to amino acid sequences in the protein database. *Anal. Chem.*, **67**, 1426–1436 (1995).
52. Spahr, C.S., Susin, S.A., Bures, E.J., Robinson, J.H., Davis, M.T., Mcginley, M.D., Kroemer, G. and Patterson, S.D., Simplification of complex peptide mixtures for proteomic analysis: reversible biotinylation of cysteinyl peptides. *Electrophoresis*, **21**, 1635–1650 (2000).
53. Valaskovic, G.A., Kelleher, N.L. and McLafferty, F.W., Attomole protein characterization by capillary electrophoresis mass spectrometry. *Science*, **273**, 1199–1202 (1996).
54. Belov, M.E., Nikolaev, E.N., Harkewicz, R., Masselon, C., Alving, K. and Smith, R.D., Ion discrimination during ion accumulation in a quadrupole interface external to a Fourier transform ion cyclotron resonance mass spectrometer. *Int. J. Mass Spectrom.*, **208**, 205–225 (2001).
55. Sannes-Lowery, K., Griffey, R.H., Kruppa, G.H., Speir, J.P. and Hofstadler, S.A., Multipole storage assisted dissociation, a novel in-source dissociation technique for electrospray ionization generated ion. *Rapid Commun. Mass Spectrom.*, **12**, 1957–1961 (1998).
56. Tolmachev, A.V., Udseth, H.R. and Smith, R.D., Radial stratification of ions as a function of mass to charge ratio in collisional cooling radio frequency multipoles used as ion guides or ion traps. *Rapid Commun. Mass Spectrom.*, **14**, 1907–1913 (2000).
57. Belov, M.E., Gorshkov, M.V., Alving, K. and Smith, R.D., Optimal pressure conditions for unbiased external ion accumulation in a 2-D RF-quadrupole for FTICR mass spectrometry. *Rapid Commun. Mass Spectrom.*, **15**, 1988–1996 (2001).
58. Harkewicz, R., Belov, M.E., Anderson, D.A., Pasa-Tolic, L., Masselon, C.D., Prior, D.C., Udseth, H.R. and Smith, R.D., ESI–FTICR mass spectrometry employing data-dependent external ion selection and accumulation. *J. Am. Soc. Mass Spectrom.*, **13**, 144–154 (2002).
59. Guan, S.H., Kim, H.S., Marshall, A.G., Wahl, M.C., Wood, T.D. and Xiang, X.Z., Shrink-wrapping an ion cloud for high-performance Fourier transform ion cyclotron resonance mass spectrometry. *Chem. Rev.*, **94**, 2161–2182 (1994).
60. Pasa-Tolic, L., Harkewicz, R., Anderson, G.A., Tolic, N., Shen, Y., Zhao, R., Thrall, B., Masselon, C. and Smith, R.D., Increased proteome coverage for quantitative peptide abundance measurements based upon high performance separations and DREAMS FTICR mass spectrometry. *J. Am. Soc. Mass Spectrom.* (2002), in press.
61. Smith, R.D., Anderson, G.A., Lipton, M.S., Pasa-Tolic, L., Shen, Y. and Udseth, H.R., The use of accurate mass tags for high-throughput microbial proteomics. *OMICS*, **6**, 61–90 (2002).
62. Washburn, M.P., Ulaszek, R., Deciu, C., Schieltz, D. and Yates, J.R., Analysis of quantitative proteomic data generated via multidimensional protein identification technology. *Anal. Chem.*, **74**, 1650–1657 (2002).
63. Gorshkov, M.V., Masselon, C., Anderson, G.A., Udseth, H.R. and Smith, R.D., Dynamically assisted gated trapping for FTICR mass spectrometry. *Rapid Commun. Mass Spectrom.*, **15**, 1558–1561 (2001).
64. Makarova, K.S., Aravind, L., Wolf, Y.I., Tatusov, R.L., Minton, K.W., Koonin, E.V. and Daly, M.J., Genome of the extremely radiation-resistant bacterium *Deinococcus radiodurans* viewed from the perspective of comparative genomes. *Microbiology*, **65**, 44–79 (2001).
65. White, O., Eisen, J.A., Heidelberg, J.F., Hickey, E.K., Peterson, J.D., Dodson, R.J., Haft, D.H., Gwinn, M.L., Nelson, W.C., Richardson, D.L., Moffat, K.S., Qin, H., Jiang, L., Pamphile, W., Crosby, M., Shen, M., Vamathevan, J.J., Lam, P., Mcdonald, L., Utterback, T., Zalewski, C., Makarova, K.S., Aravind, L., Daly, M.J., Minton, K.W., Fleischmann, R.D., Ketchum, K.A., Nelson, K.E., Salzberg, S., Smith, H.O., Venter, J.C. and Fraser, C.M., Genome sequence of the radioresistant bacterium *Deinococcus radiodurans* R1. *Science*, **286**, 1571–1577 (1999).

66. Venkateswaran, A., Mcfarlan, S.C., Ghosal, D., Minton, K.W., Vasilenko, A., Makarova, K., Wackett, L.P. and Daly, M.J., Physiologic determinants of radiation resistance in *Deinococcus radiodurans*. *Appl. Environ. Microbiol.*, **66**, 2620–2626 (2000).
67. Belov, M., Gorshkov, M., Udseth, H. and Smith, R., Controlled ion fragmentation in a 2-D quadrupole ion trap for external ion accumulation in ESI FTICR mass spectrometry. *J. Am. Soc. Mass Spectrom.*, **12**, 1312–1319 (2001).
68. Patel, P.H., Suzuki, M., Adman, E., Shinkai, A. and Loeb, L.A., Prokaryotic DNA polymerase I: evolution, structure, and 'base flipping' mechanism for nucleotide selection. *J. Mol. Biol.*, **308**, 823–837 (2001).
69. Sharp, P.M. and Li, W.-H., The codon adaptation index — a measure of directional synonymous codon usage bias, and its potential applications. *Nucleic. Acids Res.*, **15**, 1281–1295 (1987).
70. Karlin, S. and Mrazek, J., Predicted highly expressed genes of diverse prokaryotic genomes. *J. Bacteriol.*, **182**, 5238–5250 (2000).
71. Curnow, A.W., Tumbula, D.L., Pelaschier, J.T., Min, B. and Soll, D., Glutamyl-tRNA(Gln) amidotransferase in *Deinococcus radiodurans* may be confined to asparagine biosynthesis. *Proc. Natl Acad. Sci. USA*, **95**, 12838–12843 (1998).
72. Daly, M.J., Ouyang, L., Fuchs, P. and Minton, K.W., In vivo damage and recA-dependent repair of plasmid and chromosomal DNA in the radiation-resistant bacterium *Deinococcus radiodurans*. *J. Bacteriol.*, **176**, 3508–3517 (1994).
73. Minton, K.W., DNA repair in the extremely radioresistant bacterium *Deinococcus radiodurans*. *Mol. Microbiol.*, **1**, 9–15 (1994).
74. Moseley, B.E. and Evans, D.M., Isolation and properties of strains of *Micrococcus* (*Deinococcus*) *radiodurans* unable to excise ultraviolet light-induced pyrimidine dimers from DNA: evidence for two excision pathways. *J. Gen. Microbiol.*, **129**, 2437–2445 (1983).
75. Oyarce, A.M. and Eipper, B.A., Identification of subcellular compartments containing peptidylglycine alpha-amidating monooxygenase in rat anterior pituitary. *J. Cell Sci.*, **108**, 287–297 (1995).
76. Posewitz, M.C. and Tempst, P., Immobilized gallium(III) affinity chromatography of phosphopeptides. *Anal. Chem.*, **71**, 2883–2892 (1999).
77. Wallin, E. and Von Heijne, G., Genome-wide analysis of integral membrane proteins from eubacterial, archaean, and eukaryotic organisms. *Protein Sci.*, **7**, 1029–1038 (1998).
78. Santoni, V., Molloy, M. and Rabilloud, T., Membrane proteins and proteomics: un amour impossible? *Electrophoresis*, **21**, 1054–1070 (2000).
79. Molloy, M.P., Two-dimensional electrophoresis of membrane proteins using immobilized pH gradients. *Anal. Biochem.*, **280**, 1–10 (2000).
80. Buttner, K., Bernhardt, J., Scharf, C., Schmid, R., Mader, U., Eymann, C., Antelmann, H., Volker, A., Volker, U. and Hecker, M., A comprehensive two-dimensional map of cytosolic proteins of *Bacillus subtilis*. *Electrophoresis*, **22**, 2908–2935 (2001).
81. Pandey, A., Podtelejnikov, A.V., Blagoev, B., Bustelo, X.R., Mann, M. and Lodish, H.F., Analysis of receptor signaling pathways by mass spectrometry: identification of Vav-2 as a substrate of the epidermal and platelet-derived growth factor receptors. *Proc. Natl Acad. Sci. USA*, **97**, 179–184 (2000).
82. Ficarro, S.B., McCleland, M.L., Stukenberg, P.T., Burke, D.J., Ross, M.M., Shabanowitz, J., Hunt, D.F. and White, F.M., Phosphoproteome analysis by mass spectrometry and its application to *Saccharomyces cerevisiae*. *Nat. Biotechnol.*, **20**, 301–305 (2002).
83. Zhou, H., Watts, J.D. and Aebersold, R., A systematic approach to the analysis of protein phosphorylation. *Nat. Biotechnol.*, **19**, 375–378 (2001).

84. Oda, Y., Nagasu, T. and Chait, B., Enrichment analysis of phosphorylated proteins as a tool for probing the phosphoproteome. *Nat. Biotechnol.*, **19**, 379–382 (2001).
85. Borisov, O.V., Goshe, M.B., Conrads, T.P., Rakov, V.S., Veenstra, T.D. and Smith, R.D., Low-energy collision-induced dissociation fragmentation analysis of cysteinyl-modified peptides. *Anal. Chem.*, **74**, 2284–2292 (2002).
86. Corthals, G.L., Wasinger, V.C., Hochstrasser, D.F. and Sanchez, J.-C., The dynamic range of protein expression: a challenge for proteomic research. *Electrophoresis*, **21**, 1104–1115 (2000).

Proteome Analysis. Interpreting the Genome.
D.W. Speicher (editor)

Chapter 9

Clinical applications of proteomics

SAM M. HANASH*

Department of Pediatrics, University of Michigan, 1150 W. Medical Center Drive, MSRBI, Room A520, Ann Arbor, MI 48109, USA

1. Introduction

There is currently substantial interest in the application of proteomics to disease investigations. Despite tremendous advances in our understanding of the molecular genetics of diseases such as cancer, the development of novel diagnostics and therapeutics has lagged behind. Promising areas in disease-related proteomics include: (i) the identification of distinctive disease proteomic profiles that correlate with clinical features; (ii) the development of novel biomarkers for diagnosis and early detection; (iii) the identification of novel targets for therapeutics on the one hand, given that the vast majority of drugs target proteins; and (iv) on the other hand, to utilize proteomics to accelerate drug development and evaluation of efficacy and toxicity. The proteome of a cell or a tissue is highly dynamic and changes much more readily in response to external factors than the genome or transcriptome. To capture all the proteomes that a cell or a tissue may manifest in health and disease represents a substantial challenge for which the current available technologies have a limited reach. Nevertheless, numerous biomedical investigations have been successfully tackled using proteomics, and

* Tel.: +1-734-763-9311; fax: +1-734-647-8148. E-mail: shanash@umich.edu (S.M. Hanash).

proteomics technologies are currently evolving in a manner that emphasizes sensitivity and throughput. It is not feasible in a concise review to provide a full presentation of all clinical applications of proteomics. While this review emphasizes the work done in the author's laboratory, it in no way is intended to give an exclusive view but rather represents an attempt to provide a coherent view of the clinical application of proteomics in one area and as practiced in one laboratory.

2. Correlative studies using proteomics and transcriptomics

The large-scale profiling of gene expression at the protein level in cancer has a long history that predates profiling at the RNA level, largely due to the availability of older procedures for 2D separation of protein mixtures [1]. In the past decade, mass spectrometry has facilitated the identification, through sequence database searching, of proteins separated by 2D gels or other means, at an unprecedented level of sensitivity and speed [2]. With the recent interest in global gene expression profiling at the RNA level, the extent to which the two types of analyses correlate or complement each other has been of substantial interest, as assays of gene expression through measurements of RNA may not necessarily predict protein levels. Furthermore, post-translational modifications cause a much greater diversity in protein expression than could be gleaned simply from RNA analysis. Additional diversity in expression stems from the various locations in which proteins reside in cells and tissue and the dynamic nature of their translocation between various cellular compartments.

While it has been frequently advanced that measures of abundance at the protein and RNA levels for the same genes correlate poorly with each other, there is currently only very limited and rather sketchy data for interpreting the significance of such poor correlations. In studies of different tumor types in the author's laboratory, the availability of mRNA expression data from DNA microarrays for the same tumor samples where protein 2D gel data was available has permitted more thorough examination of RNA/protein relationships. In cases where the identity of the protein spot is known, such investigations can answer the question of how well mRNA levels for a protein predict that protein's abundance. In cases where protein spots have not yet been identified or were identified with low confidence, such correlations can lead to or confirm hypothetical spot identifications. More generally, one can search for larger groups of proteins and mRNA whose abundance is controlled by some common mechanism. In related studies of lung adenocarcinomas, mRNA and protein expression were compared for a cohort of genes in the same tumors [3]. The abundance of 165 protein spots representing 98 individual genes was analyzed in 76 lung adenocarcinomas and nine non-neoplastic lung tissues using 2D gels. For the same 85 samples, mRNA levels were determined using oligonucleotide microarrays, allowing a comparative analysis of mRNA levels and protein expression among the 165 protein spots. Twenty-eight of the 165 protein spots (17%) or 21 of 98 genes (21.4%) had a statistically significant correlation between protein and mRNA expression ($r > 0.2445$; $p < 0.05$); however, among all 165 proteins, the correlation coefficient values (r) ranged from 0.467 to 0.442. Correlation coefficient values were not related to protein abundance. No significant correlation between mRNA and protein expression was found ($r - 0.025$) if the average levels of mRNA or protein among all samples were applied across the 165 protein spots (98 genes). The mRNA/protein correlation coefficient also varied among proteins with multiple isoforms, indicating potentially separate isoform-specific mechanisms for the regulation of protein abundance. Interestingly, among the 21 genes with a significant correlation between mRNA and protein, five genes differed significantly between stage I and stage III lung adenocarcinomas. Using a quantitative analysis of mRNA and protein expression within the same lung adenocarcinomas, we showed that only a subset of the proteins exhibited a significant correlation with mRNA abundance.

Aside from the aspect of protein–RNA correlations, the informational value and complementary contributions of proteomic and transcriptomic approaches are of interest. Two independent approaches were utilized to analyze the expression profile of human $CD14^+$ blood monocytes and their derived dendritic cells (DCs) because the process of differentiation and maturation of DCs remains poorly understood and is highly relevant to immunotherapy [4]. Analysis of gene expression changes at the RNA level, using oligonucleotide microarrays complementary to 6300 human genes, showed that approximately 40% of the genes were expressed in DCs. A total of 255 genes (4%) were found to be regulated during DC differentiation or maturation. Most of these genes were not previously associated with DCs and included genes encoding secreted proteins as well as genes involved in cell adhesion, signaling, and lipid metabolism. 2D gel electrophoresis was used for protein analysis of the same cell populations. A total of 900 distinct protein spots were included, 4% of which exhibited quantitative changes during DC differentiation and maturation. Differentially expressed proteins were identified by mass spectrometry and found to represent proteins with Ca^{2+} binding, fatty acid binding or chaperone activities as well as proteins involved in cell motility. In addition, proteomic analysis provided an assessment of post-translational modifications. The chaperone protein, calreticulin, was found to undergo cleavage, yielding a novel form. This study showed that combined oligonucleotide microarray and proteomic approaches can uncover novel genes associated with DC differentiation and maturation and can detect post-translational modifications of specific proteins associated with these processes.

It is obvious that at the present time, essentially all the genes in the genome can be interrogated for their expression at the RNA level using DNA microarrays.

While proteomic technologies do not currently approach this level of comprehensiveness, it is clear that as the sensitivity and throughput of proteomics tools increase, the potential of proteomics for profiling gene expression and uncovering biologically and clinically relevant patterns may far exceed that of DNA microarrays.

3. Disease marker identification using proteomics

A comprehensive strategy for identification of disease markers using proteomics includes analysis of normal and disease tissue as well as analysis of serum and other biological fluids. Proteomic analysis of tissues and cell populations uniquely contributes an understanding of both protein post-translational modifications and the distribution of protein gene products in subcellular compartments. Currently, many groups are interested in profiling biological fluids using proteomic technologies from antibody microarrays to 'industrial-scale' proteomics, a term which has referred to the in-depth identification of proteins in liter quantities of samples such as serum or plasma. These studies are illustrated here by the analysis of bronchoalveolar lavage (BAL) fluid to identify disease markers and the analysis of serum to identify cancer markers. A pertinent source of information on this subject is a recent special issue of the journal *Disease Markers* (Volume 17, #4, 2001), which is dedicated to proteomics.

Changes in secreted lung tissue factors, notably surfactants, have been relied upon to diagnose, treat, and monitor lung diseases. BAL fluid has a highly complex protein content due in part to the diversity of its cellular origins. Proteome analysis of BAL fluid to search for new lung disease marker proteins has been undertaken. The first 2D gel study of BAL was published more than 20 years ago [5] and resulted in the identification of 23 serum-like proteins in BAL. Subsequent studies encompassing a greater number of polypeptides uncovered disease-related protein changes in idiopathic pulmonary fibrosis and other diseases. Some of these proteins may turn out to be disease markers as well as reflect differences in protein patterns between smokers and non-smokers [6]. Studies of BAL fluid in cystic fibrosis have uncovered changes including high levels of lipocalin-1 [7]. A more exhaustive identification of protein in BAL fluid leading to the construction of a map was undertaken using immobilized pH gradients for the first-dimension separation in a 2D gel system [8] (Fig. 1). The map has included over 1200 silver-stained spots, 900 of which were identified by matching with human plasma protein and other published 2D gel maps. A large number of proteins, some novel, exhibited disease-related differences. Further studies are likely to result in a battery of markers of relevance for diagnosis and for guiding therapy.

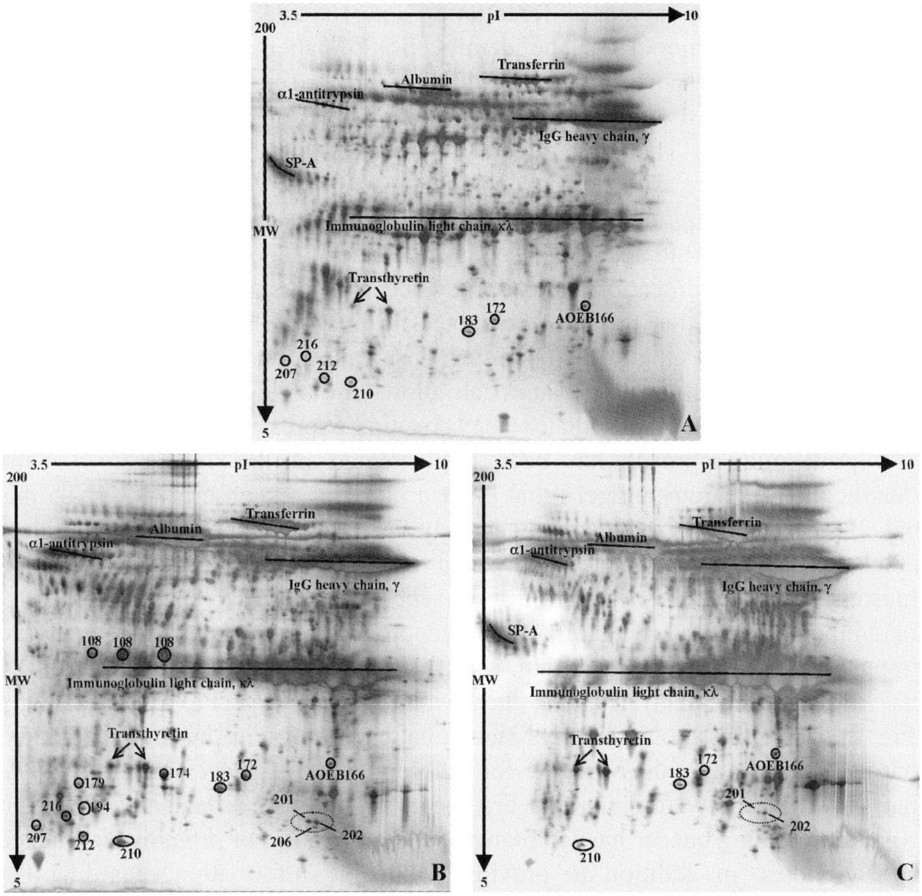

Fig. 1. Silver-stained 2D gels of human BAL fluids from (A) a healthy subject (B) a patient with idiopathic pulmonary fibrosis, and (C) a patient with sarcoidosis. Circles are placed around proteins upregulated in idopathic pulmonary fibrosis (Reprinted with permission from: Noel-Georis, I., et al., *Dis. Markers*, **17(4)**, 271–284 (2001)).

An important objective of the lung cancer effort in this laboratory is the identification of novel markers for early detection. In addition to direct analysis of protein expression in tumors, a highly productive approach for marker identification has been the analysis of serum for tumor proteins that induce a serological response in the form of autoantibodies. There is increasing evidence for an immune response to cancer in humans, demonstrated in part by the identification of autoantibodies against a number of intracellular and surface antigens detectable in sera from patients with different cancer types [9–12]. The majority of tumor-derived antigens that have been identified as eliciting a humoral response in lung cancer, as in other tumor types, are not the products of mutated genes. They include differentiation antigens and other proteins that are overexpressed in tumors [13]. The identification of panels of tumor antigens that elicit an antibody response may have utility in cancer screening, diagnosis, or in establishing prognosis. Such antigens may also have utility in immunotherapy against the disease. There are several approaches for the detection of tumor antigens that induce an immune response. A number of antigens have been detected by screening expression libraries with patient sera [9–11,14–16]. The merits of the proteomic approach used in the author's laboratory is that it allows proteins in their modification states as they occur in cells to be analyzed for their antigenicity. Given that proteins are subject to post-translational modifications, antibodies to epitopes that result from such post-translational modifications can be detected. Additionally, the 2D approach allows the serial serum samples to be analyzed much more readily than when screening expression libraries.

A proteomic approach has been implemented in this laboratory for the identification of tumor antigens that elicit a humoral response [17–21]. To this end, 2D gels were utilized to simultaneously separate several thousand individual cellular proteins from tumor tissue or tumor cell lines. Separated proteins are transferred onto membranes. Sera from cancer patients are screened individually, for antibodies that react against separated proteins, by Western blot analysis. Proteins that specifically react with sera from cancer patients are identified by mass spectrometric analysis and further evaluated with respect to their specificity (Fig. 2). Using this proteomic approach, a battery of proteins has been identified that induce autoantibodies specific for different types of cancer including some that show specificity for lung cancer. The availability of a database of protein expression in lung cancer has facilitated the identification of proteins that induce autoantibodies, in addition to providing valuable information regarding the expression pattern of such protein antigens in different tumor types and cell lines. One such antigen we have identified in lung cancer is protein PGP 9.5 [18]. Circulating PGP 9.5 antigen was also detected in sera from some patients with lung cancer, without detectable PGP 9.5 autoantibodies.

In another study, sera from 54 newly diagnosed patients with lung cancer, 60 patients with other cancers, and 61 non-cancer controls were analyzed for autoantibodies to lung tumor proteins. Sera from 60% of patients with lung adenocarcinoma and 33% of patients with squamous cell lung carcinoma but none of the non-cancer controls exhibited IgG-based reactivity against proteins identified as glycosylated annexins I and/or II. Immunohistochemical analysis showed that annexin I was diffusely expressed in neoplastic cells in lung tumor tissues, whereas annexin II was predominant at the cell surface. Interestingly, annexin I is a target of autoantibodies in autoimmune diseases such as systemic lupus erythematosus [22,23] and rheumatoid arthritis [24]. Annexin II, specifically, has not been previously implicated as a target of autoantibodies in any disorders.

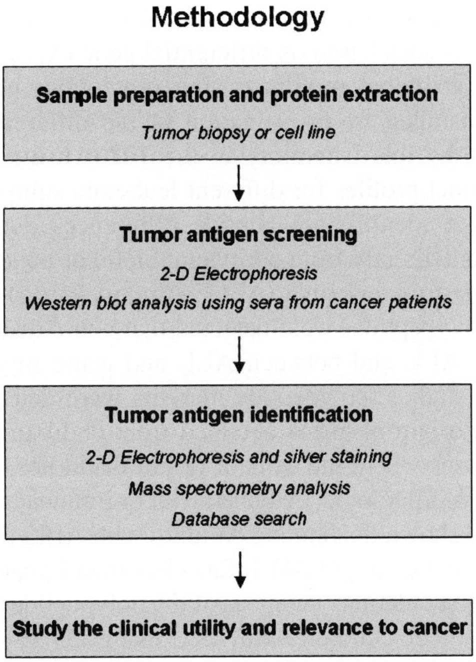

Fig. 2. Schema for discovery and identification of tumor antigens using proteomics.

4. Disease tissue analysis using proteomics

Tissues from numerous diseases other than cancer have been investigated using proteomics. A full review of the studies undertaken and resulting findings are beyond the scope of this chapter. However, several insightful reviews have been published [25–28].

A molecular analysis study of normal and diseased tissues invariably has to address issues pertaining to tissue heterogeneity and sampling. Given that tissues are made of numerous components, bulk tissue has the potential for substantial heterogeneity stemming from numerous cell types and stromal elements. This is particularly problematic for tumor tissue. For example, prostatic tissue consists of epithelial and stromal cells with the epithelial component representing only a small percentage. Prostatic cancer, particularly, shows substantial heterogeneity with respect to extent of stroma and benign prostatic epithelium [29]. Laser capture microdissection has been developed as a method to procure pure populations of cells from specific microscopic regions of tissue sections under direct visualization [30]. While ideally it would be desirable to obtain pure populations of cells from tissues for molecular analysis, because of tissue heterogeneity, in practice such an approach may be difficult to implement because of the current requirement for large amounts of protein for analysis.

With the recent interest in the use of microarrays to molecularly characterize normal and diseased tissues based on differential gene expression, it is of interest to note that efforts at global profiling using proteomics go back at least two decades. This is illustrated by investigation of the different subtypes of acute lymphoid leukemias by this laboratory in the 1980s using proteomics and the identification of distinct profiles for different leukemia subtypes [31]. This study utilized 2D PAGE to identify polypeptide differences due to lineage among lymphoblasts obtained directly from peripheral blood or bone marrow of children with acute lymphoblastoid leukemia (ALL). Among some 400 polypeptides that were analyzed, 12 polypeptides were detected that could distinguish between the major subgroups of ALL and between ALL and acute myelogenous leukemia (AML). In separate studies, polypeptide markers were detected in 2D gels that indicated a myeloid origin of blasts obtained from children who presented with acute leukemia in cases where the cells of origin could not be determined at the time of diagnosis by morphologic, cytochemical, or immune marker analysis [32]. Clinical studies have shown that among children with ALL, age at diagnosis is an important prognostic indicator [33,34]. Infants less than 1 year of age at the time of diagnosis had the worst outcome. Analysis of the polypeptide patterns of leukemic cells of infants and older children with ALL using 2D PAGE showed distinct levels of a polypeptide, designated L3 between infants and older children with otherwise similar cell surface markers [35].

More recently, proteins with restricted expression in different types of cancer have been identified by numerous groups. In a recent study of breast, ovary and lung tumors, 20 differentially expressed proteins were identified [36] and in a prior study, 16 polypeptides were found to be associated with different histopathological features of lung cancer [37,38]. A large number of studies involving lung cancer have been independently performed in the author's laboratory. At the protein level, these studies have resulted in over 1000 samples related to lung cancer, which have been processed using 2D gels and for which information has been recorded in the Lung Protein Database. While lung adenocarcinomas represent a major portion of the lung cancer database, other lung tumor types including squamous cell carcinomas and small cell lung cancers are represented, as are control lung tissues. Other 2D patterns were produced from studies of cell lines that have been manipulated by transfection or by treatment with specific agents, as well as patterns produced after different cell fractionation schemes. In an initial study of 25 adenocarcinomas of the lung, 12 small cell lung cancers, and 16 squamous cell tumors [39], an initial analysis of protein 2D patterns uncovered a group of 52 protein spots that differed in average integrated intensity between the three groups. Performing simple two-sample T-tests gave P-values of less than 0.05 for the 52 spots for at least one of the pairs of groups. Most of the spots differed between small cell and the remaining two diagnostic groups, with 47 spots differing significantly between small cell and adenocarcinoma groups and 44 between small cell and squamous ($P < 0.05$). Between the adenocarcinoma and squamous groups 12 spots with difference of this significance were found. Of the 52 spots, 39 were identified by either N-terminal sequencing and/or mass spectrometry of spot digests. Small cell lung cancers were characterized by higher average amounts for some proteins associated with cell proliferation such as PCNA and Op18 [40–43], particularly the once-phosphorylated form of Op18, as well as protein products of the UCHL1, RBP1, CRABP2, KRT15, and TUBB genes among others. Squamous cell and adenocarcinoma samples had greater amounts of the S100 proteins, S100A8, S100A9, and S100A11, as well as larger average amounts of both the unphosphorylated and phosphorylated 27 kDa heat shock protein (HSPB1). These two groups also had larger amounts of several protein spots detected on these gels that did not occur in similar gels made from cell lines and were thought to be cleavage products from proteins present in cells or plasma surrounding the tumor cells (e.g. cleaved albumin). The number of protein spots that differed between lung adenocarcinomas and squamous tumors were fewer than the number of proteins that distinguished between small cell lung cancer and the other two lung cancer types.

Tumor profiling studies are beginning to emerge that do not involve 2D gels. As a model to better understand how patterns of protein expression shape the tissue microenvironment, Knezevic et al. analyzed protein expression in tissue derived from squamous cell carcinoma of the oral cavity through an antibody microarray approach for high-throughput proteomic analysis [44]. Utilizing laser capture microdissection to procure total protein from specific microscopic cellular populations, they demonstrated that quantitative, and potentially qualitative, differences in expression patterns of multiple proteins within epithelial cells reproducibly correlate with oral cavity tumor progression. Differential expression of multiple proteins was found in stromal cells surrounding and adjacent to regions of diseased epithelium that directly correlated with tumor progression of the epithelium. Most of the proteins identified in both cell types were involved in signal transduction pathways. They hypothesized therefore that extensive molecular communications involving complex cellular signaling between epithelium and stroma play a key role in driving oral cavity cancer progression.

A reverse phase protein array approach that immobilizes the whole repertoire of a tissue's proteins has been developed [45]. A high degree of sensitivity, precision, and linearity was achieved, making it possible to quantify the phosphorylated status of signal proteins in human tissue cell subpopulations. Using this approach, Paweletz et al. have longitudinally analyzed the state of pro-survival checkpoint proteins at the microscopic transition stage from patient-matched, histologically- normal, prostate epithelium to prostate intraepithelial neoplasia and then to invasive prostate cancer. Cancer progression was associated with increased phosphorylation of Akt, suppression of apoptosis pathways, and decreased phosphorylation of ERK. At the transition from histologically normal epithelium to intraepithelial neoplasia, a statistically significant surge in phosphorylated Akt and a concomitant suppression of downstream apoptosis pathways which precedes the transition into invasive carcinoma were observed.

A recent development in mass spectrometry allows profiling of the major proteins in fresh tissue sections [46]. With this approach, fresh tissue sections are sampled and blotted onto a polyethylene membrane for protein transfer and then subsequently analyzed by matrix-assisted laser desorption/ionization-mass spectrometry (MALDI-MS). Using this technology, Chaurand et al. [46] compared the protein expression of normal and cancerous mouse colon tissue obtained from the same animal. Several protein signals specific to cancerous tissue were observed. A protein extract obtained from the tumors was fractionated by high-performance liquid chromatography and the individual fractions analyzed by MALDI-MS. The fractions containing the targeted proteins were subjected to trypsin digestion. The resulting tryptic peptides were sequenced by tandem mass spectrometry, and based on the recovered partial amino acid sequences, three of the tumor-specific protein markers were identified as calgranulin A, calgranulin B and calgizzarin.

5. Protein microarrays as a novel technology for disease investigations

There is a need to develop technologies for proteomics that allow the systematic analysis of thousands of proteins in parallel for basic biological research for developing a better understanding of disease processes and for the identification of novel therapeutic targets. Determination of selective interactions between proteins and many other biomolecules, including other proteins, antibodies, drugs and various small ligands are essential for target discovery and validation. Thus, the development of procedures and standardized assays for the simultaneous analysis of protein-directed interactions in a rapid, inexpensive and low-sample volume format have substantial merit. Recently, several protein microarray methodologies have been developed for the high-throughput investigation of proteins. Some approaches incorporate recombinant proteins obtained using cDNA expression libraries [47–49] and phage-display libraries [50]. Other approaches array antibodies to specific proteins [51–58]. In some cases, whole tissue-derived samples have been directly arrayed onto slides to assess reactivity of total protein lysates [45,59,60]. A limitation of these methodologies is that proteins undergo numerous post-translational modifications — phosphorylations,

Fig. 3. Heat-map of average PGP antibody signal divided by average of nine human serum samples. Bright red colors are >2, bright green is <0.5. 1D fractions are columns, 2D fractions rows, except that pairs of columns are from replicate wells dotted to the slides (e.g. 5.1 and 5.2).

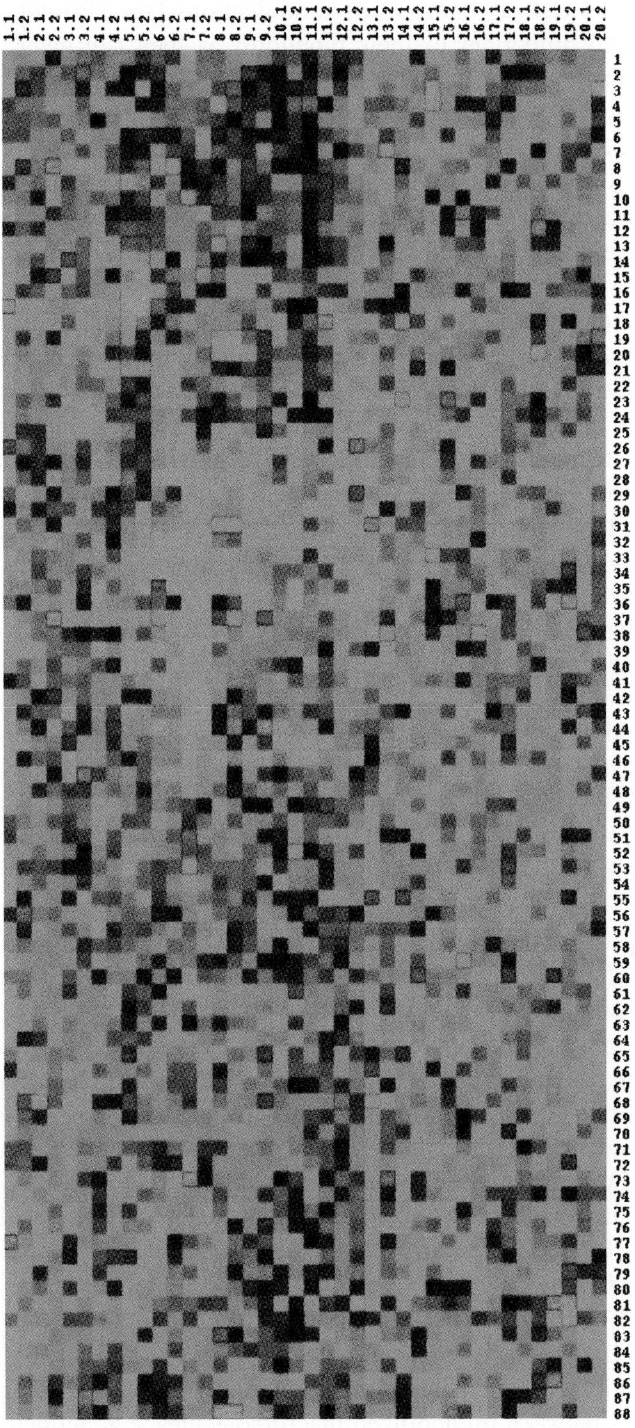
1.1
1.2
2.1
2.2
3.1
3.2
4.1
4.2
5.1
5.2
6.1
6.2
7.1
7.2
8.1
8.2
9.1
9.2
10.1
10.2
11.1
11.2
12.1
12.2
13.1
13.2
14.1
14.2
15.1
15.2
16.1
16.2
17.1
17.2
18.1
18.2
19.1
19.2
20.1
20.2
1
2
3
4
5
6
7
8
9
10
11
12
13
14
15
16
17
18
19
20
21
22
23
24
25
26
27
28
29
30
31
32
33
34
35
36
37
38
39
40
41
42
43
44
45
46
47
48
49
50
51
52
53
54
55
56
57
58
59
60
61
62
63
64
65
66
67
68
69
70
71
72
73
74
75
76
77
78
79
80
81
82
83
84
85
86
87
88

glycosylations as well as many other modifications (see Chapters 5 and 6) — which are highly important to their functions, as they can determine activity, stability, localization, and turnover. However, these modifications are generally not captured using either recombinant proteins or antibodies that recognize a single form of the protein.

The author's laboratory has investigated several approaches to the liquid-based separation of cell and tissue lysates in order to obtain protein fractions with reduced complexity or pure individual proteins [61]. The separation products can be arrayed in a manner that allows probing of cell and tissue-derived proteins to uncover specific targets. For example, using a combination of anion exchange and reverse phase LC, some 1200 individual protein fractions have been obtained that have been utilized to produce microarrays for interrogating cancer cell proteomes. Fractions that react with specific probes are well within the reach of mass spectrometric techniques for identification of their constituent proteins, and chromatographic and gel-based separation techniques for resolving their individual protein constituents (Fig. 3). The LC procedures allow sufficient protein amounts to be resolved for the construction of large numbers of microarrays from a given cell or tissue source. Different types of microarray surface binding chemistry have been analyzed for their potential to achieve the highest level of sensitivity in different assays. Individual fractions from whole cell lysates were arrayed onto glass slides and tested as probes to detect the binding of specific antibodies to their corresponding antigens in arrayed lysate fractions. The binding chemistry of the amine-coated slides gave poor results in this laboratory's experience. A slight improvement was achieved with the use of slides that had been treated with an aldehyde-containing silane reagent. The same assays on nitrocellulose membrane covered slides showed that adjacent fractions containing lower levels of antigen could also be detected, whereas the aldehyde microarrays were not able to display them. Antibody/antigen interactions were detected by measuring the fluorescence emission of the labeled antibody when attached to its corresponding antigen. The strategy used for protein microarrays was based on the technology developed for DNA microarrays [62,63]. Each slide was hybridized with a different primary antibody against a specific antigen. The reactive fractions were highly concordant with fractions shown to be reactive by Western blot analysis. The intensity of the fluorescent signal varied in an inter-dots fashion, reflecting likewise the different proportions of the specific detected protein, in relation with the total amount of protein in each fraction. The analysis showed a high signal-to-noise ratio, which provides excellent sensitivity and reproducible detection. The findings were also concordant with dot-blot controls, run in parallel and developed with the ECL chemiluminescence technology.

One of the limitations of using nitrocellulose slides is the inability to control protein orientation during the immobilization process. The necessity of optimizing the interactions between immobilized macromolecules, e.g. antigens, antibodies, peptides, and their corresponding target ligands in solution as well as with the solid device interface has been recognized [64,65]. Numerous procedures are currently available for oriented immobilization: ionic interaction, specific covalent binding, apoenzyme reconstituted on the surface that binds to a prosthetic group, receptor/ ligand interactions, specific affinity motifs engineered on the surface of the protein, etc. In most cases, it has been shown that optimal binding of protein to solid supports requires hydrophilic spacers [66]. The coupling of technologies for protein separation with techniques for orientational control would permit different surfaces of the proteins to interact with other proteins or ligands and enhance efficiency. Protein microarrays of different types are likely to become commercially available for assays of broad sets of proteins and may well rival DNA microarrays as a tool for global expression analysis.

6. Summary

It is clear that while some progress has been made in disease proteomics, the field is still in its infancy. There will be a need to employ and integrate multiple technologies, as it is unlikely that one technology will address all facets of proteomics. There is also a need to begin an organized international effort in proteomics that brings together investigators in the public as well as private sectors, given the enormity of the task of identifying and characterizing all the proteins in the human proteome [67]. Beyond technological innovations that are required to increase throughput and sensitivity, there is a substantial need to develop informatics tools that allow the mining of proteomic data that will be generated at an unprecedented scale.

References

1. Hanash, S.M., Biomedical applications of two-dimensional electrophoresis using immobilized pH gradients: current status. *Electrophoresis*, **21(6)**, 1202–1209 (2000).
2. Patterson, S., Mass spectrometry and proteomics. *Physiol. Genomics*, **2**, 59–65 (2000).
3. Chen, G., Gharib, T.G., Huang, C.C., Thomas, D.G., Shedden, K.A., Taylor, J.M., Kardia, S.L., Misek, D.E., Giordano, T.J., Iannettoni, M.D., Orringer, M.B., Hanash, S.M. and Beer, D.G., Proteomic analysis of lung adenocarcinoma: identification of a highly expressed set of proteins in tumors. *Clin. Cancer Res.*, **8(7)**, 2290–2305 (2002).
4. Le Naour, F., Hohenkirk, L., Grolleau, A., Misek, D.E., Lescure, P., Geiger, J.D., Hanash, S.M. and Beretta, L., Profiling changes in gene expression during differentiation and maturation of monocyte-derived dendritic cells using both oligonucleotide microarrays and proteomics. *J. Biol. Chem.*, **276(21)**, 17920–17931 (2001).
5. Bell, D.Y., Haseman, J.A., Spock, A., McLennan, G. and Hook, G.E., Plasma proteins of the bronchoalveolar surface of the lungs of smokers and nonsmokers. *Am. Rev. Respir. Dis.*, **124**, 72–79 (1981).

6. Westermeier, R., Postel, W., Wester, J. and Gorg, A., High-resolution two-dimensional electrophoresis with isoelectric focusing in immobilized pH gradients. *J. Biochem. Biophys. Methods*, **8**, 321–330 (1983).
7. Redhl, B., Wojnar, P., Ellemunter, H. and Feichtinger, H., Identification of a lipocalin in mucosal glands of the human tracheobronchial tree and its enhanced secretion in cystic fibrosis. *Lab. Invest.*, **78**, 1121–1129 (1998).
8. Noel-Georis, I., Bernard, A., Falmagne, P. and Wattiez, R., Proteomics as the tool to search for lung disease markers in bronchoalveolar lavage. *Dis. Markers*, **17(4)**, 271–284 (2001).
9. Yamamoto, A., Shimizu, E., Ogura, T. and Sone, S., Detection of auto-antibodies against L-myc oncogene products in sera from lung cancer patients. *Int. J. Cancer*, **22**, 283–289 (1996).
10. Stockert, E., Jager, E., Chen, Y.T., Scanlan, J.J., Gout, I., Arnad, M., Knuth, A. and Old, L.J., A survey of the humoral immune response of cancer patients to a panel of human tumor antigens. *J. Exp. Med.*, **187**, 1349–1354 (1998).
11. Gure, A.O., Altorki, N.K., Stockert, E., Scanlan, M.J., Old, L.J. and Chen, Y.T., Human lung cancer antigens recognized by autologous antibodies: definition of a novel cDNA derived from the tumor suppressor gene locus on chromosome 3p21.3. *Cancer Res.*, **58**, 1034–1341 (1998).
12. Soussi, T., p53 antibodies in the sera of patients with various types of cancer: a review. *Cancer Res.*, **60**, 1777–1788 (2000).
13. Yamamoto, A., Shimizu, E., Takeuchi, E., Houchi, H., Doi, H., Bando, H., Ogura, T. and Sone, S., Infrequent presence of anti-c-Myc antibodies and absence of c-Myce oncoprotein in sera from lung cancer patients. *Oncology*, **56(2)**, 129–133 (1999).
14. Gourevitch, M.M., von Mensdorff-Pouilly, S., Litvinov, S.V., Kenemans, P., van Kamp, G.J., Verstraeten, A.A. and Hilgers, J., Polymorphic epithelial mucin (MUC-1)-containing circulating immune complexes in carcinoma patients. *Br. J. Cancer*, **72**, 934–938 (1995).
15. Soussi, T., The humoral response to the tumor-suppressor gene product p53 in human cancer: implications for diagnosis and therapy. *Immunol. Today*, **17**, 354–356 (1996).
16. Old, L.J. and Chen, Y.T., New paths in human cancer serology. *J. Exp. Med.*, **187**, 1163–1167 (1998).
17. Prasannan, L., Misek, D.E., Hinderer, R., Michon, J., Geiger, J.D. and Hanash, S.M., Identification of beta-tubulin isoforms as tumor antigens in neuroblastoma. *Clin. Cancer Res.*, **6(10)**, 3949–3956 (2000).
18. Brichory, F., Beer, D., Le Naour, F., Giordano, T. and Hanash, S., Proteomics-based identification of protein gene product 9.5 as a tumor antigen that induces a humoral immune response in lung cancer. *Cancer Res.*, **61(21)**, 7908–7912 (2001).
19. Brichory, F.M., Misek, D.E., Yim, A.M., Krause, M.C., Giordano, T.J., Beer, D.G. and Hanash, S.M., An immune response manifested by the common occurrence of annexins I and II autoantibodies and high circulating levels of IL-6 in lung cancer. *Proc. Natl Acad. Sci. USA*, **98(17)**, 9824–9829 (2001).
20. Le Naour, F., Brichory, F., Misek, D.E., Brechot, C., Hanash, S.M. and Beretta, L., A distinct repertoire of autoantibodies in hepatocellular carcinoma identified by proteomic analysis. *MCP*, **1(4)**, 304–313 (2002).
21. Le Naour, F., Misek, D., Krause, M., Deneux, L., Giordano, T., Scholl, S. and Hanash, S., Proteomics-based identification of RS/DJ-1 as a novel circulating tumor antigen in breast cancer. *Clin. Cancer Res.*, **7(11)**, 3328–3335 (2001).
22. Kristensen, T., Saris, C.J., Hunter, T., Hicks, L.J., Noonan, D.J., Glenney, J.R. Jr. and Tack, B.F., Primary structure of bovine calpactin I heavy chain (p36), a major cellular substrate for retroviral protein-tyrosine kinases: homology with the human phospholipase A2 inhibitor lipocortin. *Biochemistry*, **25(16)**, 4497–4503 (1986).

23. Gure, A., Stockert, E., Scanlan, M., Keresztes, R., Jager, D., Altorki, N., Old, L. and Chen, Y., Serological identification of embryonic neural proteins as highly immunogenic tumor antigens in small cell lung cancer. *Proc. Natl Acad. Sci. USA*, **97(8)**, 4198–4203 (2000).
24. Dowlati, A., Levitan, N. and Remick, S.C., Evaluation of interleukin-6 in bronchoalveolar lavage fluid and serum of patients with lung cancer. *J. Lab. Clin. Med.*, **134(4)**, 405–409 (1999).
25. Karlsen, A.E., Sparre, T., Nielsen, K., Nerup, J. and Pociot, F., Proteome analysis — a novel approach to understand the pathogenesis of Type1 diabetes mellitus. *Dis. Markers*, **17(4)**, 205–216 (2001).
26. Banks, R.E., Dunn, M.J., Hochstrasser, D.F., Sanchez, J.C., Blackstock, W., Pappin, D.J. and Selby, P.J., Proteomics: new perspectives, new biomedical opportunities. *Lancet*, **356(9243)**, 1749–1756 (2000).
27. Rohlff, C. and Southan, C., Proteomic approaches to central nervous system disorders. *Curr. Opin. Mol. Ther.*, **4(3)**, 251–258 (2002).
28. Van Eyk, J.E., Proteomics: unraveling the complexity of heart disease and striving to change cardiology. *Curr. Opin. Mol. Ther.*, **3(6)**, 546–553 (2001).
29. Rubin, M.A., Use of laser capture microdissection, cDNA microarrays, and tissue microarrays in advancing our understanding of prostate cancer. *J. Pathol.*, **195(1)**, 66–80 (2001).
30. Emmert-Buck, M.R., Bonner, R.F., Smith, P.D., Chuaqui, R.F., Zhuang, Z., Goldstein, S.R., Weiss, R.A. and Liotta, L.A., Laser capture microdissection. *Science*, **274(5289)**, 998–1001 (1996).
31. Hanash, S.M., Baier, L.J., McCurry, L. and Schwartz, S., Lineage related polypeptide markers in acute lymphoblastic leukemia detected by two-dimensional electrophoresis. *Proc. Natl Acad. Sci. USA*, **83**, 807–811 (1986).
32. Hanash, S.M., Baier, L.J., Neel, J.V. and Niezgoda, W., Genetic analysis of thirty three platelet polypeptides detected in two-dimensional polyacrylamide gels. *Am. J. Hum. Genet.*, **38**, 352–360 (1986).
33. Reaman, G., Zeltzer, P., Bleyer, W.A., Amendola, B., Level, C., Sather, H. and Hammond, D., Acute lymphoblastic leukemia in infants less than one year of age: a cumulative experience of the Children's Cancer Study Group. *J. Clin. Oncol.*, **3(11)**, 1513–1521 (1985).
34. Crist, W., Pullen, J., Boyett, J., Falletta, J., van Eys, J., Borowitz, M., Jackson, J., Dowell, B., Frankel, L., Quddus, F., Ragab, A. and Vietti, T., Clinical and biologic features predict a poor prognosis in acute lymphoid leukemias in infants: a pediatric oncology group study. *Blood*, **67(1)**, 135–140 (1986).
35. Hanash, S.M., Kuick, R., Strahler, J.R., Richardson, B.C., Reaman, G., Stoolman, L., Hanson, C., Nichols, D. and Tueche, J., Identification of a cellular polypeptide that distinguishes between acute lymphoblastic leukemia in infants and in older children. *Blood*, **73**, 527–532 (1989).
36. Bergman, A.C., Benjamin, T., Alaiya, A., Waltham, M., Sakaguchi, K., Franzen, B., Linder, S., Bergman, T., Auer, G., Appella, E., Wirth, P.J. and Jornvall, H., Identification of gel-separated tumor marker proteins by mass spectrometry. *Electrophoresis*, **21**, 679–686 (2000).
37. Hirano, T., Franzen, B., Uryu, K., Okuzawa, K., Alaiya, A.A., Vanky, F., Rodrigues, L., Ebihara, Y., Kato, H. and Auer, G., Detection of polypeptides associated with the histopathological differentiation of primary lung carcinoma. *Br. J. Cancer*, **72**, 840–848 (1995).
38. Schmid, H.R., Schmitter, D., Blum, P., Miller, M. and Vonderschmitt, D., Lung tumor cells: a multivariate approach to cell classification using two-dimensional protein pattern. *Electrophoresis*, **16**, 1961–1968 (1995).
39. Hanash, S.M., Strahler, J.R., Neel, J.V., Hailat, N., Melhem, R., Keim, D., Zhu, X.X., Wagner, D., Gage, D.A. and Watson, J.T., Highly resolving two-dimensional gels for protein sequencing. *Proc. Natl Acad. Sci. USA*, **88**, 5709–5713 (1991).

40. Wang, Y.K., Liao, P.-C., Allison, J., Gage, D.A., Andrews, P.C., Lubman, D.M., Hanash, S.M. and Strahler, J.R., Phorbol 12-myristate 13-acetate-induced phosphorylation of Op18 in Jurkat T cells. *J. Biol. Chem.*, **268**, 14269–14277 (1993).
41. Melhem, R.F., Zhu, X.X., Hailat, N., Strahler, J. and Hanash, S.M., Characterization of the gene for a proliferation related phosphoprotein (Op18) expressed in high amounts in acute leukemia. *J. Biol. Chem.*, **266(27)**, 17747–17753 (1991).
42. Zhu, X.X., Kozarsky, K., Strahler, J.R., Eckerskorn, C., Lottspeich, F., Melhem, R., Lowe, J., Fox, D.A., Hanash, S.M. and Atweh, G.F., Molecular cloning of a novel human leukemia associated gene: evidence of conservation in animal species. *J. Biol. Chem.*, **264(24)**, 14556–14560 (1989).
43. Hanash, S.M., Strahler, J.R., Kuick, R., Chu, E.H.Y. and Nichols, D., Identification of a polypeptide associated with the malignant phenotype in acute leukemia. *J. Biol. Chem.*, **263(26)**, 12813–12815 (1988).
44. Knezevic, V., Leethanakul, C., Bichsel, V.E., Worth, J.M., Prabhu, V.V., Gutkind, J.S., Liotta, L.A., Munson, P.J., Petricoin, E.F. III and Krizman, D.B., Proteomic profiling of the cancer microenvironment by antibody arrays. *Proteomics*, **1**, 1271–1278 (2001).
45. Paweletz, C.P., Charboneau, L., Bichsel, V.E., Simone, N.L., Chen, T., Gillespie, J.W., Emmert-Buck, M.R., Roth, M.J., Petricoin, E.F. III and Liotta, L.A., Reverse phase protein microarrays which capture disease progression show activation of pro-survival pathways at the cancer invasion front. *Oncogene*, **20(16)**, 1981–1989 (2001).
46. Chaurand, P., DaGue, B.B., Pearsall, R.S., Threadgill, D.W. and Caprioli, R.W., Profiling proteins from azoxymethane-induced colon tumors at the molecular level by matrix-assisted laser desorption/ionization mass spectrometry. *Proteomics*, **1**, 1320–1326 (2001).
47. Bussow, K., Cahill, D., Nietfeld, W., Bancroft, D., Scherzinger, E., Lehrach, H. and Walter, G., A method for global protein expression and antibody screening on high-density filters of an arrayed cDNA library. *Nucleic Acids Res.*, **26**, 5007–5008 (1998).
48. Lueking, A., Horn, M., Eickhoff, H., Bussow, K., Lehrach, H. and Walter, G., Protein microarrays for gene expression and antibody screening. *Anal. Biochem.*, **270(1)**, 103–111 (1999).
49. Zhu, H., Klemic, J.F., Chang, S., Bertone, P., Casamayor, A., Klemic, K.G., Smith, D., Gerstein, M., Reed, M.A. and Snyder, M., Analysis of yeast protein kinases using protein chips. *Nat. Genet.*, **26**, 283–289 (2000).
50. DeWildt, R., Mundy, C., Gorick, B. and Tomlinson, I., Antibody arrays for high-throughput screening of antibody–antigen interactions. *Nat. Biotechnol.*, **18**, 989–994 (2001).
51. Ge, H., UPA, a universal protein array system for quantitative detection of protein–protein, protein–DNA, protein–RNA and protein–ligand interactions. *Nucleic Acids Res.*, **28(2)**, e3 (2000).
52. Rowe, C.A., Scruggs, S.B., Feldstein, M.J., Golden, J.P. and Ligler, F.S., An array immunosensor for simultaneous detection of clinical analytes. *Anal. Biochem.*, **71**, 433–439 (1999).
53. Mendoza, L.G., McQuary, P., Mongan, A., Gangadharan, R., Brignac, S. and Eggers, M., High-throughput microarray-based enzyme-linked immunosorbent assay (ELISA). *Biotechniques*, **27**, 778–780 (1999) see also pp. 782–786, 788.
54. Silzel, J.W., Cercek, B., Dodson, C., Tsay, T. and Obremski, R.J., Mass-sensing, multianalyte microarray immunoassay with imaging detection. *Clin. Chem.*, **44**, 2036–2043 (1998).
55. Arenkov, P., Kukhtin, A., Gemmell, A., Voloshchuk, S., Chupeeva, V. and Mirzabekov, A., Protein microchips: use for immunoassay and enzymatic reactions. *Anal. Biochem.*, **278**, 123–131 (2000).
56. Haab, B.B., Dunham, M.J. and Brown, P.O., Protein microarrays for highly parallel detection and quantitation of specific proteins and antibodies in complex solutions. *Genome Biol.*, **2(2)**, Research0004 (2001).

57. MacBeath, G. and Schreiber, S.L., Printing proteins as microarrays for high-throughput function determination. *Science*, **289(5485)**, 1760–1763 (2000).
58. Schweitzer, B., Wiltshire, S., Lambert, J., O'Malley, S., Kukanskis, K., Zhu, Z., Kingsmore, S.F., Lizardi, P.M. and Ward, D.C., Inaugural article: immunoassays with rolling circle DNA amplification: a versatile platform for ultrasensitive antigen detection. *Proc. Natl Acad. Sci. USA*, **97**, 10113–10119 (2000).
59. Kononen, J., Bubendorf, L., Kallioniemi, A., Barlund, M., Schraml, P., Leighton, S., Torhorst, J., Mihatsch, M.J., Sauter, G. and Kallioniemi, O.P., Tissue microarrays for high-throughput molecular profiling of tumor specimens. *Nat. Med.*, **4**, 844–847 (1998).
60. Kallioniemi, O.P., Wagner, U., Kononen, J. and Sauter, G., Tissue microarray technology for high-throughput molecular profiling of cancer. *Hum. Mol. Genet.*, **10**, 657–662 (2001).
61. Madoz-Gurpide, J., Wang, H., Misek, D.E., Brichory, F. and Hanash, S.M., Protein based microarrays: a tool for probing the proteome of cancer cells and tissues. *Proteomics*, **1(10)**, 1279–1287 (2001).
62. Schena, M., Shalon, D., Heller, R., Chai, A. and Brown, P.O., Quantitative monitoring of gene expression patterns with a complementary microarray. *Science*, **270(5235)**, 467–470 (1995).
63. DeRisi, J., Penland, L., Brown, P.O., Bittner, M.L., Meltzer, P.S., Ray, M., Chen, Y., Su, Y.A. and Trent, J.M., Use of a cDNA microarray to analyse gene expression patterns in human cancer. *Nat. Genet.*, **14(4)**, 367–370 (1996).
64. Madoz-Gurpide, J., Abad, J.M., Fernandez-Recio, J., Velez, M., Vazquez, L., Gomez-Moreno, C. and Fernandez, V.M., Modulation of electroenzymatic NADPH oxidation through oriented immobilization of ferredoxin: NADP(+) reductase onto modified gold electrodes. *J. Am. Chem. Soc.*, **122**, 9808–9817 (2000).
65. Delamarche, E., Sundarababu, G., Biebuyck, H., Michel, B., Gerber, C., Sigrist, H., Wolf, H., Ringsdorf, H., Xanthopoulos, N. and Mathieu, H.J., Immobilization of antibodies on a photoactive self-assembled monolayer on gold. *Langmuir*, **12(8)**, 1997–2006 (1996).
66. Muller, W., Ringsdorf, H., Rump, E., Wildburg, G., Zhang, X., Angermaier, L., Knoll, W., Liley, M. and Spinke, J., Attempts to mimic docking processes of the immune system: recognition-induced formation of protein multilayers. *Science*, **262**, 1706–1708 (1993).
67. Hanash, S. and Celis, J.E., The human proteome organization: a mission to advance proteome knowledge. *MCP*, **1(6)**, 413–414 (2002).

Proteome Analysis. Interpreting the Genome.
D.W. Speicher (editor)

Chapter 10

Affinity-based biosensors, microarrays and proteomics

EDOUARD NICE[a,b,*] and BRUNO CATIMEL[a]

[a] *The Ludwig Institute for Cancer Research, Melbourne Tumour Biology Branch, P.O. Box 2008, The Royal Melbourne Hospital, Parkville, Vic. 3050, Australia*
[b] *The CRC for Cellular Growth Factors, Melbourne, Australia*

* Corresponding author. The Ludwig Institute for Cancer Research, Melbourne Tumour Biology Branch, P.O. Box 2008, The Royal Melbourne Hospital, Parkville, Vic. 3050, Australia. Tel.: +61-3-9341-3135; fax: +61-3-9341-3104. E-mail: ed.nice@ludwig.edu.au (E. Nice).

1. Introduction

The emerging fields of genomics, proteomics and bioinformatics are becoming fundamental to the comprehensive investigation of biological processes and the rapid development of novel pharmaceuticals. Many of the current advances in these fields have arisen from increased sensitivity of key technologies, often accompanied by component miniaturisation to facilitate both high throughput and the handling of small samples.

The field of proteomics [1] now encompasses three major fields of research: protein profiling or proteome analysis, interaction analysis and structural genomics. The separation sciences will obviously play a fundamental and pivotal role in such studies, coupled with highly sensitive and specific downstream analytical techniques such as mass spectrometry and NMR. To date two-dimensional (2D) gel electrophoresis has been routinely used as the preferred method for the separation of complex mixtures of cellular proteins prior to downstream protein characterisation using mass spectrometry. 2D gel electrophoresis potentially provides comparative protein maps that will enable the identification of differentially regulated proteins in cells undergoing various stimuli, or in different disease states. However, it is now realised that the 2D gel approach is not a panacea and, in particular, low-abundance proteins (e.g. growth factors, receptors, membrane proteins and proteins involved in signal transduction) may be masked by the presence of 'house-keeping' proteins with similar size/charge properties or may be present at levels below the current detection limits of the analytical techniques used [2–9]. Indeed, inspection of the current databases shows that the aforementioned classes of proteins are conspicuous by their absence. Thus, e.g. a recent study has shown that 2D-MS was unable to identify proteins of low abundance (codon bias < 0.1) in the yeast proteome, even with over half the genes in yeast having a low-codon bias [5]. This suggests that the detection of low-abundance proteins would be a serious challenge for higher eukaryotes [5].

While proteomics lacks an equivalent technique to PCR to amplify proteins, second-generation functional proteomic protocols, in which specific subsets of the proteome are analysed, are beginning to realise the potential of trace enrichment chromatographic techniques to both concentrate low-level components and simultaneously remove the more abundant 'house-keeping' proteins [2,3,6, 10–19] (Fig. 1). In particular, affinity-based techniques have enormous potential in this area because of their high selectivity and relatively gentle elution conditions which help minimise protein denaturation [17,18].

The use of such affinity-based techniques, which have been used previously in the search for orphan biomolecules, has suggested to several groups the potential of using affinity-based biosensors as microaffinity purification platforms. Affinity biosensor systems have the obvious advantages of having automated sample

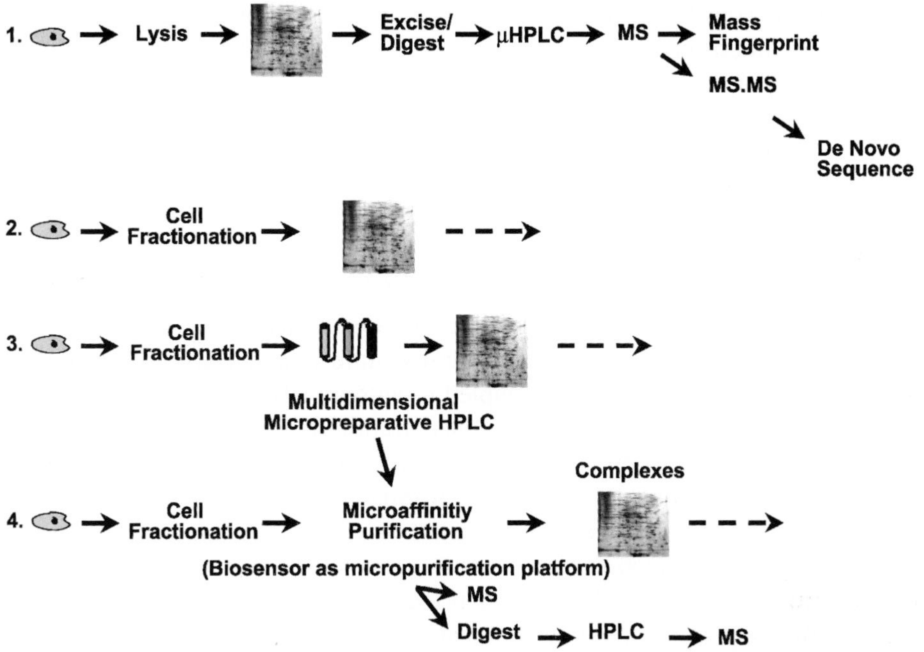

Fig. 1. Separation technologies and proteomics. Strategy 1. This represents the basic 2D gel approach where cells lysates are separated on 2D gel electrophoresis, the spots excised and digested and analysed by HPLC/MS by either mass fingerprinting or MS/MS for protein identification. Strategy 2. In this case, cell fractionation is used to reduce the complexity of the 2D map. This can allow the identification of proteins that are masked by more abundant species in Strategy 1, but does not give any significant increase in concentration for compounds present at very low levels. Strategy 3. Multi-dimensional micropreparative HPLC is used for both trace enrichment and partial purification to facilitate the identification of trace components. Strategy 4. In this example, the high selectivity afforded from affinity-based methods is used for both trace enrichment and purification. Alternatively, the microaffinity purification may be used as part of the multi-dimensional purification strategy shown in 3.

injection, online detection, direct quantitation, are compatible with small (μl) volume recovery, show reduced non-specific binding compared with chromatographic supports and enable continuous monitoring of surface stability [20]. Furthermore, the level of sensitivity and sample recovery (femtomole level) of biosensors is complementary with downstream mass spectrometric analysis [20].

A significant advantage of these biosensor techniques is that they are compatible with almost all protein interactions, and may be used to probe the formation of complexes in a manner analogous to immunoprecipitation [20]. It is now established that the large majority of proteins interact biologically to form larger complexes and function within complicated cellular pathways. As described in this chapter, biosensors may be involved in the systematic study of protein–protein interaction through the automated capture or ligand fishing of protein and protein complexes from biological sources and may provide defined material for 2D gels, microsequence analysis, mass spectrometry and database analyses (Fig. 1).

2. Biosensor technology

2.1. Instrumentation

Biosensors are analytical devices that detect molecules with high selectivity based on molecular recognition [21]. The central feature is a selective active surface consisting of a biological species coupled to an optically or electronically active medium. Biorecognition, ideally in a concentration-dependent manner, will alter the properties of the sensor surface enabling suitable transducers to convert biochemical interactions to electronic information that can be amplified and translated into a quantitative signal [21]. Many of the available instrumental biosensors are multi-function biosensors that are suitable for a range of research applications and use optical detection principals [22]. These instruments have opened new perspectives in the detection and analysis of biomolecular interactions and have become a major tool in the field of biomedical research [22–24]. Instrumental optical biosensors are either commercially available or under development from a number of manufacturers (Table 1) [23,25]. The BIAcore™ range from BIAcore, Uppsala, Sweden (http://www.biacore.com) and IAsys systems from Affinity Sensors, Cambridge, UK (http://www.affinity-sensors.com) are currently the most widely used [24]. These instruments (Fig. 2) measure biomolecular interactions in real time without labelling the interactants and allow detailed investigation of the reaction kinetics by analysis of the resultant signals. Several new instruments are also in development, including biosensor arrays, utilising grating-coupled surface plasmon resonance (SPR) instead of prism-based SPR technology [25].

Both the BIAcore and IAsys biosensors use detection principles based on optical evanescence. The BIAcore uses SPR detection [26], while the IAsys uses a waveguide technique called a prism coupler or resonant mirror [27]. These instruments continuously monitor the resonance angle and thus can detect changes in refractive index caused by changes in sensor surface mass when a ligand binds to, or dissociates from, its immobilised binding partner. Data are presented as sensorgrams that display the change in resonance units (RU, BIAcore) or angle (arc seconds, IAsys) versus time (Fig. 3).

The BIAcore biosensors are flow-based instruments that use two microsyringe pumps to give accurate pulse-free liquid flow at low flow rates (down to 1 μl/min). One pump maintains constant buffer flow over the sensor surface while the other is connected to a robotic sample dispenser, which can supply reagents or samples to

Table 1

Commercially available optical biosensors

Manufacturer	Instrument	Detection	System	Website
BIAcore AB	BIAcore 1000, 2000, 3000 BIAcore X BIAcore J S51	Surface plasmon resonance	Flow cells	www.biacore.com
	BIAcore 2D SPR		Array format	
Thermobio	IAsys IAsys + IAsys Auto +	Mirror resonance	Cuvettes	www.thermo.com
IBIS	IBIS I IBIS II	Surface plasmon resonance	Cuvettes	www.ibis-spr.nl/
Texas	Spreeta	Surface plasmon resonance	Flow cells	www-s.ti.com
Nipon laser electronics	SPR670 SPR Cellia	Surface plasmon resonance	Flow cells	www.nle-lab.co.jp/English/ ZO-HOME.htm
Analytical μ Systems	BIO-SUPPLAR 2	Surface plasmon resonance	Flow cell	www.micro-systems.de
HTS Biosystems	FLEX CHIP	Surface plasmon resonance	Array format	www.htsbiosystems.com
AVIV Instruments	PWR model 400	Plasmon waveguide resonance		www.protein-solutions.com/
Leica Microsystems Inc.	Leica	Surface plasmon resonance	Flow cell	www.leica-ead.com
Farfield Sensor Ltd	AnaLight Bio250	Interferometer Planar waveguide	Flow cell	www.farfield-sensors.co.uk
Prolinx	OCTAVE	Surface plasmon resonance	Cuvettes	www.prolinx.com
Graffinity	Plasmon Imager	Surface plasmon resonance	Array format	www.graffinity.com

(*continues*)

Table 1
Continued

Manufacturer	Instrument	Detection	System	Website
Luna Innovations	LPG sensor Fiber optique biosensor	Long-period gratings	Fiber optique	www.lunainnovations.com
ThreeFold Sensors	Prototype (label-free)	Evanescent excitation of fluorofore	Fiber optique (cartridge)	http://ic.net/~tfs/TFS Default.html
SRU biosystems	SRU BIND	Colorimetric resonant reflection	Array format\flow cells	www.srubiosystems.com
Quantech	FasTraQ	Surface plasmon resonance	Array format	www.quantechltd.com

the sensor surface via an injection port. Up to 192 samples can be accommodated using two 96-well microtitre plates. These sample plates can be independently thermostatically controlled using an external circulator. The sensor chip interfaces with a thermostatically controlled integrated fluidic cartridge (IFC) to form four parallel-flow cell channels (60 nl volume, 1.6 mm^2 surface area) on the sensor surface. In the BIAcore 2000 and 3000 (Fig. 2), samples can be passed sequentially over the sensor surfaces allowing multi-channel analysis. Samples can be recovered from these instruments, although the mechanism of recovery, which is described later in this chapter, is different [20,22]. A new instrument, the BIAcore S51, especially designed for drug discovery applications, is also available now. This biosensor possesses a novel IFC that allows simultaneous detection of up to three spots per flow cell, which enhances throughput and provides parallel referencing to control for non-specific binding. The instrument also incorporates a high quality, low-pulsation pump and an improved autosampler allowing optimal sample usage with minimal carryover. This instrument has a significantly higher throughput, allowing up to 384 samples to be analysed in 24 h, and has been designed to facilitate analysis of low molecular weight compounds down to 100 Da.

By comparison, the IAsys Auto + is a cuvette-based sensor (Fig. 2) with dual microcuvettes (10–80 μl capacity), which have integral sensor surfaces and optical coupling [20,28]. Sample and reagent delivery and sample recovery are performed using a robotic arm connected to an autosampler, syringe pump, wash station and autoaspirator, which ensures accurate and reproducible low volume transfer (minimum volume 1 μl). There is the potential to use up to 96 separate sample and recovery positions in two defined racks plus 10 reagent vials [20,28].

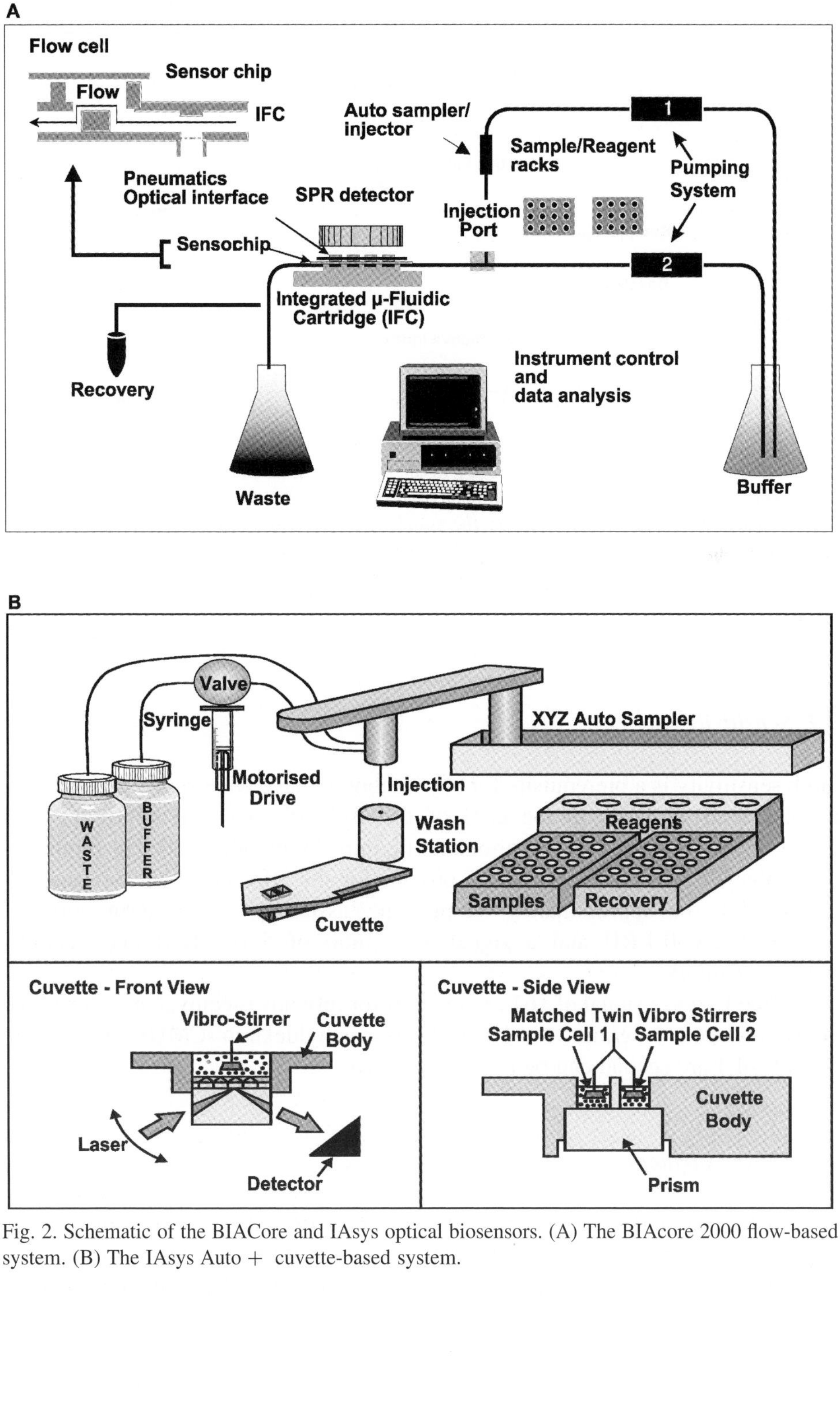

Fig. 2. Schematic of the BIACore and IAsys optical biosensors. (A) The BIAcore 2000 flow-based system. (B) The IAsys Auto + cuvette-based system.

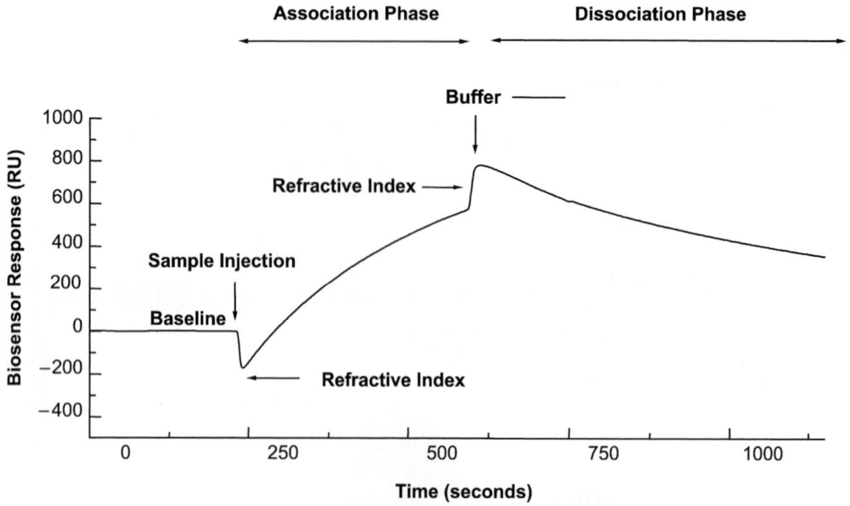

Fig. 3. Interpretation of a sensorgram. Prior to analysis, one interactant is immobilised onto the sensor surface. Upon injection, binding of analyte results in an increase in signal corresponding to the association phase. Following injection, the sample is replaced by a continuous flow of buffer, the sample concentration drops to zero and the decrease in signal now reflects dissociation of analyte from the surface-bound complex. Kinetic rate constants can be derived from the association and dissociation phases of the sensorgram.

2.2. Sensitivity

High sensitivity is a prerequisite for any method to have widespread applicability in biological analysis. In the case of the BIAcore, a signal of 1000 RU is equivalent to a surface concentration of 1 ng/mm^2 for proteins [29]. The minimum detectable surface concentration of protein for the BIAcore 3000 biosensor is estimated to be approximately 0.5 pg/mm^2, assuming a short-term noise of approximately 0.1 RU and a signal/noise ratio of 5 (R. Karlsson, personal communication).

For the IAsys, a signal of 163 arc seconds for proteins is equivalent to a surface concentration of 1 ng/mm^2 using the carboxymethyldextran (CMD) surface [30]. Signals of 1 arc second can be readily detected.

2.3. Sensor surfaces

Both BIAcore and IAsys offer a range of sensor surfaces to suit various applications (Table 2). The BIAcore sensor chips (with the exception of the HPA chip, which has a flat hydrophobic surface) consist of a glass slide coated with

Table 2

Details of IAsys and BIAcore biosensor surfaces

Surface	Chemistry	Applications
Carboxymethyldextran (CM5 BIAcore, CMD IASys)	Coupling via primary amines using NHS/EDC chemistry	Compatible with a wide range of protein and peptide applications: amino coupling of proteins and peptides Immobilisation of thiol reactive reagents for thiol coupling Immobilisation of hydrazine for coupling of oxidised carbohydrate Immobilisation for affinity capture
CMD select carboxymethyldextran (IAsys)	Large coupling capacity (16 mm^2 surface)	Ideal for ligand fishing
Dextran matrix with low degree of carboxylation (B1 BIAcore)	Coupling via primary amines using NHS/EDC chemistry	Kinetic applications requiring a low density of immobilised ligand Shows reduced non-specific binding
Short carboxymethyldextran (F1 BIAcore)	Coupling via primary amines using NHS/EDC chemistry	High molecular weight analytes and cell binding studies
Carboxylate planar surface (IAsys)	Coupling via primary amines using NHS/EDC chemistry	High molecular weight analytes and cell binding studies
Carboxyl group attached directly to the surface (C1 BIAcore)		
Amino planar surface containing primary amine) (IAsys)	Coupling directly to primary amines on the sensor surface using a homobifunctional cross-linker or glutaraldehyde	Coupling of ligands with low p*I* High molecular weight analytes and cell binding studies

(*continues*)

Table 2
Continued

Surface	Chemistry	Applications
Biotin (IAsys)	Coupling of biotinylated ligand following streptavidin immobilisation	Immobilisation of biotinylated ligands or membranes (e.g. proteins, oligonuclotides)
SA Streptavidin	Pre-immobilized streptavidin surface	Capture of biotinylated ligands (e.g. proteins, oligonuclotides)
N-(5-amino-1-carboxypentyl)iminoacetic acid surface (NTA BIAcore)	Metal chelation	Immobilisation of histidine-tagged recombinant proteins
Untreated gold surface (J1 BIAcore)	Partial hydrophobic character	To design customised surface chemistry using, e.g. self assembled monolayers, thin polymeric films
Hydrophobic (IAsys)	Hydrophobic binding of biomolecules, such as lipid monolayers	Lipids, liposomes
Long-chain alkanethiol (HPA BIAcore)		Self assembled monolayers Membrane receptor analyte interactions
Lipophilic modified surface (L1 BIAcore)	Non-covalent anchorage of lipids Formation of lipids bilayer	Immobilisation of cell membranes, liposomes

Reprinted with permission from: Catimel, B., et al., *J. Biochem. Biophys. Methods*, **49**, 289–312 (2001).

a thin (50 nm) gold film to which is attached, by an inert (alkanethiol) linker layer, a carboxymethylated matrix onto which one of the binding partners can be immobilized using a number of well-defined chemistries [20,28].

The IAsys sensor surfaces, which form the bottom of the cuvette, are available in analytical (4 mm^2 surface) or preparative (16 mm^2 surface) format. The glass surface can be derivatised to give similar functionalities to those available from BIAcore.

2.4. Surface immobilisation

Fundamental to the operation of these instruments is the ability to immobilise one of the binding partners, ideally in a defined biologically active orientation mimicking that found in vivo, on the sensor surface. Targets for immobilisation may be isolated from bulk biological samples or produced by recombinant or synthetic techniques. Since the biological specificity resides in this surface, the compounds for immobilisation should be homogeneous. A number of suitable chemistries (Table 2) have now been defined including amino (amino-terminus and ε-group of lysine residues), aldehyde and thiol coupling, thiazolidine-mediated ligation via amino-terminal serine or threonine residues [31] as well as affinity capture including biotin/streptavidin, protein A/Fc domain, metal chelation, coiled-coil interaction and antibodies directed against specific tags (e.g. glutathione *S*-transferase (GST), FLAG and Myc peptides). These methods have been described in detail in several reviews [20,28]. This range of immobilisation strategies has allowed instrumental biosensors to be used to study a wide range of biomolecular interactions between proteins and peptides, DNA, lipids, carbohydrates, cells and drugs in a range of research applications including structure–function studies, kinetic analysis, epitope mapping, clinical studies, antibody and enzyme engineering, immunoassays and vaccine development, monitoring of chromatographic fractions, ligand searching and drug discovery [20,22–24,28,32–34].

3. Biosensor applications

3.1. Biosensor-based ligand searching

Biosensor technology combined with micropreparative HPLC has been used successfully by a number of groups for the identification, purification and characterisation of ligands for orphan receptors identified by evolutionary conserved motifs using molecular biology or bioinformatics-based strategies [35–41], or monoclonal antibodies raised against complex mixture of proteins [42] whose binding partners and biological function remained unknown.

In ligand-searching studies, the purified orphan protein (e.g. recombinant forms of the extracellular binding region of the orphan receptor, monoclonal antibodies or antibody fragments) can be immobilised on the sensor surface and used to screen against a multitude of biological extracts (e.g. cell and tissue culture conditioned media, cell extracts, biological fluids) in order to detect the presence of possible ligand(s) for the target protein. Large sample 'libraries' can be rapidly screened using the robotic handling devices of automated instruments. During the initial screening, careful analysis of the data is required as large sample related bulk refractive index signals can potentially mask very small positive responses. Importantly, since the optical detection systems used do not readily distinguish between specific and non-specific (or inappropriate) binding, the selectivity of positive signals should be confirmed using competition experiments with a soluble form of the immobilised target [37,38], or some alternative assay based on an alternative detection principle. For example, an initial concentration of ligand in human placental conditioned medium of approximately 1 ng/ml (10 RU signal) was detected during the initial screening of the HEK ligand [38].

In some studies, the identification of a ligand source has been followed by expression cloning to characterise the ligand [41], while in others [36–38,41–43], the biosensor has been used as a specific affinity-based detector for monitoring fractions during chromatographic purification of the ligand of interest. The complexity of the purification scheme will obviously depend on the relative abundance of the ligand and the nature of contaminating proteins (e.g. ranging from a single receptor affinity step [36] to a seven-step purification protocol with an overall purification factor of 1.8×10^6 [38]). Frequently, a receptor-based affinity chromatographic step has been included in the purification protocol since the target protein is generally produced by recombinant techniques and is therefore available in suitable quantities [36–38,40,41]. The affinity step can also be introduced in a revised purification protocol [44], once the chromatographic characterisation of the ligand has been established [45] in order to obtain sufficient material for structure–function and kinetic studies [44–46].

3.2. Preparative biosensor ligand fishing and proteomics

The examples cited earlier clearly demonstrate the potential of biosensors as affinity detectors for the identification and purification of orphan biomolecules. It therefore seemed logical to attempt to use such sensor surfaces as microaffinity purification platforms to recover sufficient material for ligand identification using specific and sensitive downstream analyses (e.g. mass spectrometry, microsequence analysis, bioassay). The multiplexed analytical technique involving the use of BIAcore sensor chips as preparative surfaces with subsequent characterisation of bound proteins by matrix-assisted laser desorption/ionisation time-of-flight mass spectrometry (MALDI-TOF-MS) has been named biomolecular interaction analysis mass spectrometry (BIA/MS) [47]. MALDI-TOF-MS [48,49] is capable of accurately determining the molecular weight of proteins up to >100 kDa with femtomole sensitivity [50]. In a 2D approach, the biosensor analysis is used to both affinity purify the binding partners and simultaneously monitor the biomolecular interaction and determine its binding parameters, while the MALDI-TOF-MS reveals mass information on compounds bound to the sensor surface.

One strategy in BIA/MS studies requires the sensor chip to be introduced directly into the MALDI-TOF mass spectrometer [51–53]. This requires careful and expert experimental manipulations (MALDI-TOF-MS optimisation, choice of matrix and matrix application, chip cutting and insertion into the mass spectrometer) requiring in-house fabricated devices. Demountable sensor surfaces have been proposed by some manufacturers (e.g. IAsys) to facilitate introduction of the sensor surface into the mass spectrometer.

In initial studies, using a well-characterised antibody–antigen system (polyclonal anti-myotoxin *a* IgG-myotoxin a), it was shown that BIA/MS was able to analyse proteins with molecular mass up to 200 kDa at the 10–100 fmol level [47,52–54]. Proof of principle was demonstrated by injecting whole snake venom (10 μl, 1 μg/ml) over anti-myotoxin *a* IgG immobilised on all four channels of the sensor chip. The resulting signal of 100 RU, corresponding to approximately 20 fmol of protein per flow cell, was sufficient to detect myotoxin *a* signals by MALDI-MS analysis [47,54]. MALDI-TOF-MS analysis can be performed on a parallel blank control channel in order to detect compounds retained by non-specific interactions [54].

Recent studies have further demonstrated the potential of BIA/MS in ligand fishing and proteome analysis by analysing epitope-tagged proteins [55,56], bacterial toxins [57] and the detection of attomole levels of proteins in complex biological mixture (saliva, urine) [58–61]. Epitope-tagged tryptic peptides were captured from *Escherichia coli* tryptic lysates using the biosensor by affinity interactions with either immobilised monoclonal antibodies or chelated Ni^{2+} surfaces and identified using MALDI-TOF-MS [55,56]. Indirect MALDI-TOF-MS analysis was performed with the chelating sensor surface. Instead of using the sensor surface directly as a target, 1.5 μl of α-cyano-4-hydroxycinnamic acid matrix was applied to the sensor surface in order to elute retained compounds and was then applied to the MALDI target [55]. The amount of peptides captured on the chip was estimated to range between 5.5 fmol and 900 amol.

Alternatively, on-chip proteolysis of a captured ligand can be performed to liberate peptide fragments for subsequent MALDI-TOF-MS or MS/MS [56]. Pepsin cleavage of human interleukin-1 alpha (60 fmol of IL-1α captured after injection of 30 μl of biosensor running buffer containing 10 mg/ml HSA and 10^{-4} mg/ml of IL-1α) was performed on-chip and resulted in the identification of 25 peptides. Of the observed peptide masses, 17 peptides corresponded to predicted pepsin fragments of IL-1α within 0.1% mass error and 6 peptides were identified as pepsin fragments.

BIA/MS was used to detect bacterial toxin (Staphylococcal enterotoxins B (SEB)) in food samples (milk and mushroom extract) at a level of 1 ng/ml [57]. In this study, the detection of multiple toxins was evaluated using multi-affinity sensor surfaces on which multiple targets (e.g. anti-SEB IgG and anti-toxic-shock syndrome toxin-1) were immobilised on the same channel. MALDI-TOF-MS allowed detailed analysis on the number and nature of the components captured on the chip [57]. This study demonstrates the potential of BIA/MS analysis for multiple-format immunoassays, which would be compatible with high-throughput screening. A similar approach was also used in the multi-analyte assay of beta-2 macroglobulin, a potential biomarker for glomerular filtration and tubular reabsorption, and cystatin C, a measure of glomerular filtration rate, and putative marker for several inflammatory aliments, in human urine [58,59,61] and human plasma [60]. Using BIA/MS analysis, the relative amounts of cystatin C versus beta-2 macroglobulin were detectable at the 10 fmol level [58]. This corresponded to 8 pg (signal of 8 RU) of retained beta-2 macroglobulin from a diluted urine sample [62]. Similarly, a detection limit of 5 pg was obtained using urinary protein 1, which was also found to be present in plasma [60].

Another advantage of BIA/MS is the potential to detect protein variants or protein complexes directly and perform comparative mass profiling using parallel surfaces. Transthyretin ($M_r = 13732$ Da) and its cysteinylated form (+119 Da) were detected when human plasma was injected over immobilised anti-transthyretin IgG [60,62,63]. Furthermore, the complex transthyretin/retinol binding protein was captured when an anti-retinol binding protein IgG was immobilised onto the sensor surface [60,63].

Comparative mass profiling was performed by injecting a saliva extract over chelating sensor surfaces charged with different metal ions [58]. Histatin 3 and its related enzymatic breakdown products (histatins 4–10) were specifically retained on a nickel-charged surface while the calcium-binding proteins PRP-1 and PRP-3 as well as histatin 5 were specifically retained using a calcium surface. Using both the biosensor and MALDI-MS data, approximately 25 fmol of each protein was estimated to be captured on the chip.

The BIA/MS studies described earlier were performed using the BIAcore surface as a direct MALDI target. A more flexible approach is the elution and recovery of specifically bound ligand from the sensor chip prior to downstream analysis. In an elegant example of this approach, G-protein coupled receptors were isolated as cell membranes and immobilised onto a BIAcore LI hydrophobic surface (Fig. 4). This surface was then used to trap the binding partner for the receptor, which was then eluted and characterised by LC/MS/MS [64]. In an alternative approach, peptides generated by on-chip digestion were trapped after elution by a microcapillary RP pre-column connected to the BIAcore fluidics

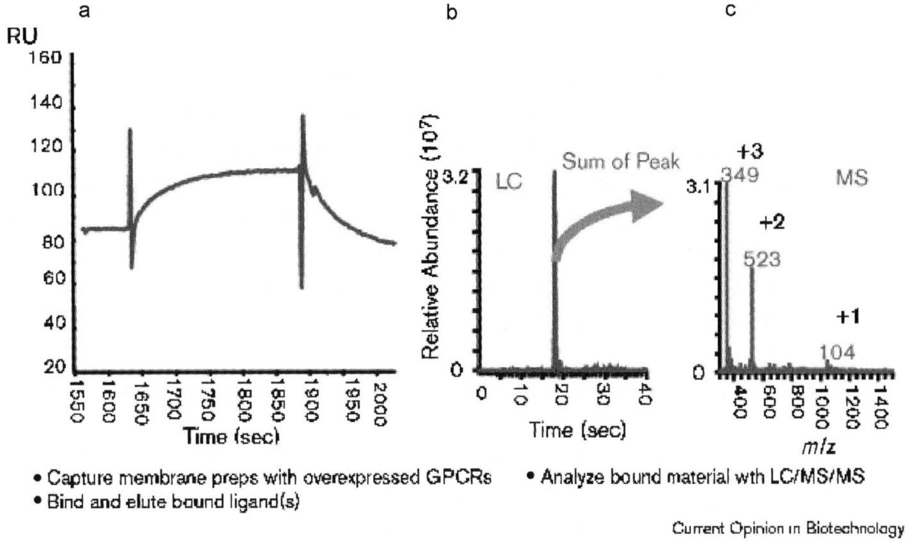

Fig. 4. Biosensor ligand fishing using an immobilised membrane receptor. (a) Sensorgram showing the binding of angiotensin peptide onto immobilised membrane containing angiotensin receptors. (b) Liquid chromatographic analysis of the peptide after elution and recovery from the sensor surface. (c) Mass spectrometry analysis of the eluted peptide (Reprinted with permission from: Williams, C., *Curr. Opin. Biotechnol.*, **11**, 42–46 (2000)).

using an in-house fabricated 'recovery port' [65,66]. The trapped peptide mixture was then analysed using HPLC–ESI/MS/MS.

The potential for direct automated elution and recovery at high concentration in small volumes (1–10 μl) has been designed into some flow-based instruments [64, 67–69]. In the original BIAcore 1000, samples could be recovered by a simple modification of the outlet tubing. This was particularly useful for the recovery of radioactively labelled proteins, which could then be analysed by SDS-PAGE and autoradiography. The integrated flow cell of the BIAcore 2000 was designed for sample recovery after the sample had passed through the IFC, although sample diffusion hampered its practical use. This problem was addressed in the design of the BIAcore 3000, where dilution effects during recovery were minimised by air segmentation and reversal of the buffer flow to return the sample back through the flow cell and collect it via the autosampler [70,71].

However, the small surface area of the BIAcore flow cells (1.6 mm^2) limits the capacity and hence the amount of material (femtomole levels) that can be recovered at each cycle. The use of the manual BIAcore PROBE, which has a surface area of 12–20 mm^2, has allowed this limitation to be overcome [68,72]. Using the BIAcore PROBE, murine IL-13 was affinity purified from concentrated cell conditioned media using an immobilised antibody: IL13Ra2-Fc [68]. The eluted murine IL-13 was found to be biologically active, and could be analysed using silver-stained SDS-PAGE gels. In-gel tryptic digestion was then performed and the resulting peptide extracts were fractionated using microcapillary RP-HPLC and sequenced directly by on-line tandem electrospray MS/MS resulting in the identification of an IL-13 peptide sequence (DPLSSVTT). In another study, purified GST was captured using an anti-GST antibody immobilised onto the BIAcore PROBE. On-probe tryptic digestion was performed, and the resulting peptides were recovered and analysed by MALDI-TOF-MS or reversed phase liquid chromatography coupled to tandem mass spectrometry (LC–MS/MS). Database searching resulted in the identification of a peptide sequence (LLLEYLEEK) arising from GST [72].

3.3. Cuvette-based biosensors as microaffinity purification platforms

In this laboratory's experience, cuvette-based biosensors (e.g. IAsys Auto +) are better suited to sample recovery than flow-based instruments [69]. The surface area of the micropreparative cuvettes (16 mm^2) increases binding capacity enabling increased ligand recovery compared with BIAcore surfaces. Furthermore, the cuvette provides a confined environment that allows long binding contact times and facilitates recovery following desorption in small (μl) eluant volumes at concomitant high concentration. This is particularly advantageous when recovering samples with appreciable off rates that may be lost during buffer change in flow-based biosensors or chromatographic systems.

In an early report [73], active peptides were recovered from a 30-peptide enkephalin library using an immobilised anti-β endorphin antibody. Human serum albumin (HSA) from an *E. coli* lysate was also purified using immobilised anti-HSA IgG and analysed post-recovery using silver-stained SDS-PAGE [74].

As proof of principle, the IAsys Auto + biosensor was used to affinity purify directly on the sensor surface, the A33 epithelial antigen present on colonic carcinoma cells, which this laboratory had originally identified using a combination of micropreparative HPLC and biosensor analysis of recovered fractions [42]. The original purification required five chromatographic stages. Attempts to purify the A33 antigen using preparative 2D-PAGE in a conventional proteomics-based approach were hampered by the similarity in molecular mass and isoelectric point between the A33 antigen and actin and cytokeratin which were major contaminants (A33: M_r 41 kDa, pI 5.0–6.0; actin: M_r 41 kDa, pI 4.0–5.3; cytokeratin: M_r 42.5 kDa, pI 5.5–6.5) [75,76].

In the preparative ligand fishing approach, the sensor surface was used directly as a micropurification platform, and the A33 integral membrane antigen was purified in situ following a single anion-exchange separation of a Triton X-100 extract of the LIM1215 colonic carcinoma cell line. The IAsys biosensor was itself used to screen the fractions from the anion-exchange column using immobilised anti-A33 $F(ab)'_2$ fragment (350 ng immobilised). Positive fractions were then pooled and the A33 ligand purified on the same surface. At this stage, SDS-PAGE analysis of the pooled fractions indicated that the A33 ligand was a trace component of a complex mixture and was not readily identifiable. After multiple (60) automated cycles of injection and recovery, microgram quantities of essentially homogeneous A33 antigen were recovered from the sensor surface (approximately 20 ng recovered per cycle) giving sufficient material for subsequent analysis using SDS-PAGE, Western blotting, micropreparative RP-HPLC and N-terminal microsequence analysis (Fig. 5) [69]. The use of on-line detection allows the surface viability to be continuously monitored throughout

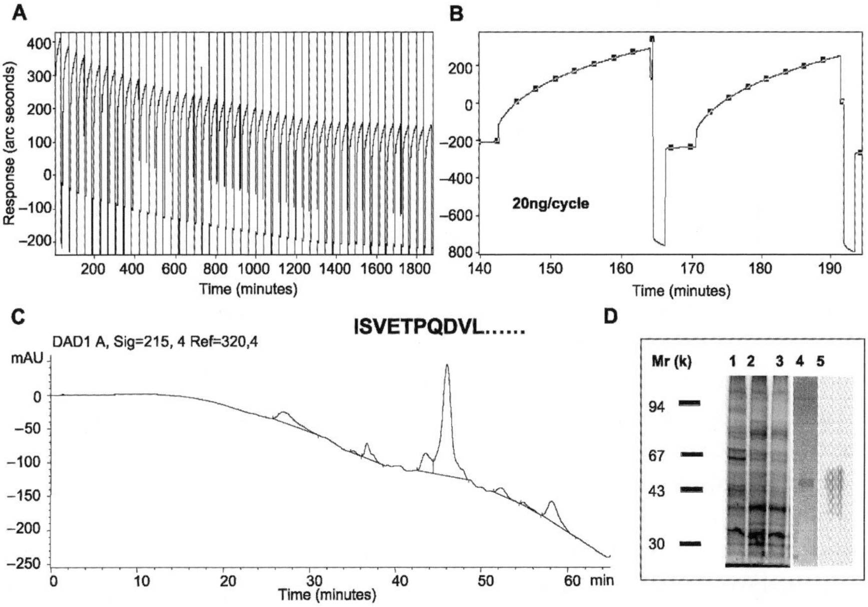

Fig. 5. Micropreparative ligand fishing using the IASys cuvette-based biosensor. (A) The A33 antigen was recovered preparatively, using a micropreparative biosensor surface, from a Triton X-114 extract of LIM1215 carcinoma cell lines following a single anion-exchange chromatography step. The anion-exchanged active fractions, identified using the biosensor in an analytical mode, were pooled and repetitively injected (50 cycles of 30 μl injection) over a micropreparative cuvette (16 mm^2) with anti-A33 Fab$'_2$ immobilised on the surface (350 ng). Bound antigen was eluted using 10 mM HCl, recovered and neutralised using 1 M Tris–HCL pH 8.0. (B) An enlarged view of two consecutive cycles of injection/recovery is shown. From the observed signal, the yield of recovery at each cycle was approximately 20 ng. (C) Analysis of the recovered material using micropreparative RP-HPLC gave a major symmetrical peak and gave the anticipated N-terminal amino acid sequence. (D) Analysis of the biosensor-recovered fraction using silver-stained gel SDS-PAGE showed an essentially homogeneous A33 antigen (M_r 43,000 protein) (lane 4), corresponding to the band recognised by the A33 mAb using Western blot analysis (lane 5). The protein complexity of the Mono Q fractions pooled as the starting material is shown for comparison (lane 1,2,3) (Reprinted with permission from: Catimel, B., et al., *J. Biochem. Biophys. Methods*, **49**, 289–312 (2001)).

the experiment and permits direct quantitation of both bound and recovered material. Furthermore, the sensor surface appears to display considerably reduced non-specific binding compared with microaffinity columns or in contrast to sensor surfaces, the latter affinity surfaces typically cannot produce homogeneous purified samples when working at low levels with trace components [38,42,69]. This is presumably due to the larger surface areas of support materials used in chromatographic columns or magnetic bead particles, which are the major cause of non-specific binding.

In an extension of these studies (Fig. 6), the cuvette-based microaffinity purification has also been used in combination with downstream MALDI-TOF-MS analysis [77]. Recombinant A33 antigen (400 nM) in 10% foetal calf serum was injected over immobilised anti-A33 IgG (400 ng). Approximately 1 pmol of antigen was recovered after seven cycles of injection/desorption. After recovery, the antigen was digested with the endoprotease Lys-C and the resulting peptide applied to a MALDI target using a C4 Zip-Tip. Ten A33-related peptides were identified, resulting in 54% coverage of the A33 protein.

4. ProteinChip mass spectrometry using SELDI

4.1. SELDI technology

The concept of derivatising the mass spectrometric probe to facilitate capture was originally proposed by Hutchens and Yip [78] and named surface-enhanced laser desorption/ionization (SELDI). This strategy allowed direct extraction and on-probe investigation of biomolecules compared with the conventional mass spectrometry approaches where complex biological material (e.g. blood, cell lysates, cell supernatant, sera) had to undergo sample preparation and purification prior to analysis. In SELDI, several previously independent technologies are now unified on a single integrated platform. Arrays of nanoscale surfaces with alternative chemical selectivity are designed to both capture components of complex mixtures and present them for detection by the laser-induced desorption/MS process. SELDI technology utilizes one of the three alternative approaches: surface-enhanced affinity captured (SEAC), surface-enhanced neat desorption (SEND) and surface-enhanced photolabile attachment and release (SEPAR) [79].

In the most commonly used form of the technology, SEAC, probes and biochips are derivatised with classic chromatographic separation functionalities and are active participants in the capture, separation, presentation, structural modification and/or amplification and detection of individual target molecule(s). The probe is first exposed to the biological source material. Washing/elution procedures are then performed on the individual surfaces, in a manner akin to stepwise gradient elution, to selectively remove bound molecules. The pattern of proteins remaining

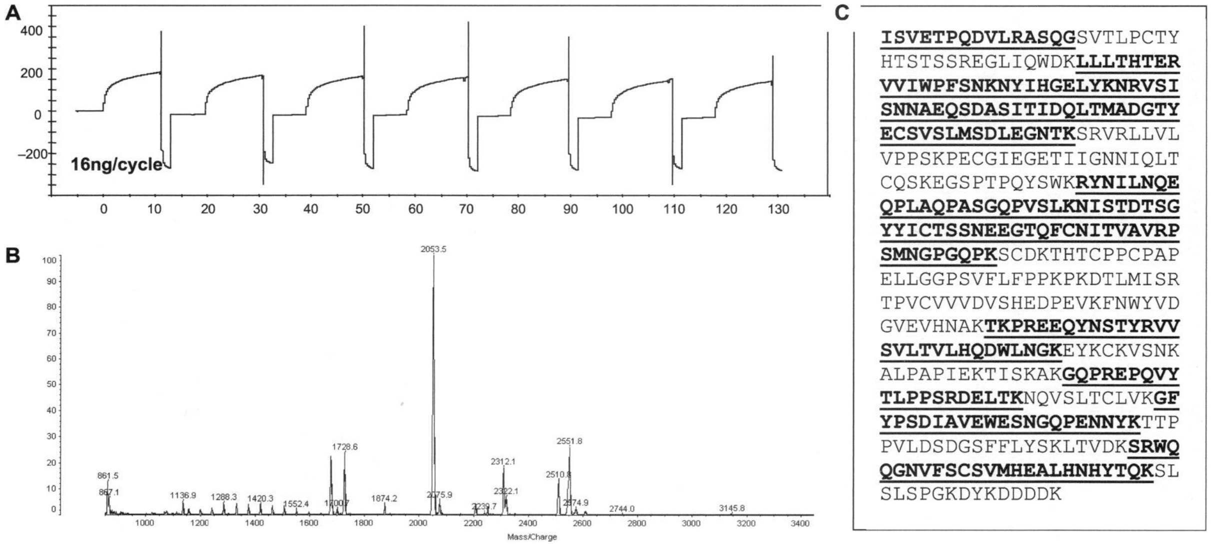

Fig. 6. Preparative biosensor fishing with downstream MS analysis. Recombinant A33 antigen in tissue culture medium was affinity purified using A33 IgG (400 ng) immobilised onto a micropreparative cuvette. (A) Seven injection cycles (25 μl, 400 nM) were performed and bound material recovered using 10 mM HCl. Each cycle yielded a recovery of 16 ng of antigen. (B) The recovered antigen was digested using Endolys C and applied to a MALDI target using a C4 Zip-Tip and the resultant spectrum is shown. (C) The peptides identified are shown on the A33 sequence (bold, underlined).

on the chip following each elution step can then be analysed by the use of TOF-MS equipped with a laser desorption ion source following matrix (energy-absorbing molecule) addition. The laser energy releases the molecules captured at each location on the biochip and initiates further separation by molecular mass (Fig. 7). In SEND technology, an energy-absorbing component is directly attached to the probe, eliminating the need of matrix addition before analysis. SEPAR is a hybrid of the two previous techniques: the capture surface also functions as an energy-absorbing platform promoting desorption and ionisation.

The first commercial SELDI system, based on SEAC technology, was introduced by Ciphergen (Ciphergen Biosystems, Palo Alto, CA, USA) around 1996. Lumicyte (www.lumicyte.com/), who has recently signed an agreement with Kratos Analytical to use their Kratos AXIMA-CFR mass spectrometer, is developing similar technologies, but has a set-up as a service company to the pharmaceutical and biotechnology industries.

ProteinChip arrays are currently available that display a number of different chromatographic selectivities: reversed phase, ion exchange, immobilised affinity capture (IMAC) or normal phase. Pre-activated surfaces with chemical groups reacting with primary amines or alcohols are also available for the immobilisation of target molecules such as receptors, antibodies, peptides or oligonucleotides [79, 80]. These various arrays can be used in series or parallel to analyse and resolve biological samples in protein subsets with common properties. Furthermore, comparison of, e.g. lysates from normal versus disease cells or tissues, or activated

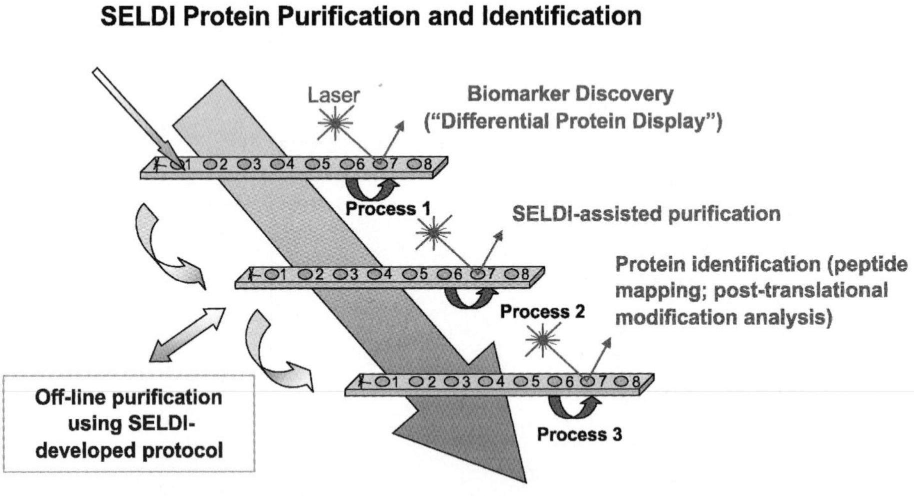

Fig. 7. Principle of the Ciphergen SELDI system. The ProteinChip technology platform can be used in a rapid stepwise process for comparative profiling to discover and validate biomarkers (process 1) followed by purification protocol development (process 2) and in situ digestion of protein for protein characterisation (process 3).

versus non-activated cells can be performed in a strategy called protein profiling or differential display. This technique can potentially identify upregulated or downregulated proteins, to define critical signalling pathways or identify the presence of biomarkers with the potential use as diagnosis, prognosis or therapeutic targets [79–81].

4.2. SELDI applications

SELDI technology has been combined with infrared laser-capture microdissection (LCM) [82] in the study of biopsy specimens. LCM allows the selective microdissection of specific cell types (e.g. neoplastic, normal cells) in defined areas of tissues and has been shown to be compatible with the isolation of sufficient purified cells for subsequent analysis of nucleic acids by performing reverse-transcriptase PCR as well as proteomic analysis using 2D gel electrophoresis, mass spectrometric and SELDI analyses [83,84]. Using this approach, protein biomarkers can be identified in the histologically defined cell population derived from heterogeneous disease tissue [85]. Comparison of the protein expression patterns from microdissected normal and tumor tissues (e.g. normal prostate, prostatic intraepithelial neoplasia and prostate cancer [86–88], head and neck cancer [85], normal squamous oesophageal epithelium and corresponding tumor cells [89], cervical intraepithelial neoplasia and cervix uteri carcinoma [90], normal and tumoral breast cells [91]) have identified differentially expressed proteins in cancer specimens. Prostate cancer cells demonstrated a downregulation of a 28 kDa protein compared to normal cells [86]. Multiple specific protein patterns were also reproducibly detected in the range from 1.5 to 30 kDa of 28 subpopulations of four prostate tumors and one control [88]. A specific 4.3 kDa protein was increased in the prostate tumor stroma and prostate tumor glands compared to normal prostate proper, transitional zone stroma and transitional zone glands [88]. In a large study set, 1500 microdissected patient-matched normal, prostatic intraepithelial neoplasia and invasive carcinoma cells were analysed using SELDI-TOF-MS [87]. A specific and reproducible phenotype was obtained for each individual cell type and the spectra for pre-malignant cells were found to represent a unique subset that was a defined mix of the normal and carcinoma protein fingerprints. In particular, two proteins were found to change concomitantly with prostatic disease progression [87]. In studies involving both tissue and body fluids (serum, urine and seminal plasma) [92,93], SELDI analysis identified known prostate cancer associated biomarkers including prostate-specific antigen (PSA), prostate-specific membrane antigen (PMSA), prostatic acid phosphatase and prostate-specific peptide, as well as detecting 30 proteins that were either over- or under-expressed in samples from prostate cancer patients. Body fluids were also analysed successfully in a number of studies investigating bladder, breast, colon, prostate, pancreatic and ovarian cancer. Serum samples from healthy controls and from patients with benign, pre-metastatic and metastatic breast disease, with pre-malignant or malignant colon disease, with PSA-negative and PSA-positive prostate cancer disease were analysed using five different protein chips (cationic, anionic, hydrophobic and IMAC with copper and nickel) following delipidation, removal of albumin and IgG and pre-fractionation using ion exchange chromatography [94]. In this study, potential cancer protein markers were identified for each cancer type. A 28.3 kDa protein correctly identified 41/41 metastatic, late and early stage breast cancers and ruled out 27/28 of the non-malignant controls. A 13.8 kDa marker was found in the sera of all the 47 patients with pre-malignant or malignant colon disease, but not in the four patients with other intestinal disorders or in the 20 normal controls. A 50.8 kDa marker was found in all the PSA-positive (36/36) and PSA-negative (28/28) prostate cancer patients but not in the 20 healthy controls.

SELDI-TOF analysis of pancreatic juice using an IMAC array loaded with copper identified that the hepatocarcinoma-intestine-pancreas/pancreatitis associated protein I (HIP/PAP-I) was overexpressed in patients with pancreatic adenocarcinoma [95]. HIP/PAP-1 levels were then quantified in pancreatic juice as well as in serum using a polyclonal anti-HIP/PAP-1 antibody immobilised via amine coupling onto a pre-activated protein array [95]. Protein patterns claimed to be capable of differentiating breast cancer patients (12) from healthy controls (15) were found by SELDI-TOF analysis of nipple aspirate fluids [96].

In another recent study, protein mass spectra were generated by SELDI-TOF analysis of serum from 50 healthy controls and 50 patients with ovarian cancer (Fig. 8) [97]. These spectra were analysed by an iterative searching algorithm that identified a proteomic pattern that discriminated cancer from non-cancer patients. This pattern was then used to correctly identify all 50 ovarian cases in a masked series of 116 serum samples including 50 women with ovarian cancer and 66 women who were healthy or who had non-malignant disorders. Of the 66 cases of non-malignant disease, 63 were classified as not cancer related. These results yielded a sensitivity of 100%, a specificity of 95% and a predictive value of 94%.

Fig. 8. Identification of a biomarker for ovarian cancer using SELDI. Serum samples from 50 unaffected controls and 50 patients with ovarian cancer were analysed using a Ciphergen system with a C16 reversed phase chip surface. Analysis of the total data set using an iterative searching algorithm revealed a pattern of *m*/*z* values that was characteristic of the cancer patients. The peak at *m*/*z* 2111 was lower in cancer patients. The top three panels show increasing magnification of the spectra from an unaffected patient. The bottom panels show the data for two cancer and two unaffected patients represented as a density plot (increased magnification in the lowest panel) (Reprinted with permission from: Petricoin, E.F., III, et al., *Lancet*, **359**, 572–577 (2002)).

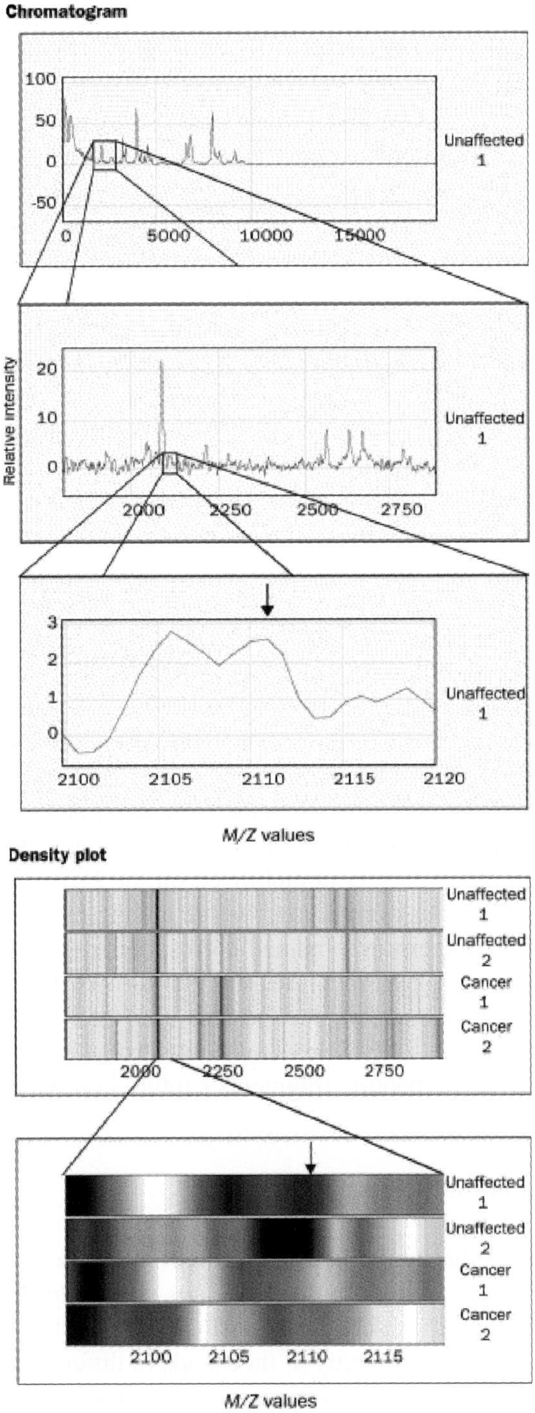
Chromatogram
100
50
0
-50
0
5000
10000
15000
Unaffected 1
Relative intensity
20
10
0
2000
2250
2500
2750
Unaffected 1
3
2
1
0
2100
2105
2110
2115
2120
Unaffected 1
M/Z values
Density plot
Unaffected 1
Unaffected 2
Cancer 1
Cancer 2
2000
2250
2500
2750
Unaffected 1
Unaffected 2
Cancer 1
Cancer 2
2100
2105
2110
2115
M/Z values

These studies demonstrate the advantages of studying protein profile patterns instead of individual proteins and the necessity of developing pattern-matching software [81,98]. An integrated approach using artificial neural networks has been designed to interrogate SELDI data [98], which identified masses that accurately predicted tumour grade in patients with high-grade astrocytoma.

SELDI technology can also be used in an immunoassay format to analyse specific proteins or biomarkers by immobilising specific antibodies on pre-activated array surfaces. PMSA was quantified in serum using anti-PSMA antibody bound to ProteinChip arrays [99,100]. Standard calibration curves can be constructed for quantitation of captured molecules and have been shown to be linear over 2–3 orders of magnitude [79]. It has been suggested that the SELDI technique is more quantitative than MALDI-MS since the surface deposition by capture is more uniform [79]. An immobilised anti-β amyloid polyclonal antibody was used to capture and identify amyloid β peptide variants secreted into the media of human HEK cells transfected with the Swedish familial mutation of β-amyloid precursor protein [101,102]. The intramembrane cleavage site of C99 amyloid precursor protein by γ-secretase was studied using a similar strategy [103]. An immunoassay was also developed for the detection of anti-microbial human β-defensin peptides (hBD1 and hBD20) secreted by human oral epithelial cells [104]. Proteins and biotinylated RNA and DNA were also immobilised as bait to study protein/protein [105–109] and protein/RNA interactions [110,111].

The SELDI platform can also be used to study post-translational modifications of proteins [80,81,112] as well as epitope mapping [113] and ligand fishing [79] using on-chip proteolytic digestion to characterise the protein. In a setting similar to the one described above for biosensor ligand fishing, the A33 colonic antigen was captured using anti-A33 IgG immobilised on the ProteinChip. The retained antigen was digested in situ using Lys-C before adding matrix and analysing the peptides. The peptide fingerprint obtained corresponded to 38% coverage of the total protein sequence. Protein identification can also be achieved using a new interface (PCI 1000) that allows the protein chips to be used with Q-TOF (Micromass) and Q-STAR (AB Sciex) mass spectrometers, allowing the direct MS/MS sequencing of the peptide fragments captured on the arrays.

5. Protein chips and microarrays

The next challenge in the study of protein/protein interactions will be to produce high-throughput protein microarray-based assays similar to those developed for DNA arrays (currently available DNA arrays contain 10^3–10^5 components) [114–118]. This will allow the study of thousands of different binding events in a parallel fashion on a genome/proteome wide scale. Until now, such large-scale studies have been performed using the yeast two-hybrid system [119]. However, this system has several limitations including problems with protein folding, non-controllable post-translational modifications and significant false negatives when protein are displayed inappropriately or when DNA domain fusions are produced in excess [120].

However, significant problems have to be resolved to create successful protein arrays, particularly in the areas of immobilisation and detection [121]. Unlike the well-defined structure of DNA, proteins have diverse and individual 3D structures that have to be retained during purification and immobilisation and there is no single immobilisation chemistry that is generally applicable [118,121]. Furthermore, functionality may also depend critically on post-translational modifications (e.g. glycosylation, phosphorylation, acetylation) [121].

5.1. Protein profiling arrays

Currently, two types of protein arrays have been defined, protein profiling arrays and protein function arrays, each requiring different strategies [121,122]. Profiling arrays measure the abundance, localisation and modification of proteins in biological samples. These arrays consist of a large number of specific, high affinity, protein ligands able to capture proteins in complex mixtures (e.g. biological fluids, cell lysates) [121,122]. Antibodies [123,124], antibody mimetics, scFV/Fab [121,122], affibodies [125,126], aptamers [127,128] and enzyme substrates [129] have all been used for this type of array.

Antibodies can be generated with high affinity and specificity for almost any protein, but the conventional hybridoma production method is not compatible with high-throughput screening. As an alternative, libraries of single chain antibodies can be displayed using bacteriophage, yeast or bacteria, allowing high-throughput screening of antibody–antigen interactions [121,122] while antibody mimetics or affibodies can also be produced by combinatorial protein engineering technology [125,126]. Plastibodies may also create novel opportunities in the development of antibody arrays [130]. These are based on molecular imprinting, where the recognition-binding site of the antibody is moulded directly in a polymer that can be used in imprinted sorbent assays [131].

A number of different supports and detection techniques have been used to generate functional immunoarrays. Glass surfaces have been used in a number of studies: one such biochip consisted of an optically flat glass plate containing 96 wells formed using hydrophobic Teflon masks [132]. Each well contained four 36-element arrays, comprising eight different antigens (immunoglobulins and a marker protein). Antibodies and marker protein were immobilised on the NHS-activated glass surface of the biochip using a continuous flow, capillary-based print head attached to a precise high-speed X–Y–Z robot with a printing capacity of 20,000 arrays per day. The assay was performed using an automatic pipettor and detection was performed using standard ELISA techniques. Arrays were quantified using a high-resolution scanning charge-coupled device (CCD) detector.

An array-based biosensor, utilising a sandwich immunoassay format, has also been fabricated on a glass support [133]. Glass microscope slides with covalently linked neutavidin were used to capture biotinylated antibodies and were utilized as a wave-guide upon which the sandwich fluoroimmunoassays were performed. In this study, 126 blind samples were analysed for the presence of three distinct classes of analytes. Antibodies were also covalently attached to polyacrylamide gel pads of the microchip using glutaraldhehyde or hydrazine chemistry [134]. Standard enzymatic assays were performed with fluorescent detection using a fluorescent microscope equipped with a CCD camera. In another example, microarrays using antibodies printed onto silanised glass have been used to quantitate PSA in 14 human serum samples using conventional ELISA techniques [135]

Anti-cytokine antibodies were coated on thiolsilane-coated and cross-linker activated glass slides divided by Teflon boundaries, forming 16 circular subarrays [136]. Within each subarray, 256 features were printed: among them, 150 represented 38 monoclonal antibodies, each specific for a cytokine and spotted in quadruplet. The remaining features were internal calibrators and controls for the signal amplification detection system. Captured cytokines were detected using biotinylated polyclonal antibodies followed by an anti-biotin antibody conjugated with the $5'$ end of an oligonucleotide primer. The amplification is performed by addition of a circular DNA molecule that hybridised to its complementary primer and, in the presence of DNA polymerase and nucleotides, rolling circle replication (RCA) occurs. The RCA product is then detected by hybridisation of multiple complementary oligonucleotides probes and measured with a conventional microarray scanning device. Using this amplification technique, 75 cytokines were measured with femtomolar sensitivity and a 3 log quantitative range. Human cytokines were also quantified using arrays in which antibodies have been printed onto silanised glass slides. Detection was performed using biotinylated detector antibody followed by streptavidin-R-phycoerythin and imaging using a ScanArray scanner [137].

Microarrays have also been constructed by conjugating antibodies and antigens to the surface of a poly-L-lysine-derivatised microscope glass slides. In one study,

---

Fig. 9. Antibody array detection of labelled antigen. The array was composed of 114 antibodies immobilised onto poly-L-lysine slides (6–12 spots of each antibody, about 1100 spots). Six protein mixtures of 115 antigens were labelled with the dye Cys5 (red fluorescence) and then mixed with a Cy3-labeled (green fluorescence) reference mixture containing each of the same 115 antigens at a constant concentration. The green to red ratio measured for each antigen spot reflects the variation in the concentration of the corresponding binding partner in the mixture. The inset highlights the anti-IgG and anti-FLAG spots and indicates the concentration of antigen applied onto the array (Reprinted with permission from: Haab, B.B., et al., *Genome Biol.*, **2**, 1–13 (2001)).

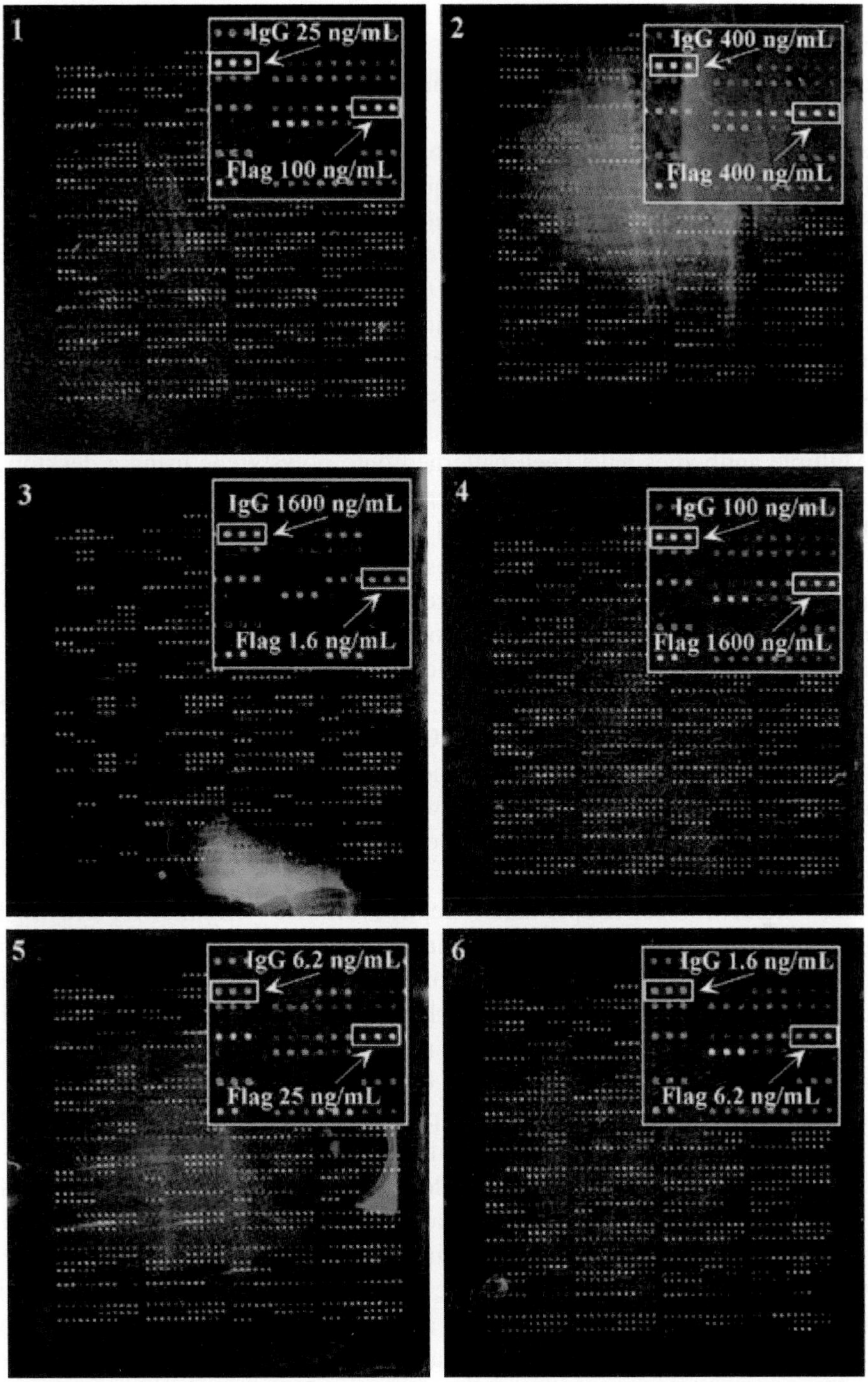
1
IgG 25 ng/mL
Flag 100 ng/mL
2
IgG 400 ng/mL
Flag 400 ng/mL
3
IgG 1600 ng/mL
Flag 1.6 ng/mL
4
IgG 100 ng/mL
Flag 1600 ng/mL
5
IgG 6.2 ng/mL
Flag 25 ng/mL
6
IgG 1.6 ng/mL
Flag 6.2 ng/mL

specificity, sensitivity and accuracy of detection of 115 antibody/antigen pairs were analysed using the array format and fluorescent detection (Fig. 9) [138]. Only 50% of the arrayed antigens and 20% of the arrayed antibodies provided specific and accurate measurements of their cognate ligands at or below concentrations of 0.34 and 1.6 μg/ml, respectively. This study emphasizes the need for further research and development with respect to accuracy in quantitation at low levels of detection and the necessity for suitable quality control of surfaces, a problem that has plagued DNA microarray slides. Using similar immunochemistry strategies, microarrays of autoantigens were used to characterise autoantibody responses in patients with autoimmune diseases [139]. A 1152-feature array containing 196 distinct biomolecules representing major autoantigens of eight distinct autoimmune diseases were attached to the surface of a poly-L-lysine-derivatised microscope glass slide. Detection was performed using fluorescent-labelled anti-human antibodies.

Multiple antibodies have been patterned onto polystyrene using a desktop jet printer in the fabrication of a mass-sensing analyte microarray immunoassay [140]. The antibodies were immobilised using the neutravidin/biotin system and detection was performed using evanescent wave excitation and fluorescent imaging detection. Other techniques to generate microarrays have included either photolithography of silane [141] or gold monolayers [142], microwells with microsphere sensors [143] and a micromolded hydrogel stamper with an aminosilylated-receiving surface [144]. Protein microarrays have also been generated by spotting crude or IMAC-purified protein expressed in liquid expression cultures onto PVDF membranes using a transfer stamp mounted onto a flat-bed spotting robot [145]. This array was used for antigen–antibody screening as well as for high-throughput screens of gene expression and receptor–ligand interactions.

Small molecules such as aptamers or peptides are also well suited for arraying for high density, high throughput, array formats. Aptamers [127,128] are nucleic acids derived from an in vitro selection and evolution process called SELEX (systematic evolution of ligands by exponential enrichment) [146] that has been engineered to display high-affinity receptors for small molecule ligands or target proteins. The SELEX protocol has been automated enabling thousands of aptamers to be produced. Furthermore, aptamers can be produced with halogenated bases that can be specifically used for photo-cross-linking to ligand [147].

Several strategies have now been developed to array peptides using miniaturised combinatorial peptide synthesis. Early array formats were achieved by parallel peptide synthesis using Fmoc chemistry in a 96-microtiter-plate format [148]. The SPOT synthesis technique also uses Fmoc chemistry but uses the hydroxyl moieties of the cellulose filter to derivatise Fmoc-alanine groups [149,150]. Peptide arrays are synthesized via the cellulose-bound alanine following deprotection. Different planar supports were also studied [151]. Applications of peptide arrays prepared using SPOT synthesis included B-cell epitope and paratope mapping, T-cell epitope mapping, enzyme–substrate recognition protein–protein, protein–DNA and protein–metal interactions [152,153].

A different chemistry has recently been used to create a peptide chip for the quantitative evaluation of protein kinase activity. This chip was prepared by the Diels-Alder mediated immobilisation of the kinase peptide substrate onto self-assembled monolayers of alkanethiolaters on gold [129]. Phosphorylation of the immobilised peptides by c-Src kinase was characterised by SPR, fluorescence and phosphorimaging while three inhibitors of the enzyme were quantitatively evaluated in an array format.

Living cells have been used to generate massively parallel arrays. In one example, the fabrication of high-density arrays using single chain antibodies produced in *E. coli* for high-throughput screening of antibody–antigen interaction was evaluated [154]. Using robotic picking and griding, 18,342 antibody clones were double spotted onto a $22 \times 22\ cm^2$ filter. This robotic system enabled the simultaneous generation of 15 identical filters, allowing screening against 15 different antigens. Detection was performed in a filter sandwich using classical ELISA techniques. A human fetal brain cDNA library was cloned into a bacterial vector that allowed expression of His6-tagged fusion proteins. Using a picking/gridding robot, the bacterial colonies were arrayed and grown in 384-well microtitre plates prior to high-density gridding on in situ filters [155]. Expressed proteins were recognised using an anti-His6 monoclonal antibody. Two candidate genes, GAPDH and HSP90alpha, were identified on high-density filters using DNA probes and antibodies.

Tissue arrays have also been developed for high-throughput profiling of tumor specimens and have been widely used in biomedical analyses (e.g. pathology, cancer research, neuropathology) [156–160]. In one study, 1000 cylindrical tissue biopsies (0.6 mm wide, 3–4 mm high) from individual tumors were distributed as a single tumor tissue microarray [156]. Sections of the microarray provided targets for parallel detection of DNA, RNA and protein targets in each specimen on the array, and consecutive sections allow the rapid analysis of hundreds of molecular markers on the same set of specimens.

5.2. Protein function arrays

The second category of protein arrays, protein function arrays, use the target proteins in order to probe their function or binding properties. Again, native conformation should be retained upon purification and immobilisation. Similar strategies are used to those described previously. Thus, proteins can be immobilised using non-covalent binding onto positively charged (aminosilane, poly-L-lysine) hydrophobic (nitrocellulose, polystyrene) surfaces. van der Walls interactions are also used to immobilise proteins onto a 3D polyacrylamide-based substrate called hydrogel (http://lifesciences.perkinelmer.com). Ready to use Fast™ glass slides are coated with a nitrocellulose polymer that binds protein non-covalently (www.arraying.com). Covalent interactions are also used to attach proteins to chemically modified surfaces (e.g. active esters, epoxy). Specific surface orientation and attachment using biomolecular interactions such as biotin/avidin, affinity tag (e.g. FLAG) or metal chelation (polyhistidine-tag) is also an attractive strategy to generate protein microarrays at high density and with a common orientation. Proteins can be expressed in vectors encoding a 15-amino acid sequence (BioTag™, www.nextgensciences.co), which is specifically biotinylated by the enzyme protein ligase. Proteins expressed using this system are then absorbed onto streptavidin-coated plastic slides.

Chemical affinity has also been successfully utilised in protein array immobilisation [161,162]. This strategy relies on the interaction between two synthetic low molecular mass molecules, phenyldiboronic acid (PDBA) and salicylhydroxamic acid (SHA). PDBA-based reagents are used to derivatise proteins at specific sites (lysine, cysteine or carbohydrate moieties of glycoproteins) for immobilisation onto SHA-coated glass slides (www.prolinxinc.com).

Several methods have been recently developed to screen biochemical and biological activity on large-scale arrays. In an early study, an array of 6144 individual yeast strains, each containing a different yeast open reading frame (ORF) was fused to GST [163]. These strains were grown in defined pools (64 pools of 96 clones) and the GST-ORFs were purified and assayed for biological affinities. Three novel ORF-associated activities were identified using this strategy. The activity of 119 different protein kinases for 17 different substrates was also assayed using a microwell-format array. The GST kinase fusion proteins were overproduced in yeast and purified using glutathione beads in a 96-well format assay. Kinase substrates were cross-linked to silicone elastomer microwells. The kinases were incubated with their substrate together with radioactively labelled ATP. After washing, phosphorylated substrates were detected using a phosphoimager. Quantitative analysis of kinase reaction was performed and novel activities of individual kinases were identified [164].

An eukaryotic proteome chip has been constructed consisting of a microarray composed of 5800 individually cloned, overproduced and purified proteins from yeast [165]. GST-His6 proteins were purified using glutathione beads in a 96-well format assay and immobilised onto aldehyde-treated microscope slides through primary amines or spotted onto nickel-coated slides. These proteome chips were evaluated by probing for several protein–protein and protein–lipid interactions: for example, the chip was probed for calmodulin- and phospholipid-binding proteins using biotinylated calmodulin in the presence of calcium and biotin-containing phosphoionoside liposomes. Novel calmodulin- and phospholipids-interacting proteins were identified. Furthermore, a common potential binding motif was identified for many of the calmodulin-binding proteins.

Table 3

Manufacturers of protein array systems

Manufacturers	Technology	Website
Affibody	Non-immunoglobulin protein scaffold arrays	www.affibody.com/
Akceli	Live cell microarrays	www.akceli.com/
Beecher Instruments	Tissue array technology	www.beecherinstruments.com
Cambridge Antibody Technology	Antibody (phage display libraries) arrays	www.cambridgeantibody.com
Ciphergen	SELDI technology	www.ciphergen.com
Clontech	Antibody microarrays	www.clontech.com
CombiMatrix Corporation	Peptide and small molecule array on semiconductor surface	http://www.combimatrix.com/
DiscernArray	Protein in situ array tag immobilisation in multi-well plate	www.discerna.co.uk
Dyax	Antibody (phage display libraries) arrays	www.dyax.com/
HTS Biosystems	Protein array surface plasmon resonance detection	www.htsbiosystems.com
Hypromatrix	Antibody arrays on membrane	www.hypromatrix.com
Lumicyte	SELDI technology	www.lumicyte.com
Large Scale Biology: Biosite	Protein markers array	www.lsbc.com
Molecularstaging	Rolling circle amplification technology detection	www.molecularstaging.com
Nextgensciences	Protein biochip arrays onto aminosilane-coated glass, hydrogel and fusion-Tag attachment	www.nextgensciences.com
Perkin Elmer Life Sciences	Hydrogel-coated slides for protein arrays	http://lifesciences.perkinelmer.com/
Phylos	Antibody-like ligand scaffolds arrays	www.phylos.com
Prolynx	Protein and peptide array: versalinx technology	www.prolinxinc.com
Proteome Systems	Gel array technology	www.proteomesystems.com/

(*continues*)

Table 3
Continued

Manufacturers	Technology	Website
Resource Center/Primary database	High-density protein arrays on PVDF filter	www.rzpd.de
Sense Functional Proteomics	COVET technology fusion-tag attachment	www.senseproteomic.com
Somalogic	Aptamer arrays	www.somalogic.com
Thermo Hybaid	Protein arrays onto gold slides coated with streptavidin	www.thermohybaid.com
Zyomix	Protein profiling biochip system Silicon-based antibody arrays	www.zyomyx.com

High-density protein arrays have been manufactured using a high-precision contact printing robot delivering nanoliter volumes of protein samples to an aldehyde-activated slide, yielding spots about 150–200 μm in diameter (1600 spots per cm^2) [120]. After immobilisation, the slides were treated with BSA to quench unreacted aldehydes. Peptides and small molecules have been printed onto ester-activated BSA slides. In this approach, a molecular layer of BSA is first attached to the slide and then activated with *N,N'*-disuccinimidyl carbonate for subsequent immobilisation of peptide. Three pairs of interacting proteins were analysed in this system. In one example, a slide containing 10,799 protein G spots and a single spot of FKBP12-rapamycin binding (FRP) domain of FKBP-rapamycin associated protein was studied. The interaction of FRP with its ligand, human immunophilin FKBP12, in the presence of rapamycin was uniquely detected using fluorescence among the large amount of protein G spots. Transient interactions that occur between enzyme and their substrates were also successfully analysed in this system, using three different kinase–substrate pairs. Recently, the analysis of the interaction between proteins and small molecules (e.g. steroid derivative/IgG interaction) was also demonstrated using this type of array.

The examples described above demonstrate that protein microarrays and chips are gradually 'coming of age' as potential high-throughput analytical tools to probe protein/protein interactions (Table 3) and have the potential to play a major role in the field of clinical proteomics [166–168]. Further developments in this field will almost certainly involve progress in the large-scale production of proteins as well as the quality control of the final purified product. The functional state of the protein (e.g. correct folding) should also be assessed [169]. Advances in detection techniques will facilitate screening of a large range of protein interactions, particularly at high sensitivity [169]. Label-free detection strategies such as optical techniques (e.g. SPR, optical spectroscopy) should further widen the range of interactions, which can be studied.

6. Summary

Recent technological advances have led to the development of a number of optical instrumental biosensors that can measure biomolecular interactions in real time. In these instruments, one of the binding partners is immobilised onto the sensor surface, ideally in a specific orientation, mimicking that found in vivo. Solutions containing possible binding partners are then introduced into the system. Detection techniques typically detect changes in mass at, or near to, the sensor surface as binding occurs. Interactions are measured in real time without the requirement for labelling and, from the shape of the binding curves, detailed analysis of the reaction kinetics is possible. If suitable desorption conditions can be identified the surface may be regenerated for further use, or retained binding partners recovered for sensitive and specific downstream analysis including bioassays or mass spectrometry. These surfaces can therefore be considered as microaffinity purification platforms that can be utilised, often as part of a multi-dimensional HPLC purification protocol, as part of a proteomics-based approach for the purification of specific binding partners and their complexes that allow the elucidation of possible signalling pathways.

There is considerable analogy between the use of biosensor chips as part of a purification protocol followed by downstream mass spectrometric analysis and the Ciphergen SELDI ProteinChip technology: in the latter case the chip itself provides the separation selectivity and MALDI-TOF-MS is used for mass readout. This technology is currently being further developed for the identification of novel biomarkers for cancer treatment and prognosis.

The sensor surfaces also represent a simple form of the protein- and peptide-chip technologies, which are currently being developed for high-throughput sample analysis. In particular many of the immobilisation chemistries that are currently being developed for biosensor analysis will be equally applicable to high-throughput chip technology.

References

1. Wilkins, M.R., Sanchez, J.C., Gooley, A.A., Appel, R.D., Humphery-Smith, I., Hochstrasser, D.F. and Williams, K.L., Progress with proteome projects: why all proteins expressed by a genome should be identified and how to do it. *Biotechnol. Genet. Engng Res.*, **13**, 19–50 (1996).
2. Gygi, S.P., Rist, B. and Aebersold, R., Measuring gene expression by quantitative proteome analysis. *Curr. Opin. Biotechnol.*, **11**, 396–401 (2000).

3. Patterson, S., Proteomics: the industrialization of protein chemistry. *Curr. Opin. Biotechnol.*, **11**, 413–418 (2000).
4. Chalmers, M.J. and Gaskell, S.J., Advances in mass spectrometry for proteome analysis. *Curr. Opin. Biotechnol.*, **11**, 384–390 (2000).
5. Gygi, S.P., Corthals, G.L., Zhang, Y., Rochon, Y. and Aebersold, R., Evaluation of two-dimensional gel electrophoresis-based proteome analysis technology. *Proc. Natl Acad. Sci. USA*, **97**, 9390–9395 (2000).
6. Washburn, M.P., Wolters, D. and Yates, J.R. III, Large-scale analysis of the yeast proteome by multidimensional protein identification technology. *Nat. Biotechnol.*, **3**, 242–247 (2001).
7. Griffin, T.J. and Aebersold, R., Advances in proteome analysis by mass spectrometry. *J. Biol. Chem.*, **276**, 45497–45500 (2001).
8. Hebestreit, H.F., Proteomics: an holistic analysis of nature's proteins. *Curr. Opin. Pharmacol.*, **1**, 513–520 (2001).
9. Santoni, V., Molloy, M. and Rabilloud, T., Membrane proteins and proteomics: un amour impossible. *Electrophoresis*, **21**, 1054–1070 (2001).
10. Opitek, G.J., Ramirez, S.M., Jorgenson, J.W. and Moseley, M.A. III, Comprehensive two-dimensional high-performance chromatography for the isolation of overexpressed proteins and proteome mapping. *Anal. Biochem.*, **258**, 349–361 (1998).
11. Link, A.J., Eng, J., Schiltz, D.M., Carmack, E., Mize, G.J., Morris, D.R., Garvick, B.M. and Yates, J.R. III, Direct analysis of protein complexes using mass spectrometry. *Nat. Biotechnol.*, **17**, 676–682 (1999).
12. Corthals, G., Wasinger, V.C., Hochstrasser, D.F. and Sanchez, J.C., The dynamic range of protein expression: a challenge for proteomic research. *Electrophoresis*, **2**, 1104–1115 (2000).
13. Richter, R., Schulz-Knappe, P., John, H. and Forssmann, W.-G., Posttranslationally processed forms of the human chemokine HCC-1. *Biochemistry*, **39**, 10799–10805 (2000).
14. Standker, L., Braulke, T., Mark, S., Mostafavi, H., Meyer, M., Honing, S., Gimenez-Gallego, G. and Forssmann, W.G., Partial IGF affinity of circulating N- and C-terminal fragments of human insulin-like growth factor binding protein-4 (IGFBP-4) and the disulfide bonding pattern of the C-terminal IGFBP-4 domain. *Biochemistry*, **39**, 5082–5088 (2000).
15. Taylor, R.S., Wu, C.C., Hays, L.G., Eng, J.K., Yates, J.R. III and Howell, K.E., Proteomic of rat liver golgi complex: minor proteins are identified through sequential fractionation. *Electrophoresis*, **21**, 3441–3459 (2000).
16. Liepke, C., Zucht, H.D., Forssmann, W.G. and Standker, L., Purification of novel peptide antibiotics from human milk. *J. Chromatogr. B*, **752**, 369–377 (2001).
17. Gavin, A.C., Bosche, M., Krause, R., Grandi, P., Marzioch, M., Bauer, A., Schultz, J., Rick, J.M., Michon, A.M., Cruciat, C.M., Remor, M., Hofert, C., Schelder, M., Brajenovic, M., Ruffner, H., Merino, A., Klein, K., Hudak, M., Dickson, D., Rudi, T., Gnau, V., Bauch, A., Bastuck, S., Huhse, B., Leutwein, C., Heurtier, M.A., Copley, R.R., Edelmann, A., Querfurth, E., Rybin, V., Drewes, G., Raida, M., Bouwmeester, T., Bork, P., Seraphin, B., Kuster, B., Neubauer, G. and Superti-Furga, G., Functional organization of the yeast proteome by systematic analysis of protein complexes. *Nature*, **415**, 141–147 (2002).
18. Ho, Y., Gruhler, A., Heilbut, A., Bader, G.D., Moore, L., Adams, S.L., Millar, A., Taylor, P., Bennett, K., Boutilier, K., Yang, L., Wolting, C., Donaldson, I., Schandorff, S., Shewnarane, J., Vo, M., Taggart, J., Goudreault, M., Muskat, B., Alfarano, C., Dewar, D., Lin, Z., Michalickova, K., Willems, A.R., Sassi, H., Nielsen, P.A., Rasmussen, K.J., Andersen, J.R., Johansen, L.E., Hansen, L.H., Jespersen, H., Podtelejnikov, A., Nielsen, E., Crawford, J., Poulsen, V., Sorensen, B.D., Matthiesen, J., Hendrickson, R.C., Gleeson, F., Pawson, T., Moran, M.F., Durocher, D., Mann, M., Hogue, C.W., Figeys, D. and Tyers, M., Systematic

identification of protein complexes in *Saccharomyces cerevisiae* by mass spectrometry. *Nature*, **415**, 180–183 (2002).
19. Kachman, M.T., Wang, H., Schwartz, D.R., Cho, K.R. and Lubman, D.M., A 2-D liquid separations/mass mapping method for interlysate comparison of ovarian cancers. *Anal. Chem.*, **74**, 1779–1791 (2002).
20. Catimel, B., Rothacker, J. and Nice, E., The use of biosensors for microaffinity purification: an integrated approach to proteomics. *J. Biochem. Biophys. Methods*, **49**, 289–312 (2001).
21. Cass, A.E.G., In: Meyers, R.A. (Ed.), *Molecular Biology and Biotechnology*. VCH, New York, 1995, pp. 110–113.
22. Nice, E.C. and Catimel, B., Instumental biosensors: new perspectives for the analysis of biomolecular interaction. *Bioessays*, **21**, 339–352 (1999).
23. Rich, R.L. and Myszka, D.G., Advances in surface plasmon resonance biosensor analysis. *Curr. Opin. Biotechnol.*, **11**, 54–61 (2000).
24. Leatherbarrow, R.J. and Edwards, P.R., Analysis of molecular recognition using optical biosensors. *Curr. Opin. Chem. Biol.*, **3**, 544–547 (1999).
25. Baird, C.L. and Muszka, D.G., Current and emerging commercial optical biosensors. *J. Mol. Recogn.*, **14**, 261–268 (2001).
26. Malmqvist, M., Biospecific interaction analysis using biosensor technology. *Nature*, **361**, 186–187 (1983).
27. Davies, R.D., Edwards, P.R., Watts, H.J., Lowe, C.R., Buckle, P.E., Yeung, D., Kinning, T.M. and Pollard-Knight, D.V., The resonant mirror: a tool for the study of biomolecular interactions. *Tech. Protein Chem. V*, 285–292 (1994).
28. Catimel, B., Domagala, T., Nerrie, N., Weinstock, J., Abud, H., Heath, J.K. and Nice, E.C., Recent applications of instrumental biosensors for protein and peptide structure–function studies. *Protein Pept. Lett.*, **6**, 319–340 (1999).
29. Stenberg, E., Persson, B., Roos, H. and Urbaniczki, J., Quantitative determination of surface concentration of protein with surface plasmon resonance using radiolabeled proteins. *J. Colloid Interface Sci.*, **43**, 513–526 (1990).
30. Edwards, P.R., Lowe, P.A. and Leatherbarrow, R.J., Ligand loading at the surface of an optical biosensor and its effect upon the kinetics of protein–protein interactions. *J. Mol. Recogn.*, **10**, 128–134 (1997).
31. Wade, J.D., Domagala, T., Rothacker, J., Catimel, B. and Nice, E.C., Use of thiazoline-mediated ligation for site specific biotinylation of mouse EGF for biosensor immobilisation. *Lett. Pept. Sci.*, **58**, 493–503 (2002).
32. Huber, A., Demartis, S. and Neri, D., The use of biosensor technology for the engineering of antibodies and enzymes. *J. Mol. Recogn.*, **12**, 198–216 (1999).
33. Rich, R.L. and Myszka, D.G., Survey of the 1999 surface plasmon resonance biosensor literature. *J. Mol. Recogn.*, **13**, 388–407 (2000).
34. Rich, R.L. and Myszka, D.G., Survey of the year 2000 commercial optical biosensor literature. *J. Mol. Recogn.*, **14**, 273–294 (2001).
35. Nice, E.C., Catimel, B., Lackmann, M., Stacker, S., Runting, A., Wilks, A., Nicola, N. and Burgess, A.W., Strategies for the identification and purification of orphan biomolecules. *Lett. Pept. Sci.*, **4**, 107–120 (1997).
36. Bartley, T.D., Hunt, R.W., Welcher, A.A., Boyle, W.J., Parker, V.P., Lindberg, R.A., Lu, H.S., Colombero, A.M., Elliott, R.L., Guthrie, B.A., Holst, P.L., Skrine, J.D., Toso, R.J., Zhang, M., Fernandez, E., Trail, G., Varnum, B., Yarden, Y., Hunter, T. and Fox, G.M., B61 is a ligand for the ECK receptor-protein tyrosine kinase. *Nature*, **368**, 558–560 (1994).
37. Stitt, T.N., Conn, G., Gore, M., Lai, C., Bruno, J., Radziejewski, C., Mattsson, K., Fisher, J., Gies, D.R., Jones, P.F., Masiakowski, P., Rian, T.E., Tobkes, N.J., Chen, D.H., DiStefano, P.S.,

Long, G.L., Basilico, C., Goldfarb, M.P., Lemke, G., Glass, D.J. and Yancopoulos, G.D., The anticoagulation factor protein S and its relative, Gas6, are ligands for the Tyro 3/Axl family of receptor tyrosine kinases. *Cell*, **80**, 661–670 (1995).

38. Lackmann, M., Bucci, T., Mann, R.J., Kravets, L.A., Viney, E., Smith, F., Moritz, R.L., Carter, W., Simpson, R.J., Nicola, N.A., Mackwell, K., Nice, E.C., Wilks, A.F. and Boyd, A.W., Purification of a ligand for the EPH-like receptor HEK using a biosensor-based affinity detection approach. *Proc. Natl Acad. Sci. USA*, **93**, 2523–2527 (1996).
39. Davis, S., Aldrich, T.H., Jones, P.F., Acheson, A., Compton, D.L., Jain, V., Ryan, T.E., Bruno, J., Radziejewski, C., Maisonpierre, P.C. and Yancopoulos, G.D., Isolation of angiopoietin-1, a ligand for the TIE2 receptor, by secretion-trap expression cloning. *Cell*, **87**, 1161–1169 (1996).
40. Sakano, S., Serizawa, R., Inada, T., Iwama, A., Itoh, A., Kato, C., Shimizu, Y., Shinkai, F., Shimizu, R., Kondo, S., Ohno, M. and Suda, T., Characterization of a ligand for receptor protein-tyrosine kinase HTK expressed in immature hematopoietic cells. *Oncogene*, **13**, 813–822 (1996).
41. Fitz, L.J., Morris, J.C., Towler, P., Long, A., Burgess, P., Greco, R., Wang, R.J., Gassaway, R., Nickbarg, E., Kovacic, S., Ciarletta, A., Giannotti, J.A., Finnerty, H., Zollner, R., Beier, D.R., Leak, L.V., Turner, K.J. and Wood, C., Characterisation of murine Flt4 ligand/VEGF-C. *Oncogene*, **15**, 613–618 (1997).
42. Catimel, B., Ritter, G., Welt, S., Old, L.J., Cohen, L., Nerrie, M.A., White, S.J., Heath, J.K., Demediuk, B., Domagala, T., Lee, F.T., Scott, A.M., Tu, G.F., Ji, H., Moritz, R.L., Simpson, R.J., Burgess, A.W. and Nice, E.C., Purification and characterization of a novel restricted antigen expressed by normal and transformed human colonic epithelium. *J. Biol. Chem.*, **271**, 25664–25670 (1996).
43. Seok, Y.J., Sondej, M., Badawi, P., Lewis, M.S., Briggs, M.C., Jaffe, H. and Peterkofsky, A., High affinity binding and allosteric regulation of *Escherichia coli* glycogen phosphorylase by the histidine phosphocarrier protein, HPr. *J. Biol. Chem.*, **272**, 26511–26521 (1997).
44. Catimel, B., Nerrie, M., Lee, F.T., Scott, A.M., Ritter, G., Welt, S., Old, L.J., Cohen, L., White, S., Heath, J.K., Hong, J., Moritz, R.L., Simpson, R.J., Burgess, A.W. and Nice, E.C., Kinetic analysis of the interaction between the monoclonal antibody A33 and its colonic epithelial antigen using an optical biosensor: a comparison of immobilization strategies. *J. Chromatogr. A*, **776**, 15–30 (1997).
45. Moritz, R.L., Ritter, G., Catimel, B., Cohen, L.S., Welt, S., Old, L.J., Burgess, A.W., Nice, E.C. and Simpson, R.J., Sequencing strategies for the human A33 antigen, a novel surface glycoprotein of human gastrointestinal epithelium. *J. Chromatogr. A*, **798**, 91–101 (1997).
46. Ritter, G., Cohen, L.S., Nice, E.C., Catimel, B., Burgess, A.W., Moritz, R.L., Ji, H., Heath, J.K., White, S., Welt, S., Old, L.J. and Simpson, R.J., Characterisation of post-translational modifications of human A33 antigen, a novel palmitoylated surface glycoprotein of human gastrointestinal epithelium. *Biochem. Biophys. Res. Commun.*, **236**, 682–686 (1997).
47. Krone, J.R., Nelson, R.W., Dogruel, D., Williams, P. and Granzow, R., BIA/MS interfacing biomolecular interaction analysis with mass spectrometry. *Anal. Biochem.*, **244**, 124–132 (1997).
48. Hillenkamp, F., Karas, M., Beavis, R.C. and Chait, B.T., Matrix-assisted laser desorption/ionization mass spectrometry of biopolymers. *Anal. Chem.*, **63**, 193A–1203A (1991).
49. Karas, M. and Hillenkamp, F., Laser desorption ionization of proteins with molecular masses exceeding 10,000 daltons. *Anal. Chem.*, **60**, 2299–2301 (1998).
50. Jardine, L., Electrospray ionization mass spectrometry of biomolecules. *Nature*, **345**, 747–748 (1990).

51. Nedelkov, D. and Nelson, R.W., Practical considerations in BIA/MS: optimising the biosensor-mass spectrometry interface. *J. Mol. Recogn.*, **13**, 140–145 (2000).
52. Nelson, R.W. and Krone, J.R., Surface plasmon resonance biomolecular interaction analysis mass spectrometry. 1. Chip based analysis. *Anal. Chem.*, **69**, 4363–4368 (1997).
53. Nelson, R.W., Krone, J.R. and Jansson, Ö., Surface plasmon resonance biomolecular interaction analysis mass spectrometry. 2. Fiber optic-based analysis. *Anal. Chem.*, **69**, 4369–4374 (1997).
54. Nelson, R.W. and Krone, J.R., Advances in surface plasmon resonance biomolecular interaction analysis mass spectrometry (BIA/MS). *J. Mol. Recogn.*, **12**, 77–93 (1999).
55. Nelson, R.W., Jarvik, J.W., Taillon, B.E. and Tubbs, K.A., BIA/MS of epitope-tagged peptides directly from *E. coli* lysate: multiplex detection and protein identification at low-femtomole to subfemtomole levels. *Anal. Chem.*, **71**, 2858–2865 (1999).
56. Nelson, R.W., Nedelkov, D. and Tubbs, K.A., Biosensor chip mass spectrometry: a chip-based proteomics approach. *Electrophoresis*, **21**, 1155–1163 (2000).
57. Nedelkov, D., Rassly, A. and Nelson, R.W., Multitoxin biosensor-mass spectrometry analysis: a new approach for rapid, real-time, sensitive analysis of staphylococcal toxins in food. *Int. J. Food Microbiol.*, **60**, 1–13 (2000).
58. Nelson, R.W., Nedelkov, D. and Tubbs, K.A., Biomolecular interaction analysis mass spectrometry. BIA/MS can detect and characterize proteins in complex biological fluids at the low- to subfemtomole level. *Anal. Chem.*, **72**, 404A–411A (2000).
59. Nedelkov, D. and Nelson, R.W., Exploring the limit of detection in biomolecular interaction analysis mass spectrometry (BIA/MS): detection of attomole amounts of native protein present in complex biological mixture. *Anal. Chim. Acta*, **423**, 1–7 (2000).
60. Nedelkov, D. and Nelson, R.W., Analysis of native proteins from biological fluids by biomolecular interaction analysis mass spectrometry (BIA/MS): exploring the limit of detection, identification of non-specific binding and detection of multi-protein complexes. *Biosens. Bioelectron.*, **16**, 1071–1078 (2001).
61. Nedelkov, D. and Nelson, R.W., Analysis of human urine protein biomarkers via biomolecular interaction analysis mass spectrometry. *Am. J. Kidney Dis.*, **38**, 481–487 (2001).
62. Nedelkov, D. and Nelson, R.W., Biomolecular interaction analysis mass spectrometry: a multiplexed proteomic approach. *PharmaGenomics*, 28–33 (2001).
63. Nedelkov, D. and Nelson, R.W., Delineation of in vivo assembled multiprotein complexes via biomolecular interaction analysis mass spectrometry. *Proteomics*, **1**, 1441–1446 (2001).
64. Williams, C., Biotechnology match making: screening orphan ligands and receptors. *Curr. Opin. Biotechnol.*, **11**, 42–46 (2000).
65. Natsume, T., Nakayama, H., Jansson, Ö., Isobe, T., Takio, K. and Mikoshiba, K., Combination of biomolecular interaction analysis and mass spectrometric amino acid sequencing. *Anal. Chem.*, **72**, 4193–4198 (2000).
66. Natsume, T., Nakayama, H. and Toshiaki, I., BIA-MS-MS: biomolecular interaction analysis for functional proteomics. *Trends Biochem.*, **19**, 528–533 (2001).
67. Sönksen, C.P., Nordhoff, E., Jansson, Ö., Malmqvist, M. and Roepstorff, P., Combining MALDI mass spectrometry and biomolecular interaction analysis using a biomolecular interaction analysis instrument. *Anal. Chem.*, **70**, 2731–2736 (1998).
68. Fitz, L., Cook, S., Nickbarg, E., Wang, J.H. and Wood, C.R., Accelerating ligand identification. *BIA J.*, **5**, 23–25 (1998).
69. Catimel, B., Weinstock, J., Nerrie, M., Domagala, T. and Nice, E.C., Micropreparative ligand fishing with a cuvette-based optical mirror resonance biosensor. *J. Chromatogr. A*, **869**, 261–273 (2000).

70. Roepstorff, P. and Sönksen, C.P., A powerful combination: BIAcore 3000 and MALDI-TOF. *BIA J.*, **6**, 9–11 (1999).
71. Zhukov, A., Suckau, D. and Buijs, B., From isolation to identification. Using surface plasmon resonance-mass spectrometry in proteomics. *Pharmagenomics*, **2**, 18–28 (2002).
72. Williams, C. and Addona, T.A., The integration of SPR biosensors with mass spectrometry: possible application for proteome analysis. *Trends Biotechnol.*, **18**, 45–48 (2000).
73. Cao, B., Urban, J., Vaisar, T., Shen, R.Y.W. and Kahn, M., Detecting and identifying active compounds from a combinatorial library using IAsys and electrospray mass spectrometry. *Technol. Protein Sci. Chem.*, **8**, 177–184 (1997).
74. Lowe, P.A., Tristan, J.H., Clark, A., Davies, R.J., Edwards, P.R., Kinning, T. and Yeung, D., New approaches for the analysis of molecular recognition using the IAsys evanescent wave biosensor. *J. Mol. Recogn.*, **11**, 194–199 (1998).
75. Ji, H., Moritz, R.L., Ritter, G., Catimel, B., Nice, E.C., Heath, J.K., White, S., Welt, S., Old, L.J., Burgess, A.W. and Simpson, R.J., Electrophoretic analysis of the novel antigen for the colon-specific monoclonal antibody A33. *Electrophoresis*, **18**, 614–621 (1997).
76. Simpson, R.J., Connolly, L.M., Eddes, J.S., Pereira, J.J., Moritz, R.L. and Reid, G.E., Proteomic analysis of the human colon carcinoma cell line (LIM 1215): development of a membrane protein database. *Electrophoresis*, **21**, 1707–1732 (2000).
77. Catimel, B., Rotacker, J. and Nice, E.C., Biosensors and proteomics. *Life Sci.*, **14**, 24–30 (2002).
78. Hutchens, T.W. and Yip, T.T., New desorption strategies for the mass spectrometric analysis of macromolecules. *Rapid Commun. Mass Spectrom.*, **7**, 576–580 (1993).
79. Merchant, M. and Weinberger, S.R., Recent advancements in surface-enhanced laser desorption/ionisation-time of flight-mass spectrometry. *Electrophoresis*, **21**, 1164–1167 (2000).
80. Fung, E.T., Thulasiraman, V., Weinberger, S.R. and Dalmaso, E.A., Protein biochips for differential profiling. *Curr. Opin. Biotechnol.*, **12**, 65–69 (2001).
81. Weinberger, S.C., Dalmaso, E.A. and Fung, E.T., Current achievement using ProteinChip Array technology. *Curr. Opin. Chem. Biol.*, **6**, 86–91 (2001).
82. Emmert-Buck, M.R., Bonner, R.F., Smith, P.D., Chuaqui, R.F., Zhuang, Z., Goldstein, S.R., Weiss, R.A. and Liotta, L.A., Laser capture microdissection. *Science*, **274**, 998–1001 (1996).
83. Banks, R.E., Dunn, M.J., Forbes, M.A., Stanley, A., Pappin, D., Naven, T., Gough, M., Harnden, P. and Selby, P.J., The potential use of laser capture microdissection to selectively obtain distinct populations of cells for proteomic analysis — preliminary findings. *Electrophoresis*, **20**, 689–700 (2000).
84. von Eggeling, F., Davies, H., Lomas, L., Fiedler, W., Junker, K., Claussen, U. and Ernst, G., Tissue-specific microdissection coupled with ProteinChip array technologies: applications in cancer research. *Biotechniques*, **29**, 1066–1070 (2000).
85. Simone, N.L., Paweletz, C.P., Charbonneau, L., Petricoin, E.F. and Liotta, L.A., Laser capture microdissection: beyond functional genomics to proteomics. *Mol. Diagn.*, **5**, 301–307 (2000).
86. Paweletz, C.P., Gillespie, J.W., Ornstein, D.K., Simone, N.L., Brown, M.R., Cole, K.A., Wang, Q.H., Huang, J., Hu, N., Yip, T.T., Rich, W.E., Kohn, E.C., Linehan, W.M., Weber, T., Taylor, P., Emmert-Buck, M.R., Liotta, L.A. and Petricoin, E.F., Rapid protein display profiling of cancer progression directly from human tissue using a protein biochip. *Drug. Dev. Res.*, **49**, 34–42 (2000).
87. Paweletz, C.P., Liotta, L.A. and Petricoin, E.F. III, New technologies for biomarker analysis of prostate cancer progression: laser capture microdissection and tissue proteomics. *Urology*, **57**, 160–163 (2001).
88. Wellmann, A., Wollscheid, V., Lu, H., Ma, Z.L., Albers, P., Schutze, K., Rohde, V., Behrens, P., Dreschers, S., Ko, Y. and Wernert, N., Analysis of microdissected prostate tissue with

ProteinChip arrays — a way to new insights into carcinogenis and to diagnostic tools. *Int. J. Mol. Med.*, **9**, 341–437 (2002).
89. Emmert-Buck, M.R., Gillespie, J.W., Paweletz, C.P., Ornstein, D.K., Basrur, V., Appella, E., Wang, Q.H., Huang, J., Hu, N., Taylor, P. and Petricoin, E.F. III, An approach to proteomic analysis of human tumors. *Mol. Carcinog.*, **27**, 158–165 (2000).
90. von Eggeling, F., Junker, K., Fiedler, W., Wollscheid, V., Durst, M., Claussen, U. and Ernst, G., Mass spectrometry meets chip technology: a new proteomic tool in cancer research? *Electrophoresis*, **22**, 2898–2902 (2001).
91. Wulfkuhle, J.D., McLean, K.C., Paweletz, C.P., Sroi, D.C., Trock, B.J., Steeg, P. and Petricoin, E.F. III, New approach to proteomic analysis of breast cancer. *Proteomics*, **1**, 1205–1215 (2001).
92. Wright, J.L. Jr., Cazares, L.H., Leung, S.M., Nasim, S., Adam, B.L., Yip, T.T., Schellhammer, P.F., Gong, L. and Vlahou, A., ProteinChip surface enhanced laser desorption/ionization (SELDI) mass spectrometry: a novel protein biochip technology for detection of prostate biomarker in complex protein mixtures. *Prostate Cancer Prostatic Dis.*, **2**, 264–276 (2000).
93. Adams, B.L., Davis, J.W., Cazares, L.H., Schellhammer, P.F. and Wright, G.L. Jr., Identifying the signature of prostate cancer in seminal plasma by SELDI affinity mass spectrometry. *Proc. Am. Assoc. Cancer Res.*, **41**, 564 (2000).
94. Watkins, B., Szaro, R., Ball, S., Knubovets, T., Briggman, J., Hlavaty, J.J., Kusinitz, F., Stieg, A. and Wu, Y.I., Detection of early-stage cancer by serum analysis. *Am. Lab.*, 32–36 (2001).
95. Rosty, C., Christa, L., Kuzdzalk, S., Baldwin, W.M., Zahurak, M.L., Carnot, F., Chan, D.W., Canto, M., Lillemoe, K.D., Cameron, J.L., Yeao, C.J., Hruban, R.H. and Goggins, M., Identification of hepatocarcinoma-intestine-pancreas/pancreatitis associated protein I as a biomarker for pancreatic ductal adenocarcinoma by protein biochip technology. *Cancer Res.*, **62**, 1868–1875 (2002).
96. Paweletz, C.P., Trock, B., Pennanen, M., Tsangaris, T., Magnant, C., Liottaand, L.A. and Petricoin, E.F. III, Proteomic patterns of nipple aspirate fluids obtained by SELDI-TOF: potential for new marker to aid in the diagnosis of breast cancer. *Dis. Markers*, **17**, 301–307 (2001).
97. Petricoin, E.F. III, Adekani, A.M., Hittr, B.A., Levine, P.J., Fusaro, V.A., Steinberg, S.M., Mills, G.B., Simone, C., Fishman, D.A., Kohn, E.C. and Liotta, L.A., Use of proteomic patterns in serum to identify ovarian cancer. *Lancet*, **359**, 572–577 (2002).
98. Ball, G., Mian, S., Holding, F., Allibone, R.O., Lowe, J., Ali, S., Li, G., McCardle, S., Ellis, I.O., Creaser, C. and Rees, R.C., An integrated approach utilizing artificial neural networks and SELDI mass spectrometry for the classification of human tumors and rapid identification of potential biomarkers. *Bioinformatics*, **18**, 395–404 (2002).
99. Xiao, Z., Beckett, M.L. and Wright, G.L. Jr., Generation of a bacculovirus recombinant prostate-specific membrane antigen and its use in the development of a novel protein biochip quantitative imunoassay. *Protein Exp. Purif.*, **19**, 12–21 (2000).
100. Xiao, Z., Adam, B.L., Cazares, L.H., Clements, M.A., Davis, J.W., Schellhammer, P.F., Dalmasso, E.A. and Wright, G.L. Jr., Quantitation of serum prostate-specific membrane antigen by a novel protein biochip immunoassay discriminates benign from malignant prostate disease. *Cancer Res.*, **61**, 6029–6033 (2001).
101. Davies, H., Lomas, L. and Austen, B., Profiling of amyloid beta peptide variants using SELDI Protein chip arrays. *Biotechniques*, **27**, 1258–1261 (1999).
102. Austen, B.M., Frears, E.R. and Davies, H., The use of SELDI proteinchip arrays to monitor production of Alzheimer's betaamyloid in transfected cells. *J. Pept. Sci.*, **6**, 459–469 (2000).

103. Lichtenthaler, S.F., Beher, D., Grimm, H.S., Wang, R., Shearman, M.S., Masters, C.L. and Beyreuther, K., The intramembrane cleavage site of the amyloid precursor protein depends on the length of its transmembrane domain. *Proc. Natl Acad. Sci. USA*, **99**, 1365–1370 (2002).
104. Diamond, D.L., Kimball, J.R., Krisanaprakornkit, S., Ganz, T. and Dale, B.A., Detection of beta-defensins secreted by human oral epithelial cells. *J. Immunol. Methods*, **256**, 65–76 (2001).
105. Tassi, E., Al-Attar, A., Aigner, A., Swift, M.R., McDonnell, K., Karavanov, A. and Wellstein, A., Enhancement of fibroblast growth factor (FGF) activity by an FGF-binding protein. *J. Biol. Chem.*, **276**, 40247–40253 (2001).
106. Schenone, M.M., Warder, S.E., Martin, J.A., Prorok, M. and Castellino, F.J., An internal histidine residue from the bacterial surface protein, PAM, mediates its binding to the kringle-2 domain of human plasminogen. *J. Pept. Res.*, **56**, 438–445 (2000).
107. Stoica, G.E., Kuo, A., Aigner, A., Sunitha, I., Souttou, B., Malerczyk, C., Caughey, D.J., Wen, D., Karavanov, A., Riegel, A.T. and Wellstein, A., Identification of anaplastic lymphoma kinase as a receptor for the growth factor pleiotrophin. *J. Biol. Chem.*, **276**, 16772–16779 (2001).
108. Amaar, Y.G., Thompson, G.R., Linkhart, T.A., Chen, S.T., Baylink, D.J. and Mohan, S., Insulin-like growth factor-binding protein 5 (IGFBP-5) interacts with a four and a half LIM protein 2 (FHL2). *J. Biol. Chem.*, **277**, 12053–12060 (2002).
109. Steyn, A.J., Collins, D.M., Hondalus, M.K., Jacobs, W.R. Jr., Kawakami, R.P. and Bloom, B.R., *Mycobacterium tuberculosis* WhiB3 interacts with RpoV to affect host survival but is dispensable for in vivo growth. *Proc. Natl Acad. Sci. USA*, **99**, 3147–3152 (2002).
110. Adilakshmi, T. and Laine, R.O., Ribosomal protein S25 mRNA partners with MTF-1 and La to provide a p53-mediated mechanism for survival or death. *J. Biol. Chem.*, **277**, 4147–4151 (2002).
111. Forde, C.E., Gonzales, A.D., Smessaert, J.M., Murphy, G.A., Shields, S.J., Fitch, J.P. and McCutchen-Maloney, S.L., A rapid method to capture and screen for transcription factors by SELDI mass spectrometry. *Biochem. Biophys. Res. Commun.*, **290**, 1328–1335 (2002).
112. Issaq, H.J., Veenstra, T.D., Conrads, T.P. and Felschow, D., The SELDI-TOF MS approach to proteomics: protein profiling and biomarker identification. *Biochem. Biophys. Res. Commun.*, **292**, 587–592 (2002).
113. Spencer, D.I., Robson, L., Purdy, D., Whitelegg, N.R., Michael, N.P., Bhatia, J., Sharma, S.K., Rees, A.R., Minton, N.P., Begentm, R.H. and Chester, K.A., A strategy for mapping and neutralizing conformational immunogenic sites on protein therapeutics. *Proteomics*, **2**, 271–279 (2002).
114. Walter, G., Bussov, K., Cagill, D., Lueking, A. and Lerach, H., Protein arrays for gene expression and molecular interaction screening. *Curr. Opin. Biotechnol.*, **3**, 298–302 (2000).
115. Cahill, D.J., Protein arrays: a high-throughput solution for proteome research? *Proteomics: A Trends Guide*, 47–51 (2000).
116. Emili, A.Q. and Cagney, G., Large-scale functional protein analysis using peptide or protein arrays. *Nat. Biotechnol.*, **18**, 393–397 (2000).
117. Zhu, H. and Snyder, M., Protein chip technology. *Curr. Opin. Chem. Biol.*, **7**, 55–63 (2003).
118. Templin, M.F., Stoll, D., Schrenk, M., Traub, P.C., Vohringer, C.F. and Joos, T.O., Protein microarray technology. *Trends Biotechnol.*, **20**, 160–166 (2002).
119. Uetz, P., Giot, L., Cagney, G., Mansfield, T.A., Judson, R.S., Knight, J.R., Lockshon, D., Narayan, V., Srinivasan, M., Pochart, P., Qureshi-Emili, A., Li, Y., Godwin, B., Conover, D., Kalbfleisch, T., Vijayadamodar, G., Yang, M., Johnston, M., Fields, S. and Rothberg, J.M.,

A comprehensive analysis of protein–protein interactions in *Saccharomyces cerevisiae*. *Nature*, **403**, 623–627 (2000).

120. MacBeath, G. and Schreiber, S.L., Printing proteins as microarrays for high-throughput function determination. *Science*, **289**, 1760–1763 (2000).
121. Kodakek, T., Protein microarrays: prospect and problems. *Chem. Biol.*, **8**, 105–115 (2001).
122. Macbeath, G., Proteomics come to the surface. *Science*, **19**, 828–829 (2001).
123. Borrebaek, C.A.K., Antibodies in diagnostics — from immunoassays to protein chips. *Immunol. Today*, **8**, 379–382 (2000).
124. Goldmann, R.D., Antibodies: indispensable tools for biomedical research. *Trends Biochem. Sci.*, **25**, 593–595 (2000).
125. Gunneriusson, E., Nord, K., Uhlén, M. and Nygren, P.A., Affinity maturation of a Taq DNA polymerase specific affibody by helix shuffling. *Protein Engng*, **12**, 873–878 (1999).
126. Gunneriusson, E., Samuelson, P., Ringdahl, J., Grönlund, H., Nygren, P.A. and Stahl, S., Staphylococcal surface-display of immunoglobulin A (IgA) and IgE specific in vitro-selected binding proteins (affibodies) based on *Staphylococcus aureus* protein A. *Appl. Environ. Microbiol.*, **65**, 4134–4140 (1999).
127. Brody, E.N., Willis, M.C., Smith, J.D., Jayasena, S., Zichi, D. and Gold, L., The use of aptamers in large arrays for molecular diagnostics. *Mol. Diagn.*, **4**, 381–388 (1999).
128. Green, L.S., Bell, C. and Janjic, N., Aptamers as reagents for high-throughput screening. *Biotechniques*, **30**, 1094–1096 (2001).
129. Houseman, B.T., Huh, J.H., Kron, S.J. and Mrksich, M., Peptide chips for the quantitative evaluation of protein kinase activity. *Nat. Biotechnol.*, **20**, 270–274 (2002).
130. Haupt, K. and Mosbach, K., Plastic antibodies: developments and application. *Trends Biotechnol.*, **16**, 468–479 (1998).
131. Vlatakis, G., Andersson, L.I., Muller, R. and Mosbach, K., Drug assay using antibody mimics made by molecular imprinting. *Nature*, **361**, 645–647 (1993).
132. Mendoza, L.G., McQuary, P., Mongan, A., Gangadharan, R., Brignac, S. and Eggers, M., High-throughput microarray-based enzyme-linked immunosorbent assay (ELISA). *BioTechniques*, **27**, 778–788 (1999).
133. Rowe, C.A., Tender, L.M., Feldstein, M.J., Golden, J.P., Scruggs, S.B., Macgraith, B.D., Cras, J.J. and Ligler, F.S., Array biosensor for simultaneous identification of bacterial, viral and protein analyte. *Anal. Chem.*, **71**, 3846–3852 (1999).
134. Arenkov, P., Kukhtin, A., Gemmell, A., Voloshchuk, S., Chupeeva, V. and Mirzabekov, A., Protein microchips: use for immunoassay and enzymatic reactions. *Anal. Biochem.*, **278**, 123–131 (2000).
135. Wiese, R., Belosludtev, Y., Powdrill, T., Thompson, P. and Hogan, M., Simultaneous multianalyte ELISA on microarray platform. *Clin. Chem.*, **47**, 1451–1457 (2001).
136. Schweitzer, B., Roberts, S., Grimwade, B., Shao, W., Wang, M., Fu, Q., Shu, Q., Laroche, I., Zhou, Z., Tchernev, V.T., Christiansen, J., Velleca, M. and Kingsmore, F., Multipexed protein profiling on microassays by rolling-circle amplification. *Proc. Natl. Acad. Sci. USA*, **97**, 10113–10119 (2000).
137. Tam, S.W., Wiese, R., Lee, S., Gilmore, J. and Kumble, K.D., Simultaneous analysis of eight human Th1/Th2 cytokines using microarrays. *J. Immunol. Methods*, **161**, 157–165 (2000).
138. Haab, B.B., Dunham, M.J. and Brown, P.O., Protein microarrays for highly parallel detection and quantification of proteins and antibodies in complex solutions. *Genome Biol.*, **2**, 1–13 (2001).
139. Robinson, W.H., DiGennaro, C., Hueber, W., Haab, B.B., Kamachi, M., Dean, E.J., Fournel, S., Fong, D., Genovese, M.C., de Vegvar, H.E., Skriner, K., Hirschberg, D.L., Morris, R.I., Muller, S., Pruijn, G.J., van Venrooij, W.J., Smolen, J.S., Brown, P.O., Steinman, L. and

Utz, P.J., Autoantigen microarrays for multiplex characterization of autoantibody responses. *Nature Med.*, **8**, 295–301 (2002).

140. Silzel, J.W., Cercek, B., Dodson, C., Tsai, T. and Obremski, R.J., Mass-sensing, multianalyte microarray immunoassay with imaging detection. *Clin. Chem.*, **44**, 2636–2643 (1998).
141. Mooney, J.F., Hunt, A.J., McIntosh, J.R., Liberko, C.A., Walba, D.M. and Rogers, C.T., Patterning of functional antibodies and other proteins by photolithography of silane monolayers. *Proc. Natl Acad. Sci. USA*, **93**, 12287–12291 (1996).
142. Jones, W.W., Kenseth, J.R., Porte, M.D., Mosher, C.L. and Henderson, E., Microminiaturized immunoassays using atomic force microscopy and compositionally patterned antigen arrays. *Anal. Chem.*, **70**, 1233–1241 (1998).
143. Michael, K.L., Taylor, L.C., Schultz, S.L. and Walt, D.R., Randomly ordered addressable high-density optical sensor arrays. *Anal. Chem.*, **70**, 1242–1248 (1998).
144. Martin, B.D., Gaber, B.P., Patterson, C.H. and Turner, D.C., Direct protein microarray fabrication using an hydrogel stamper. *Langmuir*, **14**, 3971–3975 (1998).
145. Lueking, A., Horn, M., Eickhoff, H., Bussov, K., Lehrach, H. and Walter, G., Protein microarrays for gene expression and antibody screening. *Anal. Biochem.*, **170**, 103–111 (1999).
146. Tuerk, C. and Gold, L., Systematic evolution of ligands by exponential enrichment: RNA ligands to bacteriophage T4 DNA polymerase. *Science*, **249**, 505–550 (1990).
147. Body, E.N. and Gold, L., Aptamers as therapeutic and diagnostics agents. *J. Biotechnol.*, **74**, 5–13 (2000).
148. Geysen, H.M., Meloen, R.H. and Barteling, S.J., Use of peptide synthesis to probe viral antigen for epitopes to a resolution of a single amino acid. *Proc. Natl Acad. USA*, **81**, 3998–4002 (1984).
149. Kramer, A. and Schneider-Mergener, J., Synthesis and screening of peptide libraries on single resins and continuous cellulose membrane supports. *Methods Mol. Biol.*, **87**, 25–39 (1998).
150. Kramer, A., Reineke, U., Dong, L., Hoffmann, B., Hoffmuller, U., Winkler, D., Volkmer-Engert, R. and Schneider-Mergener, J., Spot synthesis: observations and optimizations. *J. Pept. Res.*, **54**, 319–327 (1999).
151. Wenschuh, H., Volkmer-Engert, R., Schmidt, M., Schulz, M., Schneider-Mergener, J. and Reineke, U., Coherent membrane supports for parallel microsynthesis and screening of bioactive peptides. *Biopolymers*, **55**, 188–206 (2000).
152. Kramer, A., Volkmer-Engert, R., Malin, R., Reinekeand, U. and Schneider-Mergener, J., Simultaneous examples for the identification of protein, metal and DNA binding peptide mixtures. Synthesis of peptide libraries on single resin and continuous cellulose membrane supports: examples for the identification of protein, metal and DNA binding peptide mixtures. *Pept. Res.*, **6**, 314–319 (1993).
153. Reineke, U., Volkmer-Engert, R. and Scheider-Mergener, J., Applications of peptide arrays prepared by the spot technology. *Curr. Opin. Biotechnol.*, **12**, 59–64 (2001).
154. deWildt, R.M.T., Mundy, C.R., Gorick, B.D. and Tomlinson, I.M., Antibody arrays for high-throughput screening of antibody–antigen interactions. *Nat. Biotechnol.*, **18**, 989–994 (2000).
155. Bussov, K., Cahill, D., Nietfeld, W., Banvroft, D., Scherzinger, E., Lehrach, H. and Walter, G., A method for global protein expression and antibody screening on high-density filter of an array cDNA library. *Nucleic Acids Res.*, **26**, 5007–5008 (1998).
156. Kononen, J., Bubendorf, L., Kallioniemi, A., Barlund, M., Schraml, P., Leighton, S., Torhorst, J., Mihatsch, M.J., Sauter, G. and Kallioniemi, O.P., Tissue microarrays for high-throughput molecular profiling of tumor specimen. *Nat. Med.*, **4**, 844–847 (1998).

157. Barlund, M., Forozan, F., Kononen, J., Bubendorf, L., Chen, Y., Bittner, M.L., Torhorst, J., Haas, P., Bucher, C., Sauter, G., Kallioniemi, O.P. and Kallioniemi, A., Detecting activation of ribosomal protein S6 kinase by complementary DNA and tissue microarray analysis. *J. Clin. Oncol.*, **15**, 2417–2428 (2002).
158. Hoos, A., Tissue microarray profiling of cancer specimens and cell lines: opportunities and limitations. *Lab. Invest.*, **81**, 1331–1338 (2001).
159. Wang, H., Wang, H., Zhang, W. and Fuller, G.N., Tissue microarrays: applications in neuropathology research, diagnosis, and education. *Brain Pathol.*, **12**, 95–107 (2002).
160. Terris, B. and Bralet, M.P., Tissue microarrays, or the advent of chips in pathology. *Ann. Pathol.*, 69–72 (2002).
161. Hughes, K., Lucas, D., Stolowitz, M. and Wiley, J., Novel affinity tools for protein immobilisation. Implications for proteomics. *Am. Biotechnol. Lab.*, **1**, 36–38 (2001).
162. Stolowitz, M., Ahlem, C., Hughes, K., Kaiser, R., Kesicki, E., Li, G., Lund, K., Torkelson, S. and Wiley, J., Phenyldiboronic acid–salicylhydroxamic acid bioconjugates I: a novel boronic complex for protein immobilisation. *Bioconjug. Chem.*, **12**, 229–239 (2001).
163. Martzen, M.R., McCraith, S.M., Spinelli, S.L., Torres, F.M., Fields, S., Grayhack, E.J. and Phizicky, E.M., A biochemical genomics approach for identifying genes by the activity of their products. *Science*, **286**, 1153–1155 (1999).
164. Zhu, H., Klemic, J.F., Chang, S., Bertone, P., Casamayor, A., Klemic, K.C., Smith, D., Gerstein, M., Reed, M.A. and Snyder, M., Analysis of yeast protein kinases using protein chips. *Nat. Genet.*, **26**, 283–289 (2000).
165. Zhu, H., Bilgin, M., Bangham, R., Hall, D., Casamayor, A., Bertone, P., Lan, N., Jansen, R., Didlingmaier, S., Houfek, T., Mitchell, T., Miller, P., Dean, R.A., Gerstein, M. and Snyder, M., Global analysis of protein activities using proteome chips. *Science*, **293**, 2101–2105 (2001).
166. Liotta, L.A., Espina, V., Mehta, A.I., Calvert, V., Rosenblatt, K., Geho, D., Munson, P.J., Young, L., Wulfkuhle, J. and Petricoin, E.F., Protein microarrays: meeting analytical challenges for clinical applications. *Cancer Cell*, **3**, 317–325 (2003).
167. Service, R.F., Recruiting genes, proteins for a revolution in diagnostics. *Science*, **300**, 236–239 (2003).
168. Tyers, M. and Mann, M., From genomics to proteomics. *Nature*, **422**, 193–197 (2003).
169. Cutler, P., Protein arrays: the current state-of-the-art. *Proteomics*, **3**, 3–18 (2003).

Proteome Analysis. Interpreting the Genome.
D.W. Speicher (editor)

Chapter 11

Protein expression library resources for proteome studies

JOSHUA LABAER* and GERALD MARSISCHKY

Institute of Proteomics, Harvard Medical School, Department of Biological Chemistry and Molecular Pharmacology, 250 Longwood Avenue, Boston, MA 02115, USA

1. Introduction

With the completion of draft sequences of the human [1] and mouse genomes [2] and the completed genome sequences of several of the major research organisms [4–6], the stage is being set for the development of genome-scale tools that will be essential to the development of the next phase of biological experimentation.

* Corresponding author. Tel.: +1-617-324-0827; fax: +1-617-324-0824. E-mail: jlabaer@hms.harvard.edu (J. LaBaer).

A large number of genes have been identified in each of these projects, e.g. approximately 30,000 genes each of human and mouse genomes. Still, there is only an approximate understanding of their gene content (particularly for human and mouse) due to such factors as imperfect gene prediction software. This problem is being addressed by techniques such as mapping available cognate EST clone sequences to the genome sequences. An improved version of this approach involves mapping sequenced full-length cDNA clones (e.g. sequences from RefSeq [7]) instead. This approach will be most useful when the number of sequenced full-length cDNA clones available approach full representation of a given genome.

As large collections of full-length cDNA clones are assembled, a new kind of genome-scale tool is becoming evident: a reduced set of high-value, sequence-validated clones in an arrayed format that is highly representative of the complete gene content of an organism (Table 1).

1.1. Public full-length cDNA clone projects

There are several public projects worldwide that have the goal of creating representative full-length cDNA clone collections, notably the NIH Mammalian Gene Collection (MGC) project [8] and the Mouse Gene Encyclopedia Project sponsored by the RIKEN Institute in Japan [9–11]. These projects have taken the approach of sifting through large numbers of cDNA clones, e.g. from tissue-specific libraries, to identify unique full-length cDNA clones. There are, additionally, independent efforts to create ORF clone collections for use as expression tools. These include the FLEXGene effort led by the Institute of Proteomics at Harvard Medical School (HIP) [12], the *Caenorhabditis elegans* ORFeome Project [13], and the Ressourcenzentrum für Genomforschung (RZPD) ORF clone collection [14].

1.2. Cloning formats

These clone collections are available at this time as either cDNA clones or ORF clones. While cDNA clones are the most straightforward to create, they have important limitations as experimental tools. The full-length coding sequence of a cDNA clone cannot be expressed as a fusion protein because of the 5′ and 3′ untranslated sequences that surround it. ORF clones, while harder to create, allow for expression of the ORF as a fusion protein. However, use of this form entails the targeted selection of each ORF to be cloned, generation of the appropriate reagents for that clone, and tracking of the entire cloning process. For instance, ORF clones require PCR amplification of the coding sequence with ORF-specific primers,

Table 1
Full-length cDNA clone collections

	Organism	Expression clones	Recombinational cloning
Mammalian Gene Collection (MGS)	Human	No	No
http://mgc.nci.nih.gov/	Mouse	No	No
RIKEN	Mouse	No	No
http://genome.gsc.riken.go.jp/index.html/			
Harvard Institute of Proteomics (HIP)	Human	Yes	Yes
http://www.hip.harvard.edu/	Yeast	Yes	Yes
	Pseudomonas	Yes	Yes
Ressourcenzentrum für Genomforschung (RZPD)	Human	Yes	Yes
http://www.rzpd.de/			
Kazusa DNA Research Institute	Human	No	No
http://www.kazusa.or.jp/en/human/			
SSP *Arabidopsis* ORFeome Collection	*Arabidopsis*	Yes	Yes
http://signal.salk.edu/SSP/index.html			
C. elegans ORFeome Project	*C. elegans*	Yes	Yes
http://worfdb.dfci.harvard.edu			
Berkeley *Drosophila* Genome Project	*Drosophila*	No	No
http://www.fruitfly.org/DGC/index.html			
Yale Genome Analysis Center	Yeast	Yes	No
http://ygac.med.yale.edu/			
Yeast Resource Center	Yeast	Yes	No
(University of Washington)			
http://depts.washington.edu/yeastrc/			

which must be generated with information from a sequence source such as RefSeq. In addition, it is technically difficult to reliably amplify ORFs from templates such as first strand cDNA in a high-throughput setting. Nevertheless, these approaches are not mutually exclusive since it is relatively simple to create ORF clones by amplifying them from sequence-validated cDNA clones. In either case, cDNA clones are typically created in vectors that offer a single promoter to control their expression (or no promoter in the vector at all for some cDNA libraries), and so are only suitable for use in a narrow range of experiments.

Although it is clearly possible to tailor experiments that utilize cDNA or ORF clones in a single format, a flexible cloning format that allows controlled expression in different experimental contexts would be preferable. This is possible if the ORF clones are created using recombinational cloning methods. With this approach, an ORF clone resource can be completely sequence validated and then used to create many different validated expression libraries. The availability of highly efficient systems from commercial suppliers, such as the Gateway cloning system from Invitrogen and the Creator cloning system from Clontech, as well as the Univector cloning system developed by the lab of Stephen Elledge [15], makes practical the use of recombinational cloning in a high-throughput setting.

1.3. Site-specific recombination-based cloning systems

Each of the cloning systems described in Section 1.2 makes use of site-specific DNA recombination. The Gateway system uses Phage Lambda att site-mediated recombination to transfer DNA between plasmids [13], while the Creator and the Univector systems use Cre-Lox recombination [15,16]. For each of these systems, a master clone is created from a PCR product with recombination sites at its ends in an in vitro recombination reaction with a donor vector (see Fig. 1a). Similarly, after validation of the cloned ORF, transfer from the master clone is also accomplished by site-specific recombination *in vitro* with an expression vector appropriately modified with recombination sites (see Fig. 1b).

2. Clone collections

2.1. Human

2.1.1. MGC

The NIH Mammalian Gene Collection (see http://mgc.nci.nih.gov) was established with the goal of identifying and sequencing at least one full-length cDNA clone for each gene in the human and mouse genomes and to make these clones available to researchers without restriction [8].

In order to accomplish this, more than 100 cDNA libraries were created using a variety of methods from a diverse set of tissues and cell lines. Libraries with a high fraction of full-length sequences, as determined from end sequencing of a limited number of clones, were screened by $5'$ end sequencing. To identify unique full-length ORF clones, the $5'$ end-read sequences were screened by several methods: (i) comparison with RefSeq and other sequence databases (Protein Information Resource, Protein Database, and SwissProt); (ii) comparison with predicted genes from the human genome sequence using GENOMESCAN [17]; and (iii) use of a new program (HSCAN) that identifies the transition from non-coding to coding sequence. The full inserts of candidate clones were then sequenced by a variety of methods to yield high-quality sequence, with an average error rate of <1 per 50,000 bp.

This effort has resulted in a sequenced, non-redundant set of full-length clones representing 9530 human and 6368 mouse genes. Candidate full-length clones for an additional 7800 human and 3500 mouse genes have also been identified.

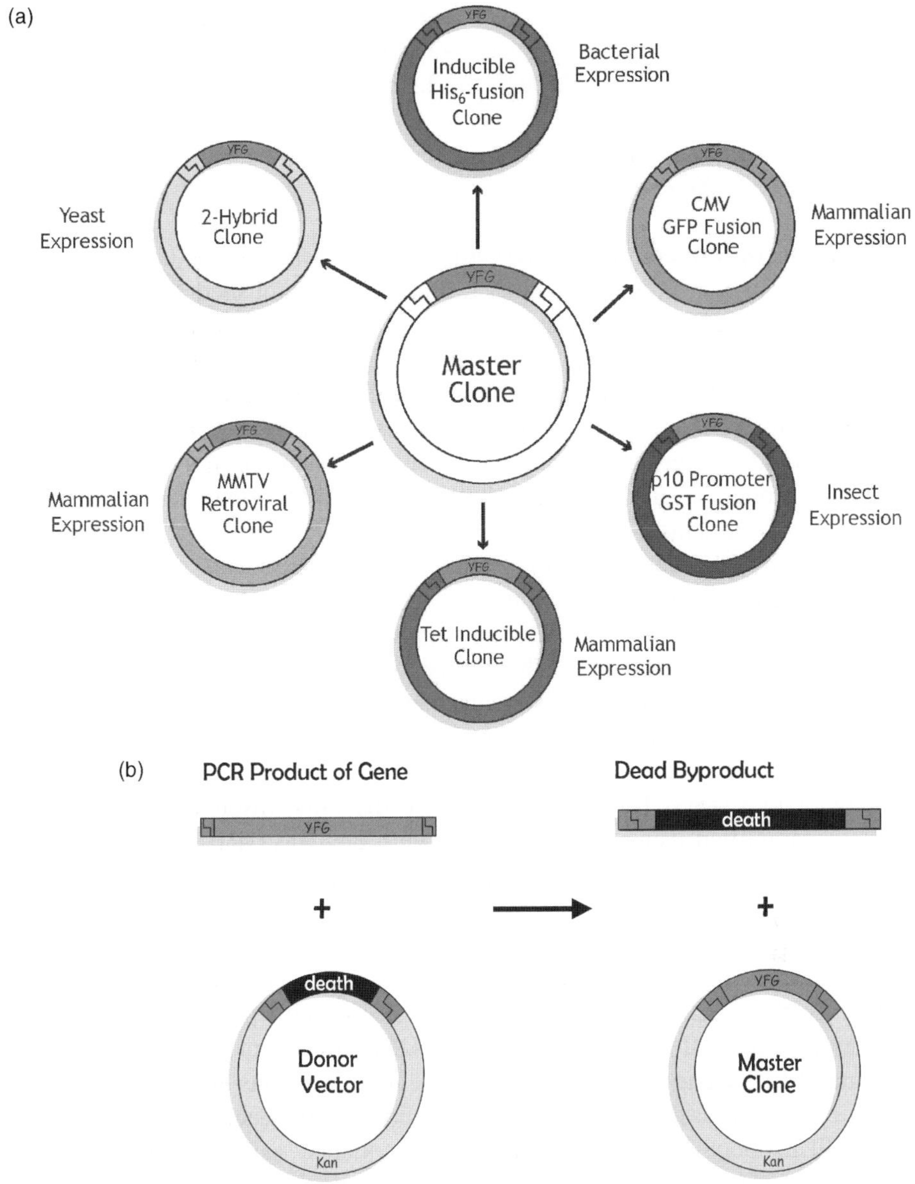

Fig. 1. (a) Creation of master clones using site-specific recombination. The product of an in vitro recombination reaction which includes the recombination site-modified ORF PCR product, a donor plasmid, and a recombinase is transformed into *E. coli*. Only correctly recombined master clones are viable on selective plates. (b) Transfer of ORF sequences to expression vectors using site-specific recombination. ORF DNA is transferred from master clones by in vitro recombination with an expression vector appropriately modified with a recombination site/counter selection cassette. Only *E. coli* transformants of correctly recombined clones are viable on selective medium. Recombinational cloning allows the creation of multiple expression libraries from a master clone collection.

The MGC clones include ~1300 human and ~1100 mouse ORFs for which there was no full-length sequence previously available [18].

MGC clones are available through the I.M.A.G.E. Consortium (http://image.llnl.gov/) and its distributors.

2.1.2. Human FLEXGene

The Institute of Proteomics at Harvard Medical School (see http://www.hip.harvard.edu/), which is led by Joshua LaBaer, was established with the goal of creating the FLEXGene (full-length expression) Resource: a complete collection of sequence-validated clones in a recombinational cloning format and representing full-length ORFs for all the human genes, as well as comprehensive collections of ORF clones for selected major model organisms and important human pathogens [12]. A fundamental principle underlying the FLEXGene initiative is that all the full-length clones in the Resource will be available to all scientists in the academic, governmental, and commercial sectors.

The Human FLEXGene ORF clones have been created using recombinational cloning systems (Creator from Clontech and Gateway from Invitrogen), and as such they can be transferred into any expression vector modified with the appropriate recombination sites. The ability to reutilize sequence-validated ORF clones in many expression formats means that the Human FLEXGene ORF clone Resource will be an invaluable tool for the entire research community.

The PCR amplification of ORFs to create Human FLEXGene clones is done in either of two ways. In the first approach, the ORF inserts of human MGC clones are amplified using low-cycle PCR with cognate ORF primers such that the amplified coding sequence is flanked with recombination sites. In the second approach, first strand cDNA is prepared from normal tissue (typically brain and placenta) and used for PCR amplification of ORFs with primers designed using the RefSeq sequence for the targeted ORF. In either case, the amplified PCR product is then gel purified and captured using either the Creator or the Gateway cloning system to create a master ORF clone. For each ORF, versions of Human FLEXGene clones with and without stop codons are created to allow the option of expression with C-terminal affinity tags. Each ORF clone in the Human FLEXGene Resource is then fully sequence validated. The observed frequency of sequence discrepancy with respect to the target sequence is <1 per 5000 bp.

At this point, the HIP Human FLEXGene cloning effort has produced full-length ORF clones representing ~5000 genes with ~2500 fully sequenced, and will eventually have FLEXGene clones that encompass the complete human MGC. A particular focus of HIP at the moment is the creation of sets of FLEXGene clones representing interesting gene classes. For example, HIP has assembled a collection of ~230 fully sequenced Human FLEXGene clones for protein kinases. HIP is also nearing completion of a set of Human FLEXGene clones for 1000 genes related to breast cancer. Data regarding the status and availability of Human FLEXGene clones is available through the HIP FLEXGene Database (http://www.hip.harvard.edu/). Fully sequenced Human FLEXGene clones for ~1100 genes are available from Open Biosystems (Huntsville, AL).

2.1.3. Ressourcenzentrum für Genomforschung

The Ressourcenzentrum für Genomforschung in Germany (see http://www.rzpd.de/) is establishing a large collection of mammalian full-length ORF clones. The collection thus far consists mainly of ORFs representing human genes. The RZPD ORF clones were made using vectors that include recombinational cloning sites flanking the cloned ORF and thus allow the transfer of the ORFs to an appropriately modified expression vector. These clones were made, for the most part, using the Gateway (Invitrogen) recombinational cloning system, and have been fully sequence validated and annotated for function. There are ~2000 full-length human ORF clones in this collection, some have also been expression verified. These clones are thus 'expression-ready' and are useful in a wide range of experimental formats [14]. The RZPD ORF expression clones are available from the Ressourcenzentrum für Genomforschung (http://www.rzpd.de/, contact orf@rzpd.de/).

2.1.4. Kazusa DNA Research Institute

The cDNA cloning effort at the Kazusa DNA Research Institute in Japan (http://www.kazusa.or.jp/en/human/) is distinct from other cloning efforts in that they are focusing on the identification of clones with large cDNA inserts (>4 kb). The Kazusa collection is thus an essential addition to the other human clone resources, which are typically under-represented with respect to large cDNA clones. Early cloning work at Kazusa DNA Research Institute was directed toward the myeloid cell line KG-1 and current cloning is centered on genes expressed in brain tissue. From these sources, they have successfully cloned and determined the complete sequences of approximately 2000 previously undiscovered cDNA species [19]. The average insert size of these novel cDNA clones is 5 kb.

The data obtained through the Kazusa cDNA project are accessible at the HUGE Database (Human Unidentified Gene-Encoded protein database, http://www.kazusa.or.jp/huge). Future efforts will be directed towards development of more comprehensive long cDNA libraries and the discovery and functional analysis of more new genes. Instructions for obtaining the Kazusa cDNA clones are available at http://www.kazusa.or.jp/huge/clone.req/index.html.

2.2. Mouse

2.2.1. MGC

As described in Section 2.1.1, the NIH Mammalian Gene Collection (see http://mgc.nci.nih.gov) effort is creating sequenced and annotated full-length mouse cDNA clones, and making them available to the research community through the I.M.A.G.E. Consortium (http://image.llnl.gov/).

2.2.2. RIKEN

The Japanese Institute of Physical and Chemical Research (RIKEN; see http://genome.gsc.riken.go.jp/index.html) has undertaken the Mouse Gene Encyclopedia Project, which they describe as a systematic approach to determine the full coding potential of the mouse genome with the goal of greatly facilitating mouse functional genomics. This work is being carried out by the RIKEN Genome Exploration Research (GER) group and involves the collection and sequencing of full-length cDNAs and the physical mapping of the corresponding genes to the mouse genome.

The RIKEN GER group has utilized a number of techniques to obtain unique full-length cDNAs with high frequency. In order to increase the representation of unique cDNAs in the collection, they applied normalization and subtractive hybridization methods to the cDNA libraries, and have prepared and characterized cDNA libraries from approximately 200 tissues and cell types. The clones contain sequence tags next to the $3'$ cloning site which identify the tissue that the cDNA clone originated from.

Clones from each of the cDNA libraries were sequenced by end reads, which were assembled into clusters representing different genes. The gene-specific clusters typically contained clones representing different splice isoforms. They next identified clones that had intact $5'$ ends and these clones were full-length sequenced. Finally, RIKEN organized an international functional annotation meeting (FANTOM, Functional Annotation of Mouse) and this body chose the annotation vocabularies to implement gene ontology (GO) terms.

The initial cloning and sequence analysis effort [20] resulted in the identification of 21,076 full-length cDNAs representing approximately 13,000 unique genes. The RIKEN GER group has continued to identify mouse full-length cDNAs, and in the latest edition of the FANTOM collection (FANTOM2), there are a total of 60,770 distinct full-length cDNA clones representing 21,744 unique protein-encoding genes in addition to a surprisingly large number of cDNA clones (11,665) which do not have protein coding potential [9–11]. The large fraction of non-coding cDNA clones observed is taken by the authors to indicate that non-coding RNA (ncRNA) transcripts comprise a substantial fraction of all transcription units. Some of these novel RNA species are likely to be of a regulatory nature, e.g. ~2400 sense/antisense pairs were found [9,10]. A recent analysis [11] of the FANTOM2 collection has also shown that there are 4280 transcription units that show no homology to transcription units known to encode protein sequence, which can be mapped to the mouse genome and are not within 10 kb of a known coding exon, and for which no CDS prediction could be made by GENSCAN. The mouse FANTOM2 collection thus may also be a useful source of clones of regulatory transcripts. Data regarding the FANTOM clone sets are available in the FANTOM and the FANTOM2 databases (see http://www.gsc.riken.go.jp/e/FANTOM/ and http://fantom2.gsc.riken.go.jp/db/, respectively).

K.K. Dnaform (Japan) is distributing FANTOM clones and clone sets under the terms and conditions of a sublicensing agreement from RIKEN (in the USA, contact fantom2-clone@postman.riken.go.jp; outside the USA, contact fantom@dnaform.co.jp).

2.3. Caenorhabditis elegans

2.3.1. C. elegans ORFeome Project

The *C. elegans* ORFeome Project is a collaborative effort that includes the laboratory of Marc Vidal (Dana Farber Cancer Institute, Boston), Invitrogen (primers), and Genome Therapeutics Corp. (sequencing effort). It is the goal of the *C. elegans* ORFeome Project to create and sequence validate ORF clones for each of the ~19,000 genes identified from the completed DNA sequence of the *C. elegans* genome [5].

Since these ORF clones have been created using the Gateway recombinational cloning system (Invitrogen) and can be combined with any expression vector modified with Gateway recombination sites, the *C. elegans* ORFs cloned by this effort will be invaluable tools for the *C. elegans* research community. They have already been used by the Vidal lab to create RNAi and 2-hybrid libraries for use in their own studies [21–23].

Individual ORF clones were constructed by PCR amplification with ORF-specific primers from pooled cDNA libraries made from different developmental stages of *C. elegans*. The ORF-specific primers used were designed using software developed by the *C. elegans* Sequencing Consortium (GeneFinder). These primers lack both ATG and Stop codons (to allow expression of both N- and C-terminal fusions), and include tails that result in the addition of Gateway recombination sites to the ends of the ORF-specific PCR products. The resulting PCR products were then used without gel purification to create Gateway Entry clones (master clones), which were kept as Entry clone transformant pools rather than as colony-purified Entry clones. Each Entry clone pool was then PCR amplified to check the insert size, and plasmid DNA from each Entry clone pool was sequence validated by two end reads. The key advantage of pooled approach is that it preserves any alternative splice forms within the Entry clone pool for a given ORF — the pools can be transferred directly to an expression vector and used in experiments. Clones for the alternative splice forms can be colony purified at a later time.

The cloning of the *C. elegans* ORFeome is now at an advanced stage, and the Worm ORFeome Collection has been declared to be at version 1.1 [24]. Of the ~19,000 ORFs for which PCRs were attempted in this effort, ~12,000 have been successfully cloned as Entry plasmid pools. Of these, ~11,000 are in-frame and ready for expression as fusion proteins. All 12,000 ORF clones are suitable for RNAi experiments. ORFeome 1.1 includes clones for ~4300 ORFs for which there were no EST data available. A future version of the *C. elegans* ORFeome Collection, ORFeome 2.1, will consist of wild-type colony-purified clones.

This data is available via the Worm ORFeome Database (WorfDB; http://worfdb.dfci.harvard.edu). The *C. elegans* ORFeome 1.1 clones are available from Open Biosystems (Huntsville, AL).

2.4. Drosophila melanogaster

2.4.1. Drosophila Gene Collection

The Berkeley *Drosophila* Genome Project is a consortium of the *Drosophila* Genome Center, funded by the National Human Genome Research Institute, National Cancer Institute, and Howard Hughes Medical Institute, through its support of work in the Gerald Rubin and Allan Spradling laboratories. The *Drosophila* Gene Collection (DGC, see http://www.fruitfly.org/DGC/index.html) is an effort by The Berkeley *Drosophila* Genome Project to identify, sequence, and annotate full-length cDNAs from within The Berkeley *Drosophila* Genome Project *Drosophila* EST clone collection.

From a total of ~250,000 EST clones from different EST libraries, analysis of end-read sequences were used to identify unique ORF clones by alignment with the annotated *D. melanogaster* genome sequence. At this point, full-insert sequencing of cDNA clones has resulted in the identification of 10,910 non-redundant full-length cDNAs or 70% of the predicted *D. melanogaster* genes [25, 26]. A small number of these have been converted to Gateway clones.

The Berkeley *Drosophila* Genome Project will make these clones available to the *Drosophila* research community (see http://www.fruitfly.org/DGC/index.html).

2.5. Saccharomyces cerevisiae

2.5.1. Fields 2-hybrid library

The laboratory of Stan Fields (University of Washington) created the first arrayed ORF expression library representing the complete coding capacity of an organism. It was used for the detection of protein–protein interactions using the 2-hybrid system [27], resulting in the identification of the first systematic protein interaction network for any organism.

The Fields 2-hybrid library was constructed using PCR products amplified from yeast genomic DNA using ORF-specific primer pairs obtained from Research Genetics. These primers were designed for each of the ~6000 ORFs identified in the *S. cerevisiae* genome sequence. The resulting PCR products were then cloned into the 2-hybrid expression plasmids using yeast gap repair recombination. This approach had an important advantage: plasmids for the 2-hybrid screen can be kept in opposite mating types until they are combined for interaction assay by mating. This library has two important limitations, however. First, the ORF clones were constructed in a single-use 2-hybrid expression vector. Second, the library was not systematically sequence validated.

The Fields 2-hybrid library is available from the National Center for Research Resources' Yeast Resource Center at the University of Washington (http://depts.washington.edu/yeastrc/).

2.5.2. Phizicky GST-fusion library

While the Fields laboratory 2-hybrid library was under construction, the laboratory of Eric Phizicky (University of Rochester) also created an yeast expression library using the Research Genetics ORF-specific primer pairs (with primers borrowed from the Fields laboratory). The Phizicky expression library was created to make possible biochemical screens for novel enzyme activities (in this case, activities involved in tRNA processing) using GST-fusion proteins that had been affinity purified from yeast [28].

As with the Fields laboratory effort, the ORF-specific PCR products were cloned into expression vectors by gap repair recombination in yeast. This library thus has the same two important limitations as the Fields 2-hybrid library: the ORF clones were constructed in a single-use expression vector and the clones received little sequence validation.

The Phizicky GST-fusion library is available as the EXClones collection from ResGen/Invitrogen (Carlsbad, CA).

2.5.3. Snyder His6-GST-fusion library

The laboratory of Michael Snyder (Yale) later constructed an yeast-arrayed ORF expression library similar to the Phizicky library [29]. The Snyder library was created for expression of affinity-tagged proteins in yeast for use in studies involving protein microarrays.

As with the Fields and Phizicky ORF libraries, the Snyder library ORFs were amplified from *S. cerevisiae* genomic DNA using ORF-specific primers and cloned into an yeast expression plasmid using gap repair recombination in yeast. In this case, the ORFs were cloned in an expression plasmid so that the cloned ORFs are expressed as N-terminal His6-GST-fusions. The Snyder library, however, was validated by single-pass end reads across the fusion junction. This allowed recloning of ORFs when defective clones were discovered. The Snyder library was also validated by determining protein levels after purification using immunodetection of the GST fusions on a protein microarray.

The Snyder His6-GST-fusion library is available from the Yale Genome Analysis Center (see http://ygac.med.yale.edu/).

2.5.4. HIP YFLEX ORF clone resource

The YFLEX project is the first major pilot project at the Institute of Proteomics at Harvard Medical School (http://www.hip.harvard.edu/), and is designed to produce a complete set of sequence-verified FLEX ORF clones from *S. cerevisiae* [12]. The HIP team, which is led by Gerald Marsischky, has been joined in this effort by the labs of Richard Kolodner (UC San Diego Medical School) and Andrew Simpson (Ludwig Institute for Cancer Research, Sao Paolo, Brazil).

Like the other yeast ORF expression libraries, the YFLEX clones were amplified from *S. cerevisiae* (S288C) genomic DNA using the Research Genetics ORF-specific primers. In this case, the ORF-specific PCR products were reamplified to add Gateway recombination sites, gel purified, and then standard Gateway cloning methods were used to create Gateway Entry clones for each ORF. Multiple isolates of each ORF were obtained, and each of these was checked by end-read sequencing to identify the best candidate clone for each ORF. At this point, there are 4200 ORFs for which clones have been validated by end-read sequencing. The cloning and sequencing of the remaining ORFs is expected to be completed by December of 2003. Another version of the YFLEX collection is planned that lacks native stop codons (which are in all other clones created with the Research Genetics ORF primers including the Fields, Phizicky, and Snyder collections described earlier), to allow for expression of the ORFs with C-terminal fusions.

Since the YFLEX ORF clones have been created using the Gateway recombinational cloning system (Invitrogen) and can be combined with any expression vector modified with Gateway recombination sites, the YFLEX clones will be invaluable tools for the yeast research community. They have already been used by the yeast group at HIP to create clones that allow galactose-inducible expression of native protein in yeast from all the YFLEX clones created to date.

The YFLEX clones will be made available to all scientists from the academic, government, and commercial sectors. Researchers interested in the YFLEX clones should contact the Institute of Proteomics (http://www.hip.harvard.edu) for information regarding their distribution.

2.6. Pseudomonas aeruginosa

2.6.1. HIP P. aeruginosa FLEX clone resource

The *Pseudomonas* FLEX project is a collaborative effort between the Institute of Proteomics at Harvard Medical School (http://www.hip.harvard.edu/, a project led by Leonardo Brizuela), led by Leonardo Brizuela, and the laboratory of Steve Lory at Harvard Medical School with funding from the Cystic Fibrosis Foundation. The goal of this effort is to produce a complete set of sequence-verified FLEX ORF clones from *P. aeruginosa.*

The *Pseudomonas* FLEX clones were amplified from *P. aeruginosa PA01* genomic DNA using ORF-specific primers designed for each of the 5070 ORFs identified in the *Pseudomonas* genome sequence [30]. The ORF-specific PCR products were reamplified to add Gateway recombination sites, and then standard Gateway cloning methods were used to create Gateway Entry clones for each ORF. Multiple isolates of each ORF were obtained, and each of these will be checked by end-read sequencing to identify the best candidate clone for each ORF. Of the ORFs, 97% were amplified and captured in the first round of cloning. At this point, there are 5067 ORFs for which Gateway Entry cloning was successful and subsets of genes have been characterized by sequencing and expression in both *Escherichia coli* and *P. aeruginosa.*

Since the *Pseudomonas* FLEX ORF clones have been created using the Gateway recombinational cloning system (Invitrogen) and can be combined with any expression vector modified with Gateway recombination sites, these clones will be essential tools for the *Pseudomonas* research community. Researchers interested in the *Pseudomonas* FLEX ORF clones should contact the Institute of Proteomics (http://www.hip.harvard.edu) for information regarding their distribution.

2.7. Arabidopsis thaliana

2.7.1. The Arabidopsis gene/ORFeome Collection

The SSP Consortium (http://signal.salk.edu/SSP/index.html), composed of the *S*alk Institute Genome Analysis Lab, the *S*tanford DNA Sequence and Technology Center, and the Plant Gene Expression Center (Albany, CA), is currently creating an *Arabidopsis* ORFeome cDNA collection using the Univector cloning system [13]. Since SSP *Arabidopsis* ORFeome clones are created using a site-specific recombinational cloning system, they can be used for expression of fusion proteins in a variety of expression vectors by using in vitro Cre-lox recombination, and will be useful in many experimental formats.

In this effort, the SSP Consortium first identified unique full-length cDNA clones from different tissue-specific cDNA libraries by end sequencing. Clones in this non-redundant collection were then fully sequenced. The coding sequence of each of the full-length cDNA clones was then PCR amplified using ORF-specific primers, and used to create a master clone in the Univector system. The resulting Univector master clones were then fully sequenced. All sequences are deposited with Genbank, and the clones deposited to the *Arabidopsis* Biological Resource Center (ABRC; Ohio State University). There are currently 9930 SSP *Arabidopsis* ORFeome clones that have been deposited with the ABRC, representing 96% of the ~10,500 clones identified by cDNA clone sequencing.

The SSP *Arabidopsis* ORFeome clones can be obtained without restriction from the ABRC (http://www.biosci.ohio-state.edu/~plantbio/facilities/abrc/abrchome.htm).

3. Summary

An important secondary outcome of the human and other genome-sequencing projects is the ongoing development of sequenced clone libraries. The cloned sequences serve not only as a tool to assist in the assignment of gene boundaries and the identification of splice isoforms, but also serve as the starting point for a new kind of genome-scale tool: an arrayed, sequence-validated, full-length ORF clone collection that is highly representative of the gene content of an organism. These types of collections are under development for human, mouse, several of the important model organisms, and an important human pathogen, *P. aeruginosa*.

Acknowledgements

We wish to acknowledge Leonardo Brizuela for helpful comments and for sharing data regarding the *P. aeruginosa* FLEXGene project.

References

1. The International Human Genome Sequencing Consortium, Initial sequencing and analysis of the human genome. *Nature*, **409**, 860–921 (2001).
2. Venter, J.C., Adams, M.D., Myers, E.W., Li, P.W., Mural, R.J., Sutton, G.G., Smith, H.O., Yandell, M., Evans, C.A., Holt, R.A., Gocayne, J.D., Amanatides, P., Ballew, R.M., Huson, D.H., Wortman, J.R., Zhang, Q., Kodira, C.D., Zheng, X.H., Chen, L., Skupski, M., Subramanian, G., Thomas, P.D., Zhang, J., Gabor Miklos, G.L., Nelson, C., Broder, S., Clark, A.G., Nadeau, J., McKusick, V.A., Zinder, N., Levine, A.J., Roberts, R.J., Simon, M., Slayman, C., Hunkapiller, M., Bolanos, R., Delcher, A., Dew, I., Fasulo, D., Flanigan, M., Florea, L., Halpern, A., Hannenhalli, S., Kravitz, S., Levy, S., Mobarry, C., Reinert, K., Remington, K., Abu-Threideh, J., Beasley, E., Biddick, K., Bonazzi, V., Brandon, R., Cargill, M., Chandramouliswaran, I., Charlab, R., Chaturvedi, K., Deng, Z., Di Francesco, V., Dunn, P., Eilbeck, K., Evangelista, C., Gabrielian, A.E., Gan, W., Ge, W., Gong, F., Gu, Z., Guan, P., Heiman, T.J., Higgins, M.E., Ji, R.R., Ke, Z., Ketchum, K.A., Lai, Z., Lei, Y., Li, Z., Li, J., Liang, Y., Lin, X., Lu, F., Merkulov, G.V., Milshina, N., Moore, H.M., Naik, A.K., Narayan, V.A., Neelam, B., Nusskern, D., Rusch, D.B., Salzberg, S., Shao, W., Shue, B., Sun, J., Wang, Z., Wang, A., Wang, X., Wang, J., Wei, M., Wides, R., Xiao, C., Yan, C., Yao, A., Ye, J., Zhan, M., Zhang, W., Zhang, H., Zhao, Q., Zheng, L., Zhong, F., Zhong, W., Zhu, S., Zhao, S., Gilbert, D., Baumhueter, S., Spier, G., Carter, C., Cravchik, A., Woodage, T., Ali, F., An, H., Awe, A., Baldwin, D., Baden, H., Barnstead, M., Barrow, I., Beeson, K., Busam, D., Carver, A., Center, A., Cheng, M.L., Curry, L., Danaher, S., Davenport, L., Desilets, R., Dietz, S., Dodson, K., Doup, L., Ferriera, S., Garg, N., Gluecksmann, A., Hart, B., Haynes, J., Haynes, C., Heiner, C., Hladun, S., Hostin, D., Houck, J., Howland, T., Ibegwam, C., Johnson, J., Kalush, F., Kline, L., Koduru, S., Love, A., Mann, F., May, D., McCawley, S., McIntosh, T., McMullen, I., Moy, M., Moy, L., Murphy, B., Nelson, K., Pfannkoch, C., Pratts, E., Puri, V., Qureshi, H., Reardon, M., Rodriguez, R., Rogers, Y.H., Romblad, D., Ruhfel, B., Scott, R., Sitter, C., Smallwood, M., Stewart, E., Strong, R., Suh, E., Thomas, R., Tint, N.N., Tse, S., Vech, C., Wang, G., Wetter, J., Williams, S., Williams, M., Windsor, S., Winn-Deen, E., Wolfe, K., Zaveri, J., Zaveri, K., Abril, J.F., Guigo, R., Campbell, M.J., Sjolander, K.V., Karlak, B., Kejariwal, A., Mi, H., Lazareva, B., Hatton, T., Narechania, A., Diemer, K., Muruganujan, A., Guo, N., Sato, S., Bafna, V., Istrail, S., Lippert, R., Schwartz, R., Walenz, B., Yooseph, S., Allen, D., Basu, A., Baxendale, J., Blick, L., Caminha, M., Carnes-Stine, J., Caulk, P., Chiang, Y.H., Coyne, M., Dahlke, C., Mays, A., Dombroski, M., Donnelly, M., Ely, D., Esparham, S., Fosler, C., Gire, H., Glanowski, S., Glasser, K., Glodek, A., Gorokhov, M., Graham, K., Gropman, B., Harris, M., Heil, J., Henderson, S., Hoover, J., Jennings, D., Jordan, C., Jordan, J., Kasha, J., Kagan, L., Kraft, C., Levitsky, A., Lewis, M., Liu, X., Lopez, J., Ma, D., Majoros, W., McDaniel, J., Murphy, S., Newman, M., Nguyen, T., Nguyen, N., Nodell, M., Pan, S., Peck, J., Peterson, M., Rowe, W., Sanders, R., Scott, J., Simpson, M., Smith, T., Sprague, A., Stockwell, T., Turner, R., Venter, E., Wang, M., Wen, M., Wu, D., Wu, M., Xia, A., Zandieh, A. and Zhu, X., The sequence of the human genome. *Science*, **291**, 1304–1351 (2001).
3. Mouse Genome Sequencing Consortium, Initial sequencing and comparative analysis of the mouse genome. *Nature*, **420**, 520–562 (2002).
4. Goffeau, A., Barrell, B.G., Bussey, H., Davis, R.W., Dujon, B., Feldmann, H., Galibert, F., Hoheisel, J.D., Jacq, C., Johnston, M., Louis, E.J., Mewes, H.W., Murakami, Y., Philippsen, P., Tettelin, H. and Oliver, S.G., Life with 6000 genes. *Science*, **274**, 563–567 (1996).
5. The *C. elegans* Sequencing Consortium, Genome sequence of the nematode *C. elegans*: a platform for investigating biology. *Science*, **282**, 2012–2018 (1998).

6. Adams, M.D., Celniker, S.E., Holt, R.A., Evans, C.A., Gocayne, J.D., Amanatides, P.G., Scherer, S.E., Li, P.W., Hoskins, R.A., Galle, R.F., George, R.A., Lewis, S.E., Richards, S., Ashburner, M., Henderson, S.N., Sutton, G.G., Wortman, J.R., Yandell, M.D., Zhang, Q., Chen, L.X., Brandon, R.C., Rogers, Y.H., Blazej, R.G., Champe, M., Pfeiffer, B.D., Wan, K.H., Doyle, C., Baxter, E.G., Helt, G., Nelson, C.R., Gabor, G.L., Abril, J.F., Agbayani, A., An, H.J., Andrews-Pfannkoch, C., Baldwin, D., Ballew, R.M., Basu, A., Baxendale, J., Bayrakataroglu, L., Beasley, E.M., Beeson, K.Y., Benos, P.V., Berman, B.P., Bhandari, D., Bolshakov, S., Borkova, D., Botchan, M.R., Bouck, J., Brokstein, P., Brottier, P., Burtis, K.C., Busam, D.A., Butler, H., Cadieu, E., Center, A., Chandra, I., Cherry, J.M., Cawley, S., Dahlke, C., Davenport, L.B., Davies, P., de Pablos, B., Delcher, A., Deng, Z., Mays, A.D., Dew, I., Dietz, S.M., Dodson, K., Doup, L.E., Downes, M., Dugan-Rocha, S., Dunkov, B.C., Dunn, P., Durbin, K.J., Evangelista, C.C., Ferraz, C., Ferriera, S., Fleischmann, W., Fosler, C., Gabrielian, A.E., Garg, N.S., Gelbart, W.M., Glasser, K., Glodek, A., Gong, F., Gorrell, J.H., Gu, Z., Guan, P., Harris, M., Harris, N.L., Harvey, D., Heiman, T.J., Hernandez, J.R., Houck, J., Hostin, D., Houston, K.A., Howland, T.J., Wei, M.H., Ibegwam, C., Jalali, M., Kalush, F., Karpen, G.H., Ke, Z., Kennison, J.A., Ketchum, K.A., Kimmel, B.E., Kodira, C.D., Kraft, C., Kravitz, S., Kulp, D., Lai, Z., Lasko, P., Lei, Y., Levitsky, A.A., Li, J., Li, Z., Liang, Y., Lin, X., Liu, X., Mattei, B., McIntosh, T.C., McLeod, M.P., McPherson, D., Merkulov, G., Milshina, N.V., Mobarry, C., Morris, J., Moshrefi, A., Mount, S.M., Moy, M., Murphy, B., Murphy, L., Muzny, D.M., Nelson, D.L., Nelson, D.R., Nelson, K.A., Nixon, K., Nusskern, D.R., Pacleb, J.M., Palazzolo, M., Pittman, G.S., Pan, S., Pollard, J., Puri, V., Reese, M.G., Reinert, K., Remington, K., Saunders, R.D., Scheeler, F., Shen, H., Shue, B.C., Siden-Kiamos, I., Simpson, M., Skupski, M.P., Smith, T., Spier, E., Spradling, A.C., Stapleton, M., Strong, R., Sun, E., Svirskas, R., Tector, C., Turner, R., Venter, E., Wang, A.H., Wang, X., Wang, Z.Y., Wassarman, D.A., Weinstock, G.M., Weissenbach, J., Williams, S.M., Woodage, T., Worley, K.C., Wu, D., Yang, S., Yao, Q.A., Ye, J., Yeh, R.F., Zaveri, J.S., Zhan, M., Zhang, G., Zhao, Q., Zheng, L., Zheng, X.H., Zhong, F.N., Zhong, W., Zhou, X., Zhu, S., Zhu, X., Smith, H.O., Gibbs, R.A., Myers, E.W., Rubin, G.M., and Venter, J.C., The genome sequence of *Drosophila melanogaster*. *Science*, **287**, 2185–2195 (2000).
7. Maglott, D.R., Katz, K.S., Sicotte, H. and Pruitt, K.D., NCBI's LocusLink and RefSeq. *Nucleic Acids Res.*, **28**, 126–128 (2000).
8. Strausberg, R.L., Feingold, E.A., Klausner, R.D. and Collins, F.S., The mammalian gene collection. *Science*, **286**, 455–457 (1999).
9. Okazaki, Y., Furuno, M., Kasukawa, T., Adachi, J., Bono, H., Kondo, S., Nikaido, I., Osato, N., Saito, R., Suzuki, H., Yamanaka, I., Kiyosawa, H., Yagi, K., Tomaru, Y., Hasegawa, Y., Nogami, A., Schonbach, C., Gojobori, T., Baldarelli, R., Hill, D.P., Bult, C., Hume, D.A., Quackenbush, J., Schriml, L.M., Kanapin, A., Matsuda, H., Batalov, S., Beisel, K.W., Blake, J.A., Bradt, D., Brusic, V., Chothia, C., Corbani, L.E., Cousins, S., Dalla, E., Dragani, T.A., Fletcher, C.F., Forrest, A., Frazer, K.S., Gaasterland, T., Gariboldi, M., Gissi, C., Godzik, A., Gough, J., Grimmond, S., Gustincich, S., Hirokawa, N., Jackson, I.J., Jarvis, E.D., Kanai, A., Kawaji, H., Kawasawa, Y., Kedzierski, R.M., King, B.L., Konagaya, A., Kurochkin, I.V., Lee, Y., Lenhard, B., Lyons, P.A., Maglott, D.R., Maltais, L., Marchionni, L., McKenzie, L., Miki, H., Nagashima, T., Numata, K., Okido, T., Pavan, W.J., Pertea, G., Pesole, G., Petrovsky, N., Pillai, R., Pontius, J.U., Qi, D., Ramachandran, S., Ravasi, T., Reed, J.C., Reed, D.J., Reid, J., Ring, B.Z., Ringwald, M., Sandelin, A., Schneider, C., Semple, C.A., Setou, M., Shimada, K., Sultana, R., Takenaka, Y., Taylor, M.S., Teasdale, R.D., Tomita, M., Verardo, R., Wagner, L., Wahlestedt, C., Wang, Y., Watanabe, Y., Wells, C., Wilming, L.G., Wynshaw-Boris, A., Yanagisawa, M., Yang, I., Yang, L., Yuan, Z., Zavolan, M., Zhu, Y., Zimmer, A., Carninci, P., Hayatsu, N., Hirozane-Kishikawa, T., Konno, H., Nakamura, M., Sakazume, N., Sato, K.,

Shiraki, T., Waki, K., Kawai, J., Aizawa, K., Arakawa, T., Fukuda, S., Hara, A., Hashizume, W., Imotani, K., Ishii, Y., Itoh, M., Kagawa, I., Miyazaki, A., Sakai, K., Sasaki, D., Shibata, K., Shinagawa, A., Yasunishi, A., Yoshino, M., Waterston, R., Lander, E.S., Rogers, J., Birney, E. and Hayashizake, Y., Analysis of the mouse transcriptome based on functional annotation of 60,770 full-length cDNAs. *Nature*, **420**, 563–573 (2002).

10. Carninci, P., Waki, K., Shiraki, T., Konno, H., Shibata, K., Itoh, M., Aizawa, K., Arakawa, T., Ishii, Y., Sasaki, D., Bono, H., Kondo, S., Sugahara, Y., Saito, R., Osato, N., Fukuda, S., Sato, K., Watahiki, A., Hirozane-Kishikawa, T., Nakamura, M., Shibata, Y., Yasunishi, A., Kikuchi, N., Yoshiki, A., Kusakabe, M., Gustincich, S., Beisel, K., Pavan, W., Aidinis, V., Nakagawara, A., Held, W.A., Iwata, H., Kono, T., Nakauchi, H., Lyons, P., Wells, C., Hume, D.A., Fagiolini, M., Hensch, T.K., Brinkmeier, M., Camper, S., Hirota, J., Mombaerts, P., Muramatsu, M., Okazaki, Y., Kawai, J. and Hayashizaki, Y., Targeting a complex transcriptome: the construction of the mouse full-length cDNA encyclopedia. *Genome Res.*, **13**, 1273–1289 (2003).
11. Numata, K., Kanai, A., Saito, R., Kondo, S., Adachi, J., Wilming, L.G., Hume, D.A., Hayashizaki, Y. and Tomita, M., Identification of putative noncoding RNAs among the RIKEN mouse full-length cDNA collection. *Genome Res.*, **13**, 1301–1306 (2003).
12. Brizuela, L., Richardson, A., Marsischky, G. and Labaer, J., The FLEXGene repository: exploiting the fruits of the genome projects by creating a needed resource to face the challenges of the post-genomic era. *Arch. Med. Res.*, **33**, 318–324 (2002).
13. Walhout, A.J., Temple, G.F., Brasch, M.A., Hartley, J.L., Lorson, M.A., van den Heuvel, S. and Vidal, M., GATEWAY recombinational cloning: application to the cloning of large numbers of open reading frames or ORFeomes. *Methods Enzymol.*, **328**, 575–592 (2000).
14. Wiemann, S., Weil, B., Wellenreuther, R., Gassenhuber, J., Glassl, S., Ansorge, W., Bocher, M., Blocker, H., Bauersachs, S., Blum, H., Lauber, J., Dusterhoft, A., Beyer, A., Kohrer, K., Strack, N., Mewes, H.W., Ottenwalder, B., Obermaier, B., Tampe, J., Heubner, D., Wambutt, R., Korn, B., Klein, M. and Poustka, A., Toward a catalog of human genes and proteins: sequencing and analysis of 500 novel complete protein coding human cDNAs. *Genome Res. Mar.*, **11**, 422–435 (2001).
15. Liu, Q., Li, M.Z., Leibham, D., Cortez, D. and Elledge, S.J., The univector plasmid-fusion system, a method for rapid construction of recombinant DNA without restriction enzymes. *Curr. Biol.*, **8**, 1300–1309 (1998).
16. http://www.clontech.com/.
17. Yeh, R.F., Lim, L.P. and Burge, C.B., Computational inference of homologous gene structures in the human genome. *Genome Res.*, **11**, 803–816 (2001).
18. Mammalian Gene Collection Program Team, Generation and initial analysis of more than 15,000 full-length human and mouse cDNA sequences. *Proc. Natl Acad. Sci. USA*, **99**, 16899–16903 (2002).
19. Kikuno, R., Nagase, T., Waki, M. and Ohara, O., HUGE: a database for human large proteins identified in the Kazusa cDNA sequencing project. *Nucleic Acids Res.*, **30**, 166–168 (2002).
20. Kawai, J., Shinagawa, A., Shibata, K., Yoshino, M., Itoh, M., Ishii, Y., Arakawa, T., Hara, A., Fukunishi, Y., Konno, H., Adachi, J., Fukuda, S., Aizawa, K., Izawa, M., Nishi, K., et al., Functional annotation of a full-length mouse cDNA collection. *Nature*, **409**, 685–690 (2001).
21. Walhout, A.J., Sordella, R., Lu, X., Hartley, J.L., Temple, G.F., Brasch, M.A., Thierry-Mieg, N. and Vidal, M., Protein interaction mapping in *C. elegans* using proteins involved in vulval development. *Science*, **287**, 116–122 (2000).
22. Davy, A., Bello, P., Thierry-Mieg, N., Vaglio, P., Hitti, J., Doucette-Stamm, L., Thierry-Mieg, D., Reboul, J., Boulton, S., Walhout, A.J., Coux, O. and Vidal, M., A protein–protein

interaction map of the *Caenorhabditis elegans* 26S proteasome. *EMBO Rep.*, **2**, 821–828 (2001).
23. Boulton, S.J., Gartner, A., Reboul, J., Vaglio, P., Dyson, N., Hill, D.E. and Vidal, M., Combined functional genomic maps of the *C. elegans* DNA damage response. *Science*, **295**, 127–131 (2002).
24. Reboul, J., Vaglio, P., Rual, J.F., Lamesch, P., Martinez, M., Armstrong, C.M., Li, S., Jacotot, L., Bertin, N., Janky, R., Moore, T., Hudson, J.R. Jr., Hartley, J.L., Brasch, M.A., Vandenhaute, J., Boulton, S., Endress, G.A., Jenna, S., Chevet, E., Papasotiropoulos, V., Tolias, P.P., Ptacek, J., Snyder, M., Huang, R., Chance, M.R., Lee, H., Doucette-Stamm, L., Hill, D.E. and Vidal, M., *C. elegans* ORFeome version 1.1: experimental verification of the genome annotation and resource for proteome-scale protein expression. *Nat. Genet.*, **34**, 35–41 (2003).
25. Rubin, G.M., Hong, L., Brokstein, P., Evans-Holm, M., Frise, E., Stapleton, M. and Harvey, D.A., A *Drosophila* complementary DNA resource. *Science*, **287**, 2222–2224 (2000).
26. Stapleton, M., Liao, G., Brokstein, P., Hong, L., Carninci, P., Shiraki, T., Hayashizaki, Y., Champe, M., Pacleb, J., Wan, K., Yu, C., Carlson, J., George, R., Celniker, S. and Rubin, G.M., The *Drosophila* gene collection: identification of putative full-length cDNAs for 70% of *D. melanogaster* genes. *Genome Res.*, **12**, 1294–1300 (2002).
27. Uetz, P., Giot, L., Cagney, G., Mansfield, T.A., Judson, R.S., Knight, J.R., Lockshon, D., Narayan, V., Srinivasan, M., Pochart, P., Qureshi-Emili, A., Li, Y., Godwin, B., Conover, D., Kalbfleisch, T., Vijayadamodar, G., Yang, M., Johnston, M., Fields, S. and Rothberg, J.M., A comprehensive analysis of protein–protein interactions in *Saccharomyces cerevisiae*. *Nature*, **403**, 623–627 (2000).
28. Martzen, M.R., McCraith, S.M., Spinelli, S.L., Torres, F.M., Fields, S., Grayhack, E.J. and Phizicky, E.M., A biochemical genomics approach for identifying genes by the activity of their products. *Science*, **286**, 1153–1155 (1999).
29. Zhu, H., Bilgin, M., Bangham, R., Hall, D., Casamayor, A., Bertone, P., Lan, N., Jansen, R., Bidlingmaier, S., Houfek, T., Mitchell, T., Miller, P., Dean, R.A., Gerstein, M. and Snyder, M., Global analysis of protein activities using proteome chips. *Science*, **293**, 2101–2105 (2001).
30. Stover, C.K., Pham, X.Q., Erwin, A.L., Mizoguchi, S.D., Warrener, P., Hickey, M.J., Brinkman, F.S., Hufnagle, W.O., Kowalik, D.J., Lagrou, M., Garber, R.L., Goltry, L., Tolentino, E., Westbrock-Wadman, S., Yuan, Y., Brody, L.L., Coulter, S.N., Folger, K.R., Kas, A., Larbig, K., Lim, R., Smith, K., Spencer, D., Wong, G.K., Wu, Z., Paulsen, I.T., Reizer, J., Saier, M.H., Hancock, R.E., Lory, S. and Olson, M.V., Complete genome sequence of *Pseudomonas aeruginosa* PA01, an opportunistic pathogen. *Nature*, **406**, 959–964 (2000).

Proteome Analysis. Interpreting the Genome.
D.W. Speicher (editor)

Chapter 12

Automation of proteome analysis

PETER JAMES*

Protein Technology, Wallenberg Laboratory II, Lund University, P.O. Box 7031, SE-220 07 Lund, Sweden

Abbreviations: CAD, collisionally activated dissociation; MS/MS, tandem mass spectrometry; MALDI TOF MS, matrix-assisted time of flight mass spectrometry; PTM, post-translational modification; 2D PAGE, two-dimensional polyacrylamide gel electrophoresis; 2D HPLC, two-dimensional high-pressure liquid chromatography.

* Tel.: +46-46-222-1496; fax: +46-46-222-1495. E-mail: peter.james@elmat.lth.se (P. James).

1. Introduction

The revolution that led to whole genome sequencing began with the development of a simple method for determining the sequence of DNA [1] some 30 years ago. There were several key steps between the process of being able to sequence a few short DNA strands and the determination of whole chromosomes. Each of these critical steps was essentially just a refinement of the basic sequencing method: the use of fluorescent dyes, capillary electrophoretic separation, etc. The critical breakthrough was the realisation and subsequent demonstration that by using the already available techniques and automating them, one could use shotgun sequencing on a genomic level [2]. Once this occurred, the next problem was to develop the software to put all the information together.

Proteomics, at least the subbranch generally called expression proteomics, i.e. the analysis and quantitation of all proteins and their modifications in a cell at a specified time, is at a similar watershed. The fundamental techniques of protein separation (2D PAGE) and peptide separation by HPLC have been established for more than a quarter of the century. In parallel, the development of new ionisation techniques, electrospray ionisation (ESI) and matrix-assisted laser desorption and ionisation (MALDI), has allowed peptides and proteins to become accessible to analysis by mass spectrometry [3,4]. The major breakthrough in the last 10 years that has paved the way to allow proteomics approaches to be truly global, has been the development of software algorithms to identify proteins on the basis of their digestion patterns or peptides from their fragmentation patterns, and the availability of sequence databases containing essentially complete genomes. The next critical step that must be taken is the development of completely automated systems capable of high-throughput analysis of protein expression in a systematic manner. Analogous to DNA sequencing, new and improved algorithms and data visualisation tools must be developed to allow the extraction of relevant data from these new experiments and help with the interpretation of huge and complex data sets.

This chapter will address the availability and status of automation for sample preparation, 2D gels, and subsequent MS analysis steps because these methods are, for 2D PAGE, currently the best established with the highest throughput. (See Chapter 10 for automation of protein chips and protein assays.) In order for proteome analysis to become a viable and widely used method, a reasonable degree of automation must be achieved to: (i) increase reproducibility to facilitate data comparison within and between laboratories; and (ii) make the process less labour intensive and increase the throughput. This chapter outlines the approaches and pitfalls in trying to automate protein identification and quantification methods for comprehensive proteome analysis.

2. Experimental design

Before launching into a discussion of automation, a general outline of the types of environments in which high-throughput techniques could be valuable will be given. The range and limitation of the current approaches is important to appreciate before assuming automation will provide a general solution. Due to the expensive nature of the equipment involved in automation, there will probably be only a limited number of truly high-throughput proteomics labs in a fashion similar to that observed with the DNA sequencing 'factories' used for the human genome project. Currently, one can identify two types of proteomics centres, those with a heavy clinical leaning, aimed at processing huge numbers of clinical specimens in order to find protein expression/modification correlations for diagnosis and monitoring treatments. The second type of proteome centre is the more problem-orientated centre aimed at understanding basic biological phenomena. For both types, the basic questions to be answered before asking how to automate the procedures are ones of experimental design. Should a protein-based (2D PAGE) or a peptide-based (ICAT and MudPIT [5,6]) or a protein chip (for a review see ref. [7] and Chapter 10) approach be used?

2.1. Experimental approaches

Despite the debates raging about their relative merits, the gel-based, peptide-based and protein chip approaches all share similar limitations. The dynamic range of protein expression in a cell is around 10^7 and in fluids such as blood serum it is probably around 10^{12}. The two main detection methods used, mass spectrometry and fluorescence, can only deal with four orders of magnitude. Thus there is no tool, chip, gel or HPLC that can cover the range of protein expression found in nature. Secondly, all the techniques are dependent on protein digestion to obtain fragments for identification or analysis of post-translational modifications (PTMs), and all proteases cease to be effective at substrate concentrations at the nanomolar level. Even antibody chips must be analysed to see if the fluorescence is due to binding of antigen protein or cross-reactivity or binding of the protein antigen plus a series of other proteins in a complex since the binding is measured under non-denaturing conditions. Thus, whichever approach one takes, one will only see a subset of the proteins that are present in the cell or serum.

2.2. Experimental design factors

One has to define a series of parameters within which the experiment will operate. The main factors are: (i) how much material is available; (ii) what coverage of protein expression and modification is required; (iii) how many samples are in the experiment; (iv) how long will the experiment last; (v) what are you looking for?

For a clinician, material is usually limiting. Tissue biopsies and material from biological specimen banks are usually extremely hard to obtain and are, hence, valuable irreplaceable material. For the clinician who may be looking for a diagnostic or prognostic tool, the need for total protein expression coverage is not critical, it is the pattern of expression that is the best indicator, not just one or two proteins. For a biologist, however, all proteins are critical when trying to define functional networks and coverage is critical. Usually the biologist has access to more material and can afford to do exhaustive pre-fractionation or multiple zoom gels to increase coverage. In content, clinicians will probably be dealing with large patient cohorts in order to extract meaningful data over a population. Hence, repetition of a single separation type is the most likely strategy and is the approach most suitable for automation. Since the proteins of interest, the markers, will not be known until the end of the experiment, i.e. once all the patient samples have been analysed, the time of the experiment is usually long and on the order of months if not years. The ability to archive gels and go back to them to excise and analyse new spots is a great advantage of this method, and it reduces the demands on the later stages of the proteome analysis pipeline, i.e. in-gel digestion and MS protein identification. For the biologist, a MudPIT-type approach in which an entire cell is digested with a protease and all the resulting peptides are analysed by multidimensional peptide HPLC coupled to mass spectrometry for on-line identification is probably preferable. However, one must note that these exhaustive analyses are extremely slow: the analysis of a single yeast experiment takes about 28 h of chromatography. The 2D gel approach (Fig. 1) allows parallelisation and a small lab can easily run 50 gels, which can correspond to up to 100–150 samples (see Chapter 3 on DIGE technology) in 2 days.

3. Sample preparation

Perhaps the single most critical point in obtaining reproducible results with gel-, HPLC- or chip-based approaches is sample preparation (Table 1). The cells or tissue must be efficiently disrupted and the contents of the cells solubilised completely in order to obtain a representative protein sample. Physical disruption

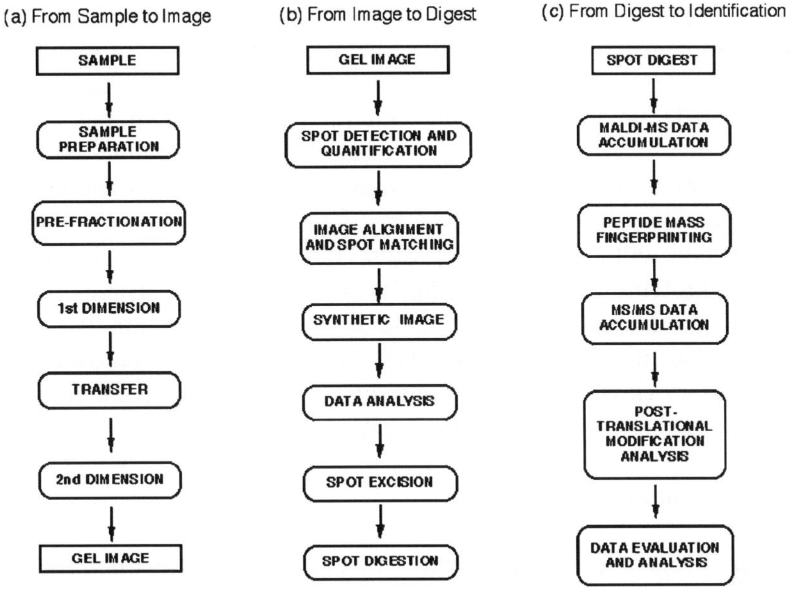

Fig. 1. Workflow charts for automated protein analysis using 2D gels: (a) depicts the workflow from sample processing to image acquisition; (b) from spot selection to protein digestion; and (c) from digest to data evaluation.

methods such as sonication, mechanically driven rapid pressure changes, homogenisation and shearing-based techniques are often used to lyse cells prior to protein extraction with a urea-based solution containing non-ionic detergents, reducing agents and a protease inhibitor cocktail. For each cell or tissue type, a specific method must be developed, though recent developments including the use

Table 1

Speed for manual vs. automated protein profile analysis

Workflow	Manual (one person per 8 h day, 5 day working week)	Automated (one system per 24 h day, 7 day working week)
Sample preparation and protein determination	50 samples	700 samples
Running 2D PAGE	10 gels	50 gels
Staining	10 gels (silver staining)	50 gels (DIGE)
Scanning	25 gels	200 gels (500 samples, DIGE)
Matching and editing	5 gels	500 samples (DIGE)
Spot cutting, digestion, spotting	150 spots	8000 spots
Protein fingerprinting (MALDI MS)	1000	1500, 8000
Peptide fingerprinting (ESI–MS/MS)	100	300, 2100

of thiourea [8], hydroxyethyl disulfide [9] and novel zwittergents [10] have improved sample solubilisation greatly. The extent of recovery of membrane and cytoskeletal proteins is very variable, with some proteins being completely solubilised while up to 10% of the cell protein remains in the pellet after extraction. Even if solubilised, many of the sparingly soluble proteins will precipitate between the first and second dimensions of a 2D gel and the strongly denaturing conditions preclude the use of antibody chips. The method of choice for these proteins will ultimately be a MudPIT/ICAT-type approach. The number of membrane proteins in a genome is fairly constant at around 30% (best estimates for bacteria up to humans). Many of these are the most important disease-related proteins because most intra and intercellular functions are controlled by cell-surface receptors as well as by cell-surface presenters, the major histocompatibility complexes and the peptides they carry.

3.1. Pre-fractionation

The extremely high degree of complexity of eukaryotic tissues often requires that a pre-fractionation step be carried out in order to reduce the complexity and allow the resolution and analysis of minor components. When dealing with a mixed cell population such as a tissue, pre-fractionation of cells using a fluorescence-activated cell sorter (FACS) [11] can allow small subpopulations to be specifically isolated, greatly increasing the sensitivity of the analysis. Similarly, pre-concentration of the proteins to be analysed can be carried out using methods orthogonal to 2D gel separation such as native PAGE or IEF [12] or by affinity pre-enrichment such as heparin chromatography for DNA binding proteins [13], immobilised metal ion affinity chromatography for phosphoproteins [14] or by antibody precipitation to select a specific protein complex. Alternatively a series of increasingly powerful solubilising buffers may be used to obtain a series of protein fractions [15] or the various cell compartments and/or organelles may be isolated [16]. All the above methods can be automated and before any large-scale study is started, a systematic study of sample preparation reproducibility should be conducted.

3.2. Protein extraction

Given the importance of this first step, it is incredible that there is not a generic protein extraction device available. DNA/RNA extraction is fairly well established but for proteomics to proceed, a fully automated device must be made available. First steps are being taken in this direction and BioSpec (Bartlesville, OK, USA) has launched a 96-well plate version of the bead-beater extraction method [17]. Since proteins are much more heterogeneous in their biophysical properties than nucleotides, it is clear that a completely generic extraction approach will be difficult to obtain. Also the huge diversity of cell types and their environments, from fragile erythrocytes that can be lysed by dilution into pure water, through hard-walled bacteria like mycobacteria to tissues such as bone and teeth, will place huge demands on any generic extraction method, if such a method should be technically feasible. Even the hardest tissues become fragile in liquid nitrogen and can easily be reduced to a powder. The powder can be more finely dispersed using a bead beater and finally, over 99% of a cell's protein content is soluble in sodium dodecyl sulphate solutions. No one has as yet put the pieces of the puzzle together into an automated sample extraction device, but it is surely only a matter of time before such tools are developed.

4. 2D PAGE

Currently, 2D PAGE is the highest resolving protein separation technique known and is capable of resolving over 10,000 spots in a single giant gel [18]. A detailed study using a prototype semi-automated instrument clearly showed the advantages of mechanical reproducibility [19], however, no automated commercial device has ever found its way to market. The standard denaturing 2D gel system has remained an essentially manual operation since it was established almost 30 years ago and as yet, there is no valid alternative method with equivalent dynamic range and resolving power. However, 2D PAGE has many drawbacks, the first being the lack of reproducibility between labs and even within labs. The move from isoelectric focussing in tube gels using ampholine buffers to focussing in immobilised pH gradient (IPG) gels and the ability to load by rehydration [20,21] went a long way to increasing reproducibility. The main problem still lies in sample preparation, and sample preparation issues will probably similarly affect reproducibility of the non-gel-based techniques being developed.

4.1. Resolution

Given the limited size of the large format standard 2D gel and the number of protein forms in a cell, completely and partially overlapping spots are a problem. This can be resolved by computational approaches but now it is becoming clear that many symmetrical spots contain two or more proteins. For prokaryotes, about 20% of all spots contain one or more proteins; for eukaryotes, the number approaches 40%. A collaborative study by Proteome Inc. and Perseptive Biosystems [22] has shown in detail the extent and nature of spot cross-contamination in 2D gels of yeast extracts. This was based solely on mass matching algorithms, and many of the spots contained three or more proteins. Although MS techniques are powerful enough to resolve such protein mixtures [23], multiple proteins per spot make interpretation of many different samples extremely difficult. This limitation can be eliminated almost totally either by using very large format gels [24] or with a series of overlapping narrow pH range gels, a technique that has been termed 'Proteome Contigs' [25] by analogy to genome mapping. The use of isotopic labelling of proteins in pairs of gels is an alternative method to deconvolute overlapping proteins in 2D gel spots [26].

4.2. Gel stability

Another hurdle that must be overcome in order to automate spot identification is that of the weak mechanical properties of the normal polyacrylamide gel matrix of the second dimension. The ideal matrix must be mechanically stable to allow the gel to be handled in automated systems without breaking, and it should also provide a clear background during protein visualisation. Mechanical stability can be increased by including pre-polymerised polyacrylamide in the polymerisation mixture though this can produce a turbid background when scanning the gel. A gel mixture with enhanced properties has been partially described [27] and is commercially available. Alternatively the gel can be covalently attached to a plastic support which has the advantage of preventing alterations in gel size during the staining procedure due to shrinkage in organic solvents and expansion upon rehydration, or dehydration upon exposure to air, which would preclude computer-driven automated spot cutting. The major drawback of commercially available plastic backings is that they are all highly fluorescent, which preclude use of dyes such as SYPRO Ruby. This is an important limitation because SYPRO Ruby is highly sensitive, has a wide dynamic range and is MS compatible. Fluorescent dyes are also ideal for automatic gel staining since only two steps are involved and they can be left indefinitely in stain, in contrast to silver staining, which is very difficult to automate. The main drawback of the fluorescent dyes is the expense. There are several non-fluorescent plastics available, however, at least in this author's experience, they are useless as gel supports because gels will not adhere to them. There are potentially many other polymer matrices that could be used and have desirable properties such as strength, solvent resistance and pH resistance, which is essential for first dimension separations at extreme pH [28,29]. However, only the cross-linker piperazine diacrylamide [30] has received any commercial attention. It is quite surprising that after almost 30 years of 2D PAGE, there is no real commercial solution to gel staining other than piling large numbers of plastic boxes on an extremely large shaker and manually changing fluids.

5. Image analysis

The next step in the workflow is gel scanning. Despite all the attention being given to the parallel nature of 2D PAGE and the potential to run hundreds of samples a day, very little thought has been given to converting the gels into computer images for subsequent analysis. Currently, digitising gel images is manual and very time consuming, roughly 20 min per large format gel. Assuming the gel has been cast on a plastic backing, handling should not be an issue and it is comparatively easy to design a system to carry out automated gel scanning though as yet nothing is commercially available.

If all the steps from sample preparation to scanning were to be automated and the gels were mechanically stable, then gel image analysis would become trivial. This not being the case, several generations of software packages have evolved such as Delta2D, Gellab II, Melanie-III, PDQuest 7, Progenesis and Z3. Currently, gel distortions must be mathematically treated, and gel images warped and transformed to allow gel matching. The analysis is a multistage process involving: spot detection, quantification, fitting and modelling, background subtraction, contrast enhancement and artefact removal, image alignment and finally gel comparison. Although these processes can all be carried out automatically in batch mode there always remains ca. 10% of the spots, which must be manually corrected and fitted due to the quality of the gel. Thus although extremely complex mathematical corrections can be made, the limiting factor is the reproducibility of sample preparation, 2D gel separation and staining, and this should be the main focus of attention for automation rather than computational 'face lifting'. One current trend has been to change the order of the mathematical analysis. Rather than identifying the spots first and using their patterns to drive the matching process, 'contour' mapping can be used. One can look at the 2D gel in three dimensions, the third being spot intensity and one gets an ordnance survey type map with contours. The first step then becomes a matching of the valleys and peaks to get the greatest overlap between images by using image warping. After this step, spot detection and fine alignment can be carried out. This approach has dramatically increased the efficiency of matching, and batch analyses can be carried out. However, even the best programs will still require at least 10–20 min of manual attention (once the system has been optimised on a specific cell type and separation) to get the final 5% of spots corrected (or the entire gel deleted if it has not run properly). However, probably the best solution practically and in terms of experimental design is the 2D-DIGE approach, which is described in Chapter 3.

6. Robotics for cutting, digestion and spotting

The most time-consuming and error-prone part of the entire 2D PAGE procedure is spot excision. As anyone knows who has worked with a pile of 10 gels and a printout of a master gel with the spots to be excised marked, it is one of the most mind and body numbing exercises. Luckily there are several commercially available spot-picking robots on the market

(Amersham Biosciences, Bio-Rad and Perkin Elmer, etc.). These devices download a spot pick list generated by the image analysis software and then pick the spots. However, this is not as trivial as it sounds.

The first problem is how to align the scanned gel image that was used for defining the spots to be excised with the real gel lying on the cutter plate. The Bio-Rad and Perkin Elmer solutions require a manual spot matching between the scanned image and the excision robot camera image. This is a slow process that requires operator intervention. An elegant solution has been devised by Amersham Biosciences. The plastic gel backing has two fluorescent sticky-backed plastic spots attached on either side of the gel. These spots act as alignment markers and are ignored by the image analysis software. The gel is placed by a robot arm onto the cutting table and a camera on the cutter arm looks for the two spots and aligns itself with them and automatically calculates the coordinates of the spots to be cut. Most of the cutters now on the market are reliable and should have success rates of 99.9% or higher. The Amersham system (the fully automated version is called the Cutter Workstation although the individual pieces (cutter, digestor, etc.) can be bought separately) places the spots into a 96-well plate, which is then removed from the cutting station and placed on the digestion station. The spots are first destained, washed and then partially dried. This is a critical step because if the spots become too dry, they easily become electrostatically charged and jump out of the wells when the metal robot arm picks up the plate. Trypsin is added by rehydration and the plates are capped and placed in an oven at 37°C for 30 min before being taken for extraction and spotting onto the MALDI plate. The system can produce 1152 samples, MALDI MS ready, from 12 gels in 24 h. A fairly comprehensive comparison of the instruments in the field is given in Table 2 for the 2D part of the procedure and in Table 3 for the mass spectrometry part. It must be emphasised that this trial was carried out with the idea of putting together a complete 2D gel-based solution. If other criteria such as absolute sensitivity, software quality, etc. were the primary criteria, then the resulting optimal system may have been quite different.

7. Protein fingerprinting by MALDI MS

Currently, the most common first-pass method of protein identification is by MALDI TOF MS analysis of the digested protein. In 1993, five groups independently proposed the idea of peptide mass fingerprinting [31–35]. The concept is that the set of peptide masses obtained by mass spectrometric analysis after digestion of a protein with a specific protease can act as a fingerprint, and this property is unique to that protein. Therefore the set of masses can be used to search a protein database in which the sequences have been replaced by the calculated fingerprint masses to find a similar pattern. As the protein database grew

Table 2

Results of a gel-based evaluation of major instrument providers

	Amersham[a] (Jan 2001)	Micromass (Dec 2000)	ABI (May 2000)	PE (Jan 2000)	Bio-Rad (Dec 2000)
Sample preparation unit(s)	0	0	0	0	0
2D gel electrophoresis unit(s)	4	0	0	4	4
Isoelectric focussing module for solutions	4	0	0	0	3
Gel staining unit	5	0	0	4	1
Gel visualisation unit	4	0	0	4	4
Gel handling unit	5	0	0	1	1
Spot excision unit	5	0	0	3	3
Sample digestion unit	4	3	3	3	0
MALDI target deposition unit	4	4	3	4	0
Sample tracking hard- and software	5	1	1	1	2
Workflow system for overall task handling	5	5	1	4	4
Simple and rapid data entry system	5	2	1	2	2
2D gel analysis software	5	0	0	3	4
Database construction and querying software	3	0	0	3	4
Data viewer linking all aspects of analysis	5	0	3	4	5
Software capable of linking to other data types such as RNA profiling, metabolomics data, etc.	4	0	3	2	5

(continues)

Table 2
continued

	Amersham[a] (Jan 2001)	Micromass (Dec 2000)	ABI (May 2000)	PE (Jan 2000)	Bio-Rad (Dec 2000)
WWW/Intranet capability for access to results	4	5	3	3	3
Overall LIMS system (laboratory information management system)	5	4	3	4	3
Compatibility with MS and MS/MS system(s)	0	5	5	5	5
Overall degree of automation	5	5	2	3	2
	81	34	28	57	55

Laboratories interested in purchasing an HT system are strongly advised to conduct a similar comparison of their currently available systems. The firms were supplied with a large amount (hundreds of milligrams) of protein extracts from a bacterium (whose genome sequence had not been released) grown under two different growth conditions. The response to the change in conditions was well defined; 35 different genes were induced. A zero score means not available; this is why some companies score low since they cannot create a fully integrated platform. The scoring system is based on relative achievement between the firms on a scale of 0–5 (maximum). Dates of test are given in parenthesis.

[a]Based on DIGE (multiple covalent fluorescence labelling prior to first dimension) and the fully automated Ettan spot cutting system (not the currently sold individual components).

Table 3

Results of an MS-based evaluation of major instrument providers

	Amersham[a] (Jan 2001)	Micromass (Dec 2000)	ABI (May 2000)	Bruker (Jan 2001)	Finnigan (Dec 2000)
MALDI TOF mass spectrometer	4	5	4	4	0
Post-source decay capability	5	3	3	4	0
Robotic system for automatic loading of MALDI plates into the MS	3	5	1	1	0
Protein identification software (MS)	3	5	3	3	0
ESI/MALDI TOF mass spectrometer	0	5	3	0	0
ESI triple quadrupole	0	3	4	0	5
ESI quadrupole ion-trap mass spectrometer	0	0	0	3	5
An HPLC system	4	4	5	5	3
An autosampler system	4	4	5	5	3
A capillary zone electrophoresis interface	0	5	5	5	5
Automated protein identification software (MS/MS)	3	5	3	3	5
Data viewer linking all aspects of analysis	4	5	3	2	2
Software capable of linking to the LIMS system	5	0	3	2	3
WWW/Intranet capability for access to results	4	5	3	2	5
Software to allow user-defined data dependent spectrum accumulation and control of external devices such as CZE or HPLC	2	4	2	3	5
The system should show a high overall degree of automation and the minimum of manual interaction	4	5	2	2	5
	45	63	49	44	46

The digests resulting from the proteins defined in the gel part were used for the MS evaluation. The dates of the tests are especially critical since new instruments would give an entirely different picture. The scoring system is based on relative achievement between the firms on a scale of 0–5 (maximum). Dates of test are given in parenthesis.

[a]Based on DIGE (multiple covalent fluorescence labelling prior to first dimension) and the fully automated Ettan spot cutting system (not the currently sold individual components).

exponentially in size, the accuracy of the methods dropped due to the huge background of noise. However, now that complete genomes are available, the protein translations have reduced the problem back in size since one only needs to search the specific genome (and perhaps a database of commonly occurring contaminants) rather than the entire one. The main problem is to ensure that the protein digests well enough to provide enough peptides to give high confidence identification. Moreover, the extent of protein coverage needs to be in the area of 90 + % to account for as many of the PTMs as possible. Currently this requires almost picomoles of proteins and at least two different protease digestions. The efficiency of protein digestion below the 100 fmol mark is extremely low. Possibly a few peptides are released which can allow protein identification by MS/MS but certainly not comprehensive coverage of all proteins in a 2D spot and their PTMs.

7.1. Automated in-gel digestion and data acquisition

Peptide mass fingerprinting can easily be adapted to a high-throughput format. Protein spots automatically excised from 2D gels can be destained, washed, dehydrated and then perfused with trypsin using a standard fluid delivery robot station (several of these are now commercially available from Applied Biosystems, Amersham, Bruker, Micromass, Perkin Elmer, etc.). Furthermore, the resultant digest can be automatically loaded onto a MALDI sample target together with matrix. Most MALDI mass spectrometers now come with sample plates capable of holding 100–10,000 samples. In order to process such a high sample volume, data accumulation must be monitored online to ensure a reasonable quality level. Fuzzy logic feedback control [36] and other algorithms have been implemented that allow sequential sample measurement. The spot is searched until a reasonable signal is found and then the laser power is optimised (reduced to the minimum possible) to obtain high-resolution spectra before a spectrum is accumulated and written to disk and the process moves on to the next sample position. Several mass spectrometer manufacturers have added robotic plate loaders for their MALDI MS instruments (Amersham, Bruker, Micromass, etc.). For example, using the Micromass Matilda MALDI mass spectrometer it is possible to load fifty 384 position plates for automated walk-away analysis, which corresponds to 19,200 samples.

7.2. Data extraction and database searching

The effect of mass accuracy was shown early on to be a critical factor for the confidence level of peptide mass searches. The development of delayed extraction MALDI MS, in which the ions are not extracted immediately after the laser impulse but are allowed 100–300 ns to equilibrate before the ions are accelerated, has dramatically improved instrument accuracy. Together with the use of 'natural' internal standards (such as known tryptic autolytic fragments) over the range 800–3000 *m/z*, mass accuracy can be kept to within 5 ppm on most current instruments.

7.3. Confidence and coverage levels

One of the fundamental problems of database searching is how to determine the confidence level of a search result. In order to resolve this, additional parameters can be included such as a molecular weight estimate, limiting the number of mismatches allowed or using a scoring system weighted according to the frequency of occurrence of a mass in the protein of a given mass range. These restrictions are not readily applicable to sequences derived from genomic or EST data since only partial or fragmented sequences (due to introns or reading-frame shifts) are represented and the accuracy of EST sequences is lower since they are obtained by single-pass DNA sequencing. The random noise element in the database search can be minimised by using two or more orthogonal data sets [37]. Data from two protease digests with differing specificities (e.g. AspN and LysC) can be combined. Alternatively a single digest can be measured in the native state and again after carrying out a chemical modification on the sample plate (such as methylation, acetylation or deuterium exchange). This greatly increases the confidence level but doubles the analysis time. The use of orthogonal data collection in a relatively high-throughput MALDI MS fingerprinting environment is to be encouraged since it reduces the load on the slower next step of MS/MS and greatly increases protein coverage for PTM analysis.

8. Peptide fingerprinting by MS/MS

Conceptually similar approaches to protein identification by peptide mass fingerprinting using the MS/MS fragmentation pattern from a single peptide have been proposed [38,39]. Tandem mass spectrometry of peptides was pioneered in the 1980s by the group of Don Hunt for low energy, triple quadrupole instrumentation [40] and Klaus Biemann [41] for high-energy, four-sector magnetic instruments. The first mass scanning stage is used to isolate a single peptide before acceleration of the ions through a region of higher pressure containing a collision gas such as argon. The second mass scanning stage is used to measure the masses of the fragments arising from the peptide as a result of collisionally activated dissociation (CAD). Post-source decay (PSD) spectra can be obtained using TOF instruments equipped with ion gates and mirrors by increasing the laser power by a factor of two over that needed to obtain a normal spectrum. A large fraction of the desorbed ions obtained by MALDI MS undergo 'delayed' fragmentation before reaching the detector as a result of multiple collisions of the peptides with the matrix during plume expansion and ion acceleration. The set of ions produced by either technique can be used to obtain sequence-related fragment masses. The set of fragment masses acts as a fingerprint for the peptide and can be used to search sequence databases for similar peptides (see Chapter 7).

8.1. Algorithms

Algorithms for searching sequence databases using uninterpreted MS/MS spectra have been developed over the past 4 years. The most widely used of these is SEQUEST, developed by John Yates and Jimmy Eng at the University of Washington, Seattle. The program was originally intended for searching protein databases using MS/MS fragmentation spectra of unmodified peptides but has been subsequently extended to allow searching with post-translationally modified peptides [42], DNA database searching [43] and PSD data [44]. Essentially, the program searches for all peptides in the database (a protease can, but does not have to be defined) that have the same mass as the parent ion (within a defined mass window), then matches the predicted MS/MS spectrum with the experimentally determined one and finally carries out a cross-correlation analysis of the best scoring peptides in order to determine the best match. The program can automatically strip all the MS/MS data from an HPLC-auto-MS/MS data file, search the databases and then produce a summary of all the individual search results. The power of this approach lies in the ability to deal with proteins present as mixtures and molar ratios of 30:1 even at the low femtomole levels [45].

8.2. New MALDI MS-based workflows

The usual workflow involves MALDI MS followed by an electrospray MS/MS step for those proteins not identified by protein fingerprinting. However, an alternative approach has recently been made possible, the use of MALDI MS/MS and the revival of PSD analysis. MALDI MS/MS has been made available, most notably with the MALDI TOF–TOF mass spectrometer from Applied Biosystems as well as modified Q-TOFs from Micromass and Sciex. These instruments provide true MS/MS analysis of peptides. Also, the development of curved field reflectrons and 'jump' voltage steps has revived PSD as a means of obtaining some sequence data in a single spectrum from peptides without true MS/MS. These are typified by the Bruker MALDI TOF (jump voltage) and the Amersham MALDI TOF (curved field) instruments. The advantage of these methods is that the entire digest can be deposited onto the sample plate and used in a first pass for protein fingerprinting and then for those not giving a positive identification for peptide fingerprinting. The main limitation is the number of laser shots required to obtain a high-quality spectrum. This competes with the need for high sensitivity. If the digest is loaded in a very small diameter spot, the peptides are more concentrated and a good spectrum is obtained more rapidly. However, one may only have enough material to obtain one or two subsequent MS/MS spectra before the spot is exhausted of material.

8.3. New ESI-based workflows

An alternative approach that is appearing is the use of high-throughput electrospray. This obviates the need for sample spotting and the digest can be desalted and then sprayed. The initial few seconds give a spectrum that can be used for protein fingerprinting, and if this is not positive, 10–20 MS/MS spectra can be accumulated within a minute for peptide fingerprinting. The advantage of electrospray is that a peptide separation can be built into the analysis, say running a very fast 1 min reversed phase HPLC gradient or a CZE separation to greatly enhance protein coverage. Commercial devices capable of high throughput are beginning to appear now, based on the use of miniaturised electrospray nozzles manufactured on silicon chips [46] (Advion Biosciences, Ithaca, NY, USA). This system is capable of running 100 samples in 2 h and since a new tip is used for each sample there is no chance of any cross-contamination.

9. The crucial elements: LIMS and data mining

Most of the basic components of automated high-throughput proteomics already exist. What is needed is a standardised format for data acquisition, storage and querying (interpretation). When dealing with large-scale projects, tasks such as tracking samples, linking samples to gels, gels to spots and spots to sequence database entries become overwhelming, and software is urgently needed to prevent chaos. The data generated in such projects are vast. For example, 2D gel images occupy around 4 MB each, HPLC–MS/MS run produces up to 40 MB of data, and sequence databases are expanding continuously and require around 400 MB. In order to deal with such vast amounts of information, parallel computing must become much cheaper and user-friendly for programming [47]. Data compression must be implemented to speed retrieval and searching [48]. Many of these tools are being developed within the commercial sector and unfortunately the platforms are frequently inaccessible to the academic researcher. A commercial platform that allows integration of home-built tools and data sharing on a common format between labs would be quite valuable.

9.1. Pre-packed solutions

Currently there are several main laboratory information management systems from the main instrument suppliers: Labworks from Perkin Elmer, Worksbase from Bio-Rad/Micromass, LWS from Amersham Biosciences, SQL from Applied Biosystems Informatics and LIMS from ThermoFinnigan. Most suppliers will perform a customised installation for an extra fee. However, it is generally preferable, unless a large bioinformatics staff is available, to obtain all instruments and software from a single source to ensure data compatibility to allow system integration (Fig. 2). Computational power is also an issue, but with systems like Red-Hat linux, access to multiprocessing becomes easier. One possible problem for some users is that whenever possible, systems involving large amounts of robotics and expensive instruments should be physically isolated from the outside world. Isolation of the system from the web prevents system breakdowns due to external activity that may upset timing on the servers or unwanted attention from 'well-wishers'.

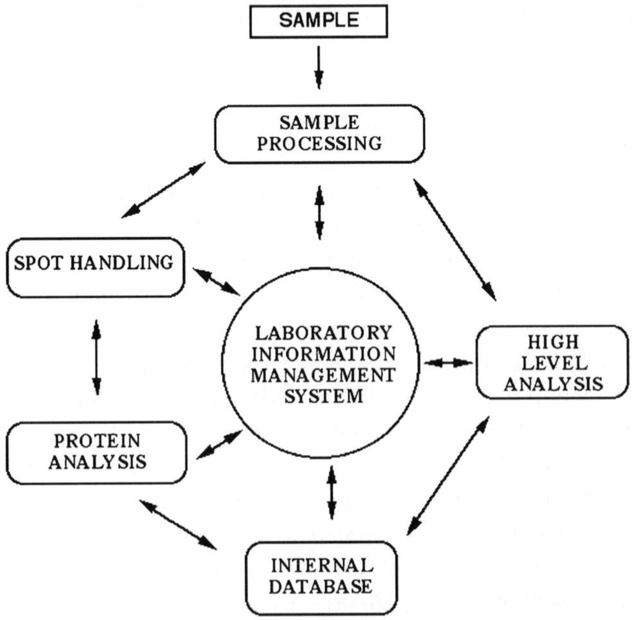

Fig. 2. System integration chart. The various modules are linked to allow process integration and data inspection. The LIMS system controls and coordinates data accumulation, data analysis and data linkage allowing sample logging and tracking, data passing between dissimilar programs and integrating the data analysis output with the databases. The high-level analysis should act as a user interface facilitating data viewing and evaluation and presenting significant links between data sets in an intuitive fashion and should allow cross-correlation of MS data and 2D patterns with a wide variety of other data such as patient files, genome database annotation, toxicity databases, gene knockout and two-hybrid experiments, etc.

9.2. Data mining

Having obtained terabytes of data, it would seem appropriate that a similar automation system be in place to handle the data processing. While this is fairly well established for processing raw data for, say, protein fingerprinting searches, higher level data processing is still a very manual process. There are a large number of bioinformatic firms appearing (and more importantly, disappearing) with a variety of solutions. Perhaps the best (but potentially very biased view based on a non-comprehensive personal survey) approach is that taken by the Genomax package from Informax (Bethesda, MD, USA). Data can be extracted from a LIMS system warehouse and then placed in a software pipeline. For example, extract all open reading frames from a genome, predict the membrane proteins, predict those membrane proteins that are involved in cell–cell contact and search an LC–MS/MS run for peptides appearing from these proteins. This could be further concatenated to join multiple LC experiments or allow the same data flow to be done with the results from a cDNA array and then to correlate the results between the DNA and protein-expression data sets. Flexibility is the key to the situation and the ability to freely plug-and-play with algorithms that are in the public domain is essential.

10. Summary

Over the past 3 years, biotechnology instrument providers have made some progress in providing a solution to the high-throughput requirements of 2D PAGE. However, perhaps due to the perceived change in the manner proteomics will be done in the future, i.e. protein chips and 2D HPLC peptide-based analysis, the limited willingness to commit necessary resources and money to develop automated tools for 2D PAGE has been somewhat of a limiting factor. Two main competitors have pulled ahead of the field and two newcomers are trying to establish themselves. The newcomers, one consortium based around Perkin Elmer and another one based around Proteome Systems, have interesting things but these need to mature soon and quickly. The more established Amersham Biosciences system and the alliance of Micromass and Bio-Rad have established comprehensive, partially automated 2D gel-based solutions. The most advanced in automation terms is the Amersham solution with the DIGE fluorescent dye and software system and a complete robotic station for multiple gel handling. However, this is only a moderate throughput solution. All other systems require a much greater degree of user intervention. In addition, there are still major gaps in the current automation solution: automation of sample preparation and gel scanning being the two most obvious gaps. It is the author's personal opinion that the field will not advance much further due to the widespread belief, however accurate or unfounded, that other techniques will take over for protein profiling. The "Emperor's new clothes" effect has already created a massive potential market for protein chip technology. However, the most urgent need for an integrated LIMS and data mining environment is taking off with small firms like Cimarron and Informax.

Acknowledgements

The author would like to thank the very generous financial support of the Wallenberg Foundation in funding the two SWEGENE proteomics centres in Lund and Gothenburg.

References

1. Sanger, F., Determination of nucleotide sequences in DNA. *Science*, **214**, 1205–1210 (1981).
2. Fleischmann, R.D., Adams, M.D., White, O., Clayton, R.A., Kirkness, W.F., Kerlavage, A.R., Bult, C.J., Tomb, J-F., Dougherty, B.A., Merrick, J.M., McKenny, K., Sutton, G., Fitzhugh, W., Fileds, C., Gocayne, J.D., Scott, J., Shirley, R., Liu, L-I., Glodek, A., Kelley, K.J.M., Weidman, J.F., Phillips, C.A., Spriggs, T., Hedblom, E., Cotton, M.D., Utterback, T.R., Hanna, M.C., Nguyen, D.T., Saudek, D.M., Brandon, R.C., Fine, L.D., Fritchman, J.L., Fuhrmann, J.L., Geoghagen, N.S.M., Gnehm, C.L., McDonald, L.A., Small, K.V., Fraser, C.M., Smith, H.O. and Venter, J.C., Whole-genome random sequencing and assembly of *Haemophilus influenzae* Rd. *Science*, **269**, 496–512 (1995).
3. Fenn, J.B., Mann, M., Meng, C.K., Wong, S.F. and Whitehouse, C.M., Electrospray ionization for mass spectrometry of large biomolecules. *Science*, **246**, 64–71 (1989).
4. Karas, M. and Hillenkamp, F., Laser desorption ionization of proteins with molecular masses exceeding 10,000 daltons. *Anal. Chem.*, **60**, 2299–2301 (1988).
5. Gygi, S.P., Rist, B., Gerber, S.A., Turecek, F., Gelb, M.H. and Aebersold, R., Quantitative analysis of complex protein mixtures using isotope-coded affinity tags. *Nat. Biotechnol.*, **17**, 994–999 (1999).
6. Washburn, M.P., Wolters, D. and Yates, J.R. III, Large-scale analysis of the yeast proteome by multidimensional protein identification technology. *Nat. Biotechnol.*, **19**, 242–247 (2001).
7. Gera, J.F., Hazbun, T.R. and Fields, S., Array-based methods for identifying protein–protein and protein–nucleic acid interactions. *Methods Enzymol.*, **350**, 499–512 (2002).
8. Rabilloud, T., Adessi, C., Giraudel, A. and Lunardi, J., Improvement of the solubilization of proteins in two-dimensional electrophoresis with immobilized pH gradients. *Electrophoresis*, **18**, 307–316 (1997).
9. Olsson, I., Larsson, K., Palmgren, R. and Bjellqvist, B., Organic disulfides as a means to generate streak-free two-dimensional maps with narrow range basic immobilized pH gradient strips as first dimension. *Proteomics*, **11**, 1630–1632 (2002).
10. Chevallet, M., Santoni, V., Poinas, A., Rouquie, D., Fuchs, A., Kieffer, S., Rossignol, M., Lunardi, J., Garin, J. and Rabilloud, T., New zwitterionic detergents improve the analysis of membrane proteins by two-dimensional electrophoresis. *Electrophoresis*, **19**, 1901–1909 (1998).

11. Madsen, P.S., Hokland, M., Ellegaard, J., Hokland, P., Ratz, G.P., Celis, A. and Celis, J.E., Major proteins in normal human lymphocyte subpopulations separated by fluorescence-activated cell sorting and analyzed by two-dimensional gel electrophoresis. *Leukemia*, **2**, 602–615 (1988).
12. Corthals, G.L., Molloy, M.P., Herbert, B.R., Williams, K.L. and Gooley, A.A., Prefractionation of protein samples prior to two-dimensional electrophoresis. *Electrophoresis*, **18**, 317–323 (1997).
13. Fountoulakis, M., Langen, H., Evers, S., Gray, C. and Takacs, B., Two-dimensional map of *Haemophilus influenzae* following protein enrichment by heparin chromatography. *Electrophoresis*, **18**, 1193–1202 (1997).
14. Porath, J., Carlsson, J., Olsson, I. and Belfrage, G., Metal chelate affinity chromatography, a new approach to protein fractionation. *Nature*, **258**, 598–599 (1975).
15. Molloy, M.P., Herbert, B.R., Walsh, B.J., Tyler, M.I., Traini, M., Sanchez, J.C., Hochstrasser, D.F., Williams, K.L. and Gooley, A.A., Extraction of membrane proteins by differential solubilization for separation using two-dimensional gel electrophoresis. *Electrophoresis*, **19**, 837–844 (1998).
16. Anderson, N.G., Preparative zonal centrifugation. *Methods Biochem. Anal.*, **15**, 271–310 (1967).
17. Hurley, S.S., Splitter, G.A. and Welch, R.A., Rapid lysis technique for mycobacterial species. *J. Clin. Microbiol.*, **25**, 2227–2229 (1987).
18. Klose, J., Large-gel 2-D electrophoresis. *Methods Mol. Biol.*, **112**, 147–172 (1999).
19. Harrington, M.G., Lee, K.H., Yun, M., Zewert, T., Bailey, J.E. and Hood, L., Mechanical precision in two-dimensional electrophoresis can improve protein spot positional reproducibility. *Appl. Theor. Electrophor.*, **3**, 347–353 (1993).
20. Bjellqvist, B., Ek, K., Righetti, P.G., Gianazza, E., Gorg, A., Westermeier, R. and Postel, W., Isoelectric focusing in immobilized pH gradients: principle, methodology and some applications. *J. Biochem. Biophys. Methods*, **6**, 317–339 (1982).
21. Sanchez, J.C., Hochstrasser, D. and Rabilloud, T., In-gel sample rehydration of immobilized pH gradient. *Methods Mol. Biol.*, **112**, 221–225 (1999).
22. Parker, K.C., Garrels, J.I., Hines, W., Butler, E.M., McKee, A.H., Patterson, D. and Martin, S., Identification of yeast proteins from two-dimensional gels: working out spot cross-contamination. *Electrophoresis*, **19**, 1920–1932 (1998).
23. Arnott, D., Henzel, W.J. and Stults, J.T., Rapid identification of comigrating gel-isolated proteins by ion trap-mass spectrometry. *Electrophoresis*, **19**, 968–980 (1998).
24. Voris, B.P. and Young, D.A., Very-high-resolution two-dimensional gel electrophoresis of proteins using giant gels. *Anal. Biochem.*, **104**, 478–484 (1980).
25. Wasinger, V.C., Bjellqvist, B. and Humphery-Smith, I., Proteomic 'contigs' of *Ochrobactrum anthropi*, application of extensive pH gradients. *Electrophoresis*, **18**, 1373–1383 (1997).
26. Münchbach, M., Quadroni, M., Miotto, G. and James, P., Quantitation and *de novo* sequencing of peptides using isotopic N-terminal tagging with a fragmentation-directing highly basic moiety. *Anal. Chem.*, **72**, 4047–4057 (2000).
27. Patton, W.F., Lopez, M.F., Barry, P. and Skea, W.M., A mechanically strong matrix for protein electrophoresis with enhanced silver staining properties. *Biotechniques*, **12**, 580–585 (1992).
28. Righetti, P.G., Chiari, M., Casale, E., Chiesa, C., Jain, T. and Shorr, R., HydroLink gel electrophoresis. *J. Biochem. Biophys. Methods*, **19**, 37–49 (1989).
29. Harrington, M.G., Lee, K.H., Bailey, J.E. and Hood, L.E., Sponge-like electrophoresis media: mechanically strong materials compatible with organic solvents, polymer solutions and two-dimensional electrophoresis. *Electrophoresis*, **15**, 187–194 (1994).

30. Hochstrasser, D.F., Harrington, M.G., Hochstrasser, A.C., Miller, M.J. and Merril, C.R., Methods for increasing the resolution of two-dimensional protein electrophoresis. *Anal. Biochem.*, **173**, 424–435 (1988).
31. Henzel, W.J., Billeci, T.M., Stults, J.T., Wong, S.C., Grimley, C. and Watanabe, C., Identifying proteins from two-dimensional gels by molecular mass searching of peptide fragments in protein sequence databases. *Proc. Natl Acad. Sci. USA*, **90**, 5011–5015 (1993).
32. James, P., Quadroni, M., Carafoli, E. and Gonnet, G., Protein identification by mass profile fingerprinting. *Biochem. Biophys. Res. Commun.*, **195**, 58–64 (1993).
33. Mann, M., Hojrup, P. and Roepstorff, P., Use of mass spectrometric molecular weight information to identify proteins in sequence databases. *Biol. Mass Spectrom.*, **22**, 338–345 (1993).
34. Pappin, D.J.C., Hojrup, P. and Bleasby, A.J., Protein identification in sequence databases. *Curr. Biol.*, **3**, 327–332 (1993).
35. Yates, J.R. III, Speicher, S., Griffin, P.R. and Hunkapiller, T., Peptide mass maps: a highly informative approach to protein identification. *Anal. Biochem.*, **214**, 397–408 (1993).
36. Jensen, O.N., Mortensen, P., Vorm, O. and Mann, M., Automation of matrix-assisted laser desorption/ionization mass spectrometry using fuzzy logic feedback control. *Anal. Chem.*, **69**, 1706–1714 (1997).
37. James, P., Quadroni, M., Carafoli, E. and Gonnet, G., Protein identification in DNA databases by peptide mass fingerprinting. *Protein Sci.*, **3**, 1347–1350 (1994).
38. Eng, J.K., McCormack, A.L. and Yates, J.R. III, Peptide identification by uninterpreted MS/MS spectra database searching. *J. Am. Soc. Mass Spectrom.*, **5**, 976–989 (1994).
39. Mann, M. and Wilm, M., Error-tolerant identification of peptides in sequence databases by peptide sequence tags. *Anal. Chem.*, **66**, 4390–4399 (1994).
40. Hunt, D.F., Yates, J.R. III, Shabanowitz, J., Winston, S. and Hauer, C.R., Protein sequencing by tandem mass spectrometry. *Proc. Natl Acad. Sci. USA*, **83**, 6233–6237 (1986).
41. Biemann, K., Sequencing of peptides by tandem mass spectrometry and high-energy collision-induced dissociation. *Methods Enzymol.*, **193**, 455–479 (1990).
42. Yates, J.R. III, Eng, J.K., McCormack, A.L. and Schieltz, D., Method to correlate tandem mass spectra of modified peptides to amino acid sequences in the protein database. *Anal. Chem.*, **67**, 1426–1436 (1995).
43. Yates, J.R. III, Eng, J.K. and McCormack, A.L., Mining genomes: correlating tandem mass spectra of modified and unmodified peptides to sequences in nucleotide databases. *Anal. Chem.*, **67**, 3202–3210 (1995).
44. Griffin, P.R., MacCoss, M.J., Eng, J.K., Blevins, R.A., Aaronson, J.S. and Yates, J.R. III, Direct database searching with MALDI-PSD spectra of peptides. *Rapid Commun. Mass Spectrom.*, **9**, 1546–1551 (1995).
45. McCormack, A.L., Schieltz, D.M., Goode, B., Yang, S., Barnes, G., Drubin, D. and Yates, J.R. III, Direct analysis and identification of proteins in mixtures by LC/MS/MS and database searching at the low-femtomole level. *Anal. Chem.*, **69**, 767–776 (1997).
46. Schultz, G.A., Corso, T.N., Prosser, S.J. and Zhang, S., A fully integrated monolithic microchip electrospray device for mass spectrometry. *Anal. Chem.*, **72**, 4058–4063 (2000).
47. Martino, R.L., Johnson, C.A., Suh, E.B., Trus, B.L. and Yap, T.K., Parallel computing in biomedical research. *Science*, **265**, 902–908 (1994).
48. Williams, H. and Zobel, J., Compression of nucleotide databases for fast searching. *Comput. Appl. Biosci.*, **13**, 549–554 (1997).

Proteome Analysis. Interpreting the Genome.
D.W. Speicher (editor)

Chapter 13

Micro- and nanotechnology for proteomics

G. MARKO-VARGA[a], J. NILSSON[b] and T. LAURELL[b,*]

[a] *Department of Analytical Chemistry, Lund University, P.O. Box 124, SE-221 00 Lund, Sweden*
[b] *Department of Electrical Measurements, Lund Institute of Technology, Lund University, P.O. Box 118, SE-221 00 Lund, Sweden*

1. Introduction

The increased focus on proteomics research following the completion of the human genome in June 2000 [1,2], has contributed to a strong demand for new bioanalytical technologies that facilitate the task of proteome mapping. More specifically, new analytical technology is greatly needed that provides simple methods for rapidly extracting and identifying proteins that may have key regulating functions in disease progresses and normal biological processes. Due to the diverse nature of proteins (hydrophilic, hydrophobic, amphiphilic, size, surface charge, etc.) it can be anticipated that no single bioanalytical technique will dominate the oncoming biotechnology development as was the case for

* Corresponding author. E-mail: thomas.laurell@elmat.lth.se (T. Laurell).

capillary electrophoresis in the mapping and sequencing of the human genome [3]. Although chromatography, gel electrophoresis and mass spectrometry are some of the cornerstones of today's approaches to protein expression analysis [4], it can be anticipated that tomorrow's methodology will not only comprise these techniques, but will also form winning concepts in smart hyphenations with new emerging analytical methodology and technology.

Biological samples frequently are very limited, whether they are biobank material, biopsies from pathogenic tissue, laser microdissections, excised 2D gel spots or capillary chromatographic fractions. Hence, there is a high priority need for new analytical protocols that can efficiently process samples of single microliters or even smaller volumes. In this quest, microfluidics has emerged as an important field in which new concepts for sample processing and transport are being developed. Also, new microfluidic technology that enables the handling of minute sample volumes without the adsorptive losses on surfaces commonly encountered in standard processes, will drive important developments within this field. Progress within the micro- and nanotechnology field has now reached a level where state-of-the-art processes and material technologies can offer a vast engineering toolbox for the manufacture of advanced micro- and nanostructures [5,6] which, if properly designed, may provide superior concepts for sample processing by performing liquid manipulation at the nanoliter to picoliter scale. Early initiatives to develop chip-integrated chemical analysis systems were proposed by Terry et al. [7,8] in 1975 with a complete gas chromatograph integrated on a 2 in. wafer. Petersen et al. [9] demonstrated the potential of using microtechnology for the development of chip-based microfluidic technology. Microtechnology has indeed been successful as illustrated by the immense impact that inkjet printing technology has had within the IT industry [10]. It can be foreseen that similar successes may be achieved in a wide range of applications within the medical and bioanalytical field. In the early nineties, the area of chip-integrated chemical analysis started to accelerate, primarily due to the fact that both the MEMS (Micro-Electro-Mechanical System) and the bioanalytical field had reached a level of maturity from which productive interdisciplinary activities were feasible. In 1990, Manz et al. coined the μTAS concept (Micro Total Analysis Systems) [11], also called Lab-On-A-Chip, and in 1994 the first μTAS conference was held in Twente, The Netherlands [12]. These conferences have emerged as the leading forum for the advancements of μTAS. The optimism in the oncoming success of new μTAS technology is also reflected in the rapidly rising number of start-up companies with a strategic base in new microfluidic concepts.

In spite of recent efforts invested in this, very few commercial successes have been seen so far, although Agilent Technologies (Palo Alto, CA, USA) and Caliper Technologies Corp. (Mountain View, CA, USA) have successfully headed commercial initiatives in glass-chip-based capillary electrophoresis for both DNA and protein analysis [13]. In fields related to proteomics, DNA-microarray technology, pioneered by Affymetrix Inc. [14], has shown the most successful commercial progress. It is anticipated that new microarray technology based on antibody probes, i.e. protein chips, will have a very strong impact in the proteomics field. Recently Zhu et al. [15] demonstrated the potential of protein chip technology by presenting a study where 119 proteins of 122 kinases in yeast were mapped on a chip, and Zhu et al. [16] performed a global protein analysis on yeast proteins using protein chips with 580 over-expressed, purified recombinant proteins. A more in-depth description of the protein chip approach to proteomics is given in Chapter 10.

In recent years, mass spectrometry and peptide mass fingerprinting has become a major strategy for protein identification in proteomics studies [17]. Although electrospray ionisation mass spectrometry initially dominated this approach to protein analysis [18], MALDI MS is an alternative and complementary technique to electrospray ionisation. MALDI MS has gained increased popularity due to its ease of use, high sensitivity and high throughput. More recently, the introduction of MALDI-based MS/MS instruments has expanded the utility of MALDI MS approaches [19]. In this context, the use of isotope labelling [20,21] has proven to be a useful strategy for quantitating proteins that are analysed, using liquid phase separation coupled online to MS/MS sequencing.

In view of μTAS developments for proteomic applications, a chip interface for MS has always been a challenge and is currently of high priority in the research community. Chip-based proteomic sample processing prior to the introduction into the mass spectrometer is of vital importance for ESI and MALDI-based MS. For the ESI MS a major effort is being invested in the development and design of robust electrospray nozzles that provide a reproducible and stable electrospray [22]. A related area of investigation is the integration of ESI-nozzles with sample processing chips [23,24]. Regarding MALDI-TOF MS, micro/nanotechnology approaches to sample transfer from sample processing chips to MALDI target plates and new designs for sample target plates are major areas for further development [25,26].

The discussion of micro- and nanotechnology based proteomics in this chapter will primarily focus on strategies for interfacing MALDI-TOF MS to microfluidic systems and μTAS.

2. Benefits of miniaturisation

As already mentioned, the increased need for handling minute sample volumes and low analyte levels is a very strong driving factor for the miniaturisation of proteomic protocols and associated analytical technology. It should be noted that miniaturisation inherently may provide analytical benefits. Parameters that may improve the overall performance of the analytical system as a result of miniaturisation are briefly listed below.

Reduced sample volumes mean that the original sample volume can be divided for multiple analysis protocols optionally providing more accurate information regarding the analyte, e.g. statistically more correct data or improved sequence coverage in peptide fingerprint mapping.

- Surface area-to-volume ratio increases, which means that as the dimensions are reduced the influence of surfaces becomes large and consequently surface-bound chemistry, e.g. immobilised chemistry, is favoured in the micro/-nanodomain as compared to bulk solution-based reactions.
- Reaction kinetics and thus the analytical throughput may be increased due to shorter distances for analyte diffusion. In the case of immobilised chemistry, e.g., an enzymatic digestion of a protein in a microchannel with the protease immobilised on the channel wall, the time for substrate diffusion from the solution in the channel to the catalytic site is reduced by two orders of magnitude as the dimension of the channel is reduced by one order of magnitude. The average diffusion distance, x, for a molecule with a diffusion coefficient, D, during a time period, t_d, is given by: $x = (2Dt_d)^{0.5}$ [27].
- As dimensions and thus volumes are reduced, evaporation becomes a major issue, e.g. small volumes in the nanoliter to picoliter range evaporate from surfaces in a few seconds or less. This may be utilised as an inherent property for sample enrichment and improved analytical readout as will be further described in this chapter.
- Reagent costs may be dramatically reduced. By reducing the length dimension of a reaction vessel by one order of magnitude, e.g. from 5 mm to 500 μm, the reagent consumption is decreased by three orders of magnitude.
- Reduced dimensions also provide improved characteristics, e.g. in separation science [28].

3. Miniaturisation in proteomics

In the currently used standard protein identification scheme, a number of steps can be identified as potential candidates for miniaturisation to improve the analytical performance with respect to sample throughput, minimum required sample volume and sensitivity. Fig. 1 is a schematic drawing of a standard process scheme for a proteomic sample. The four highlighted boxes indicate process steps that may provide improvements to the protein identification process if converted to a microscale format. Section 4 will give examples of micro- and nanotechnology developments that address these issues. Although no detailed example will be given on the benefits of miniaturising the process step of biomolecule separation in this chapter, it is evident that this is a most promising approach to improve

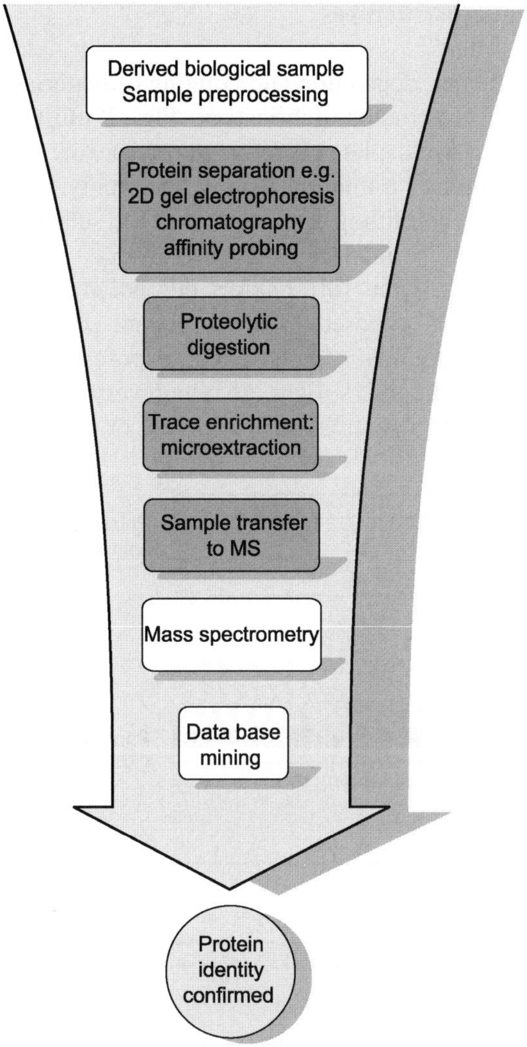

Fig. 1. Proteomic sample processing scheme. The shaded boxes indicate process steps that may provide improved overall analytical performance by incorporating micro/nanotechnology concepts.

analytical performance. The field of chromatography has historically driven the development of capillary separation technology and has frequently benefited from miniaturisation by reduced column dimensions. A good proteome analysis illustration is the multidimensional capillary separations on a large scale where a major part of the yeast proteome was mapped [29].

4. Fabrication of microstructures

Photolithography, etching, deposition and bonding of materials serve as the basis for principally all microstructure fabrication. Fig. 2 illustrates the process of etching a groove in a glass plate. Briefly, the shape of a desired structure is generated as a 2D pattern in a thin metal film, commonly chromium, on a planar glass plate, i.e. the photolithography mask. The substrate into which the microstructure is to be formed, e.g. a glass wafer, is spin coated with a photosensitive polymer. After soft baking the polymer film at an elevated temperature, the wafer is exposed in a subsequent photolithographic process where a photo mask contains a pattern that exposes desired regions of the photosensitive polymer film to UV-light. To obtain very high-resolution micro- or nanostructures, other energy sources such as X-ray or e-beam exposures may be

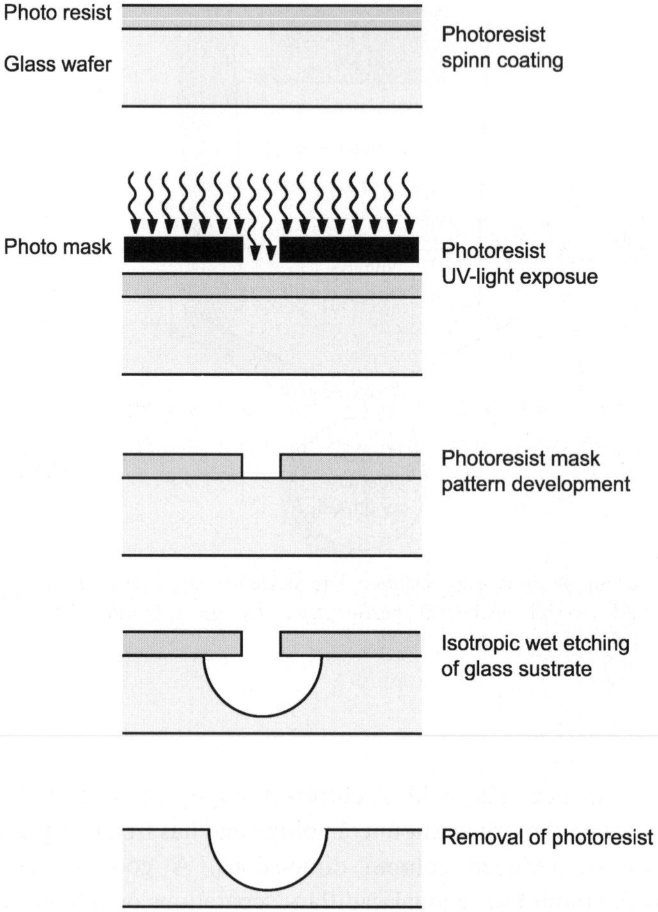

Fig. 2. Principal processing sequence for etching a microchannel in a glass wafer.

utilised. After the exposure, the polymer film is developed and the desired areas on the glass wafer are accessible for further process steps such as an isotropic etch to fabricate a capillary in the glass wafer. When processing silicon, inherent properties of the mono-crystalline material can be utilised to obtain very well defined v-grooves or vertical walls by means of anisotropic wet etchants. Fig. 3 shows an array of parallel vertical channels anisotropically wet etched in ⟨110⟩-oriented silicon. This structure has served as the base for the protease microreactors discussed later in this chapter.

High surface area microstructures are highly desirable when performing surface bound chemistry in a lab-on-a-chip format. Silicon offers favourable intrinsic properties for fabricating very high surface area structures. By means of anodic etching, porous silicon can be achieved [30]. Porous silicon can have surface area properties that match any high surface area solid phase matrix available [31]. Fig. 4 shows a scanning electron microscope image of a porous silicon surface that displays very good catalytic performance when immobilised with enzymes. The surface was obtained by anodic etching in a mixture of hydrofluoric acid and ethanol [32].

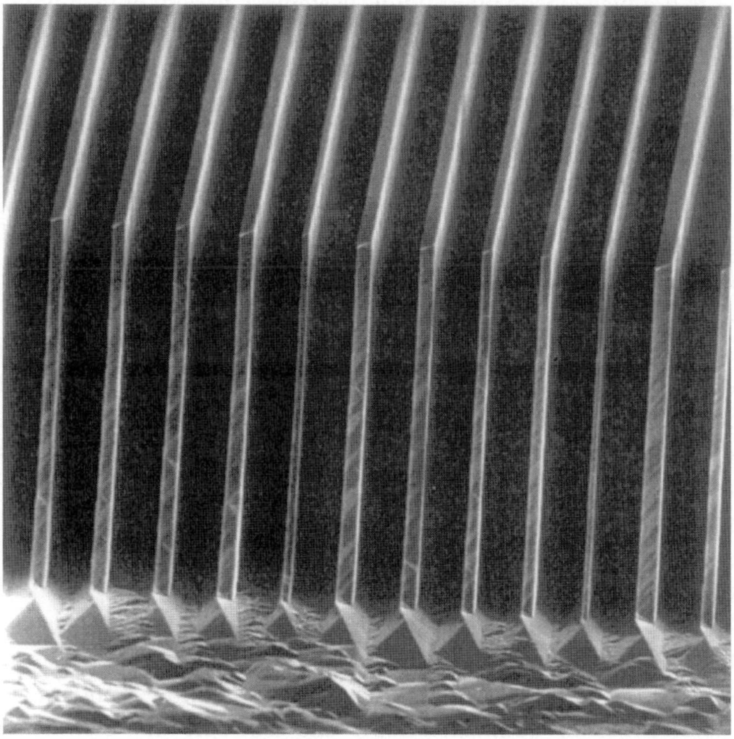

Fig. 3. Channel array with vertical walls anisotropically etched in ⟨110⟩-silicon.

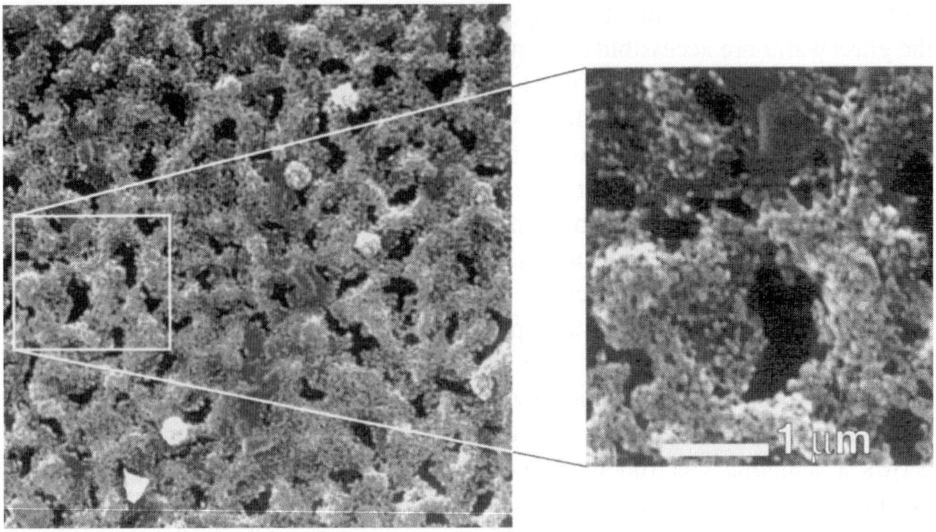

Fig. 4. Scanning electron image of a porous surface well suited for immobilising enzymes.

In order to produce closed microfluidic systems, a variety of techniques to bond microstructurable materials are available. For example, silicon microstructures are commonly sealed by means of anodic bonding of glass or direct bonding to another silicon wafer, and glass chips are commonly bonded to glass. Addition of material to the wafer being processed is done by spin or spray coating when working with polymer films, and when working with metals or semiconductor films the material is evaporated, sputtered, chemical vapour deposited or gas phase deposited.

The conditions for etching may be changed by addition of material in the substrate wafer. For example, by doping the silicon crystal with boron atoms, the etch rate of silicon in an anisotropic etch can be essentially stopped. Similarly, by doping silicon with phosphorous atoms, doped regions may be protected from anisotropic etching by applying a potential to the doped silicon. These stop etch techniques enable the formation of advanced microstructures by doping silicon in selected areas so that well-defined thin membranes of micrometer thickness can be formed over large wafer areas or non-planar structures can be developed by sacrificial etching of silicon that contains doped cavities. Fig. 5 shows an example of the process steps for fabricating the protruding droplet nozzle described in Section 4.3. The resulting nozzle is imaged in Fig. 12a.

Over the past several years, plasma etching techniques, i.e. deep reactive ion etching (DRIE), have become very popular and are now standard techniques offering possibilities of both anisotropic and isotropic dry etching with very attractive process features such as high etch rate and high aspect ratio processing

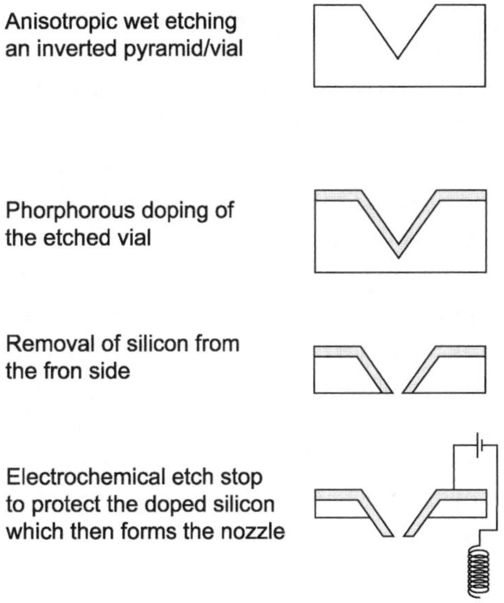

Fig. 5. Process scheme for fabricating well-defined protruding microdroplet nozzles in silicon.

[33,34]. In addition, polymer micromoulding and casting techniques are rapidly gaining ground in the area of lab-on-a-chip developments as large-scale and low-cost fabrication of disposable microfluidic structures can be accomplished [35,36].

More in-depth descriptions of the process schemes for microstructure fabrication are described by Madou [5] and Gad-el-Hak [6].

5. Microstructures for proteomics

5.1. Protein digestion on-chip

The parallel channel microstructure shown in Fig. 2 was initially developed as a microreactor for immobilised glucose oxidase in a glucose monitoring system. The enzyme was immobilised by standard protocols for coupling enzymes to silica matrices. Although the device worked well, further increase in the amount of enzyme loaded onto the microchip channels was sought. Subsequently, porous silicon technology was introduced in the authors' laboratory as a method for increasing surface area in chemically surface activated microfluidic structures. By tailoring the porous silicon morphology, it was possible to obtain a several 100-fold increase in catalytic turnover in the microchip. A thorough survey of the development of porous silicon enzyme microreactors is given by Laurell [37].

During miniaturisation efforts in the proteomics area, it became evident that immobilised protease chemistry could effectively be performed in the porous silicon microreactor already under development in the authors' research group. The initial experiments by Ekström et al. [38] reported online tryptic digestion of 1 μl samples of myoglobin, lysozyme and cytochrome c (≈500 μM). The system allowed a throughput of 100 proteins in 3.5 h. The increased speed of digestion, 200–1000 times, as compared to standard bulk solution digestion protocols is attributed to the reduced dimensions and the high surface area-to-volume ratio due to the porous silicon in the microchannels.

The stability of the microdigestion chips was found to be quite satisfactory as 400 digestions were performed over several weeks before a degradation in digestion efficiency was observed. It was also shown that improved digestion efficiency was obtained when the digestion chip temperature was increased. Table 1 shows the increase in number of identified peptides as the temperature was increased from 20°C to 60°C when digesting cytochrome c.

A detailed analysis of the porous silicon in the channels of the microdigestion chip showed that the porous layer displayed an inhomogeneous morphology from the bottom of the channels to the top. In order to maximise the performance of the microdigestion the porous surface has to have the proper morphology throughout the channel walls. By altering the chip design and the process conditions for forming the porous silicon layer, a digestion chip with a homogeneous and optimised pore morphology was obtained [39]. Fig. 6a shows a cross-section of the early version non-optimised porous silicon trypsin chip and Fig. 6b shows the optimised chip with a homogenous pore depth along the full height of the flow channels. The new chip displayed satisfactory digestion of β-casein with only 12 s exposure to the immobilised trypsin, as shown in Fig. 7.

While the digestion microchip might not offer quite the same surface area as a microcolumn with packed material, it has a lower hydrodynamic pressure drop, is less sensitive to carry-over effects and is much simpler to batch manufacture in the desired small sizes.

As an alternative to in situ fabricated porous surfaces for immobilised proteases in microdigestion protocols, trypsin-activated beads loaded into microfluidic chips

Table 1

Effects of temperature on digestion efficiency

Temperature (°C)	Number of peptides[a]
20	8
40	15
60	17

[a] Improved digestion efficiency can be obtained by placing the microdigestion chip on a temperature controlled plate. The number of identified peptides increased from 8 to 17 when rising the temperature from 20°C to 60°C.

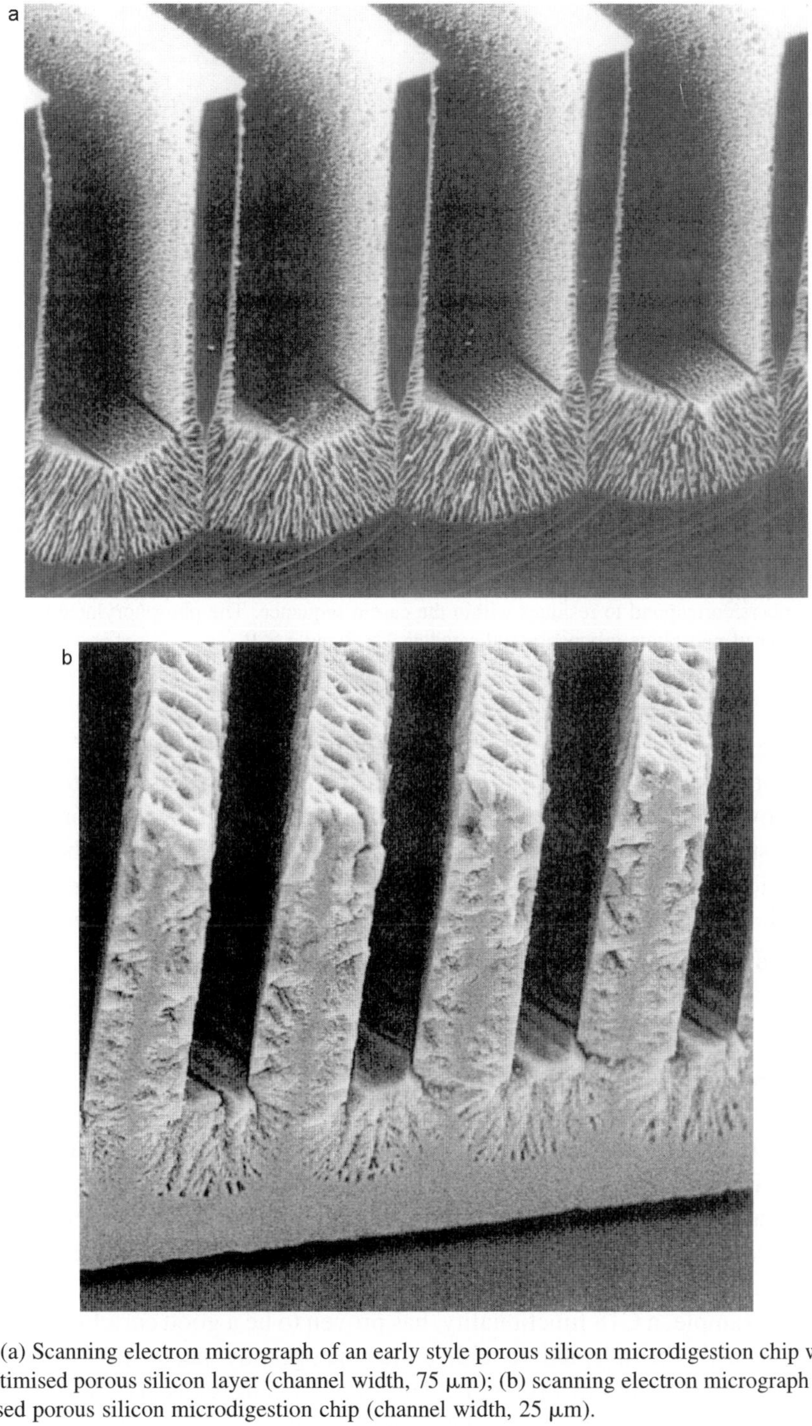

Fig. 6. (a) Scanning electron micrograph of an early style porous silicon microdigestion chip with a non-optimised porous silicon layer (channel width, 75 μm); (b) scanning electron micrograph of an optimised porous silicon microdigestion chip (channel width, 25 μm).

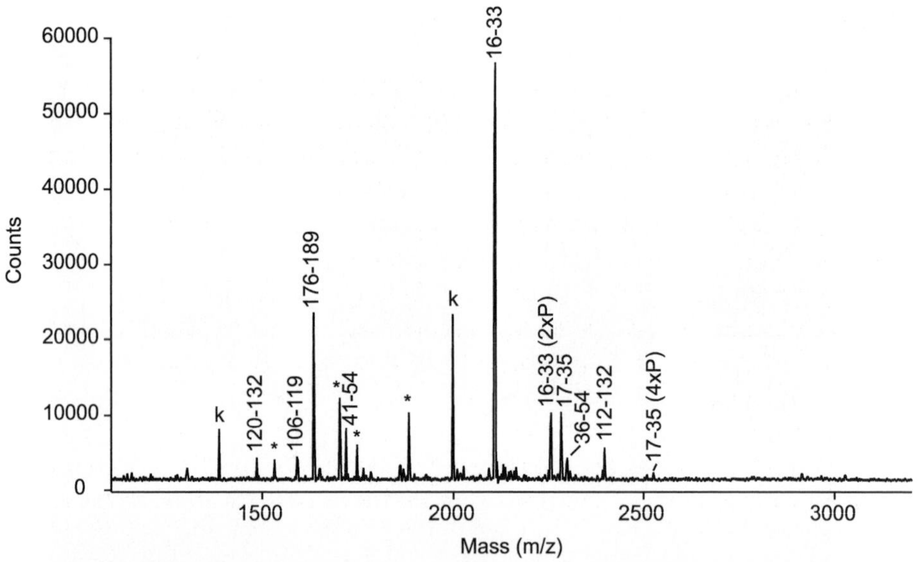

Fig. 7. MALDI-TOF mass spectra showing the peptide map of a 12 s digestion of a β-casein sample. The numbers correspond to residues within the casein sequence. The phosphorylated peptides and the number of phosphate groups are indicated as 2 × P and 4 × P.

have also been investigated. Xue et al. performed tryptic digestion of melittin in a microchip linked to ESI MS [40]. A similar investigation was reported by Wang et al. showing on-chip digestion and CE separation of the digest prior to ESI MS [41]. Ekstrom et al. also investigated digestion by trypsinated microbeads loaded into a microchip. A 1 μl BSA sample, 100 nM, was injected into the reactor and digested. The effluent was deposited on a MALDI target and identified by peptide mass fingerprint mapping [42].

5.2. Microchip solid phase enrichment

Methods to improve detection level are always attractive from an analytical point of view. Excised 2D gel spots and the resulting digests are frequently at very low concentrations, and thus new and improved protocols for sample enrichment are of great importance. The field of chromatography has again guided strategies adapted in proteomics and the use of protein and peptide adsorbing porous microbeads, with for example, a C18 functionality, has proven to be a good enrichment strategy that can easily be adapted to either chromatography needs or direct sample enrichment and clean-up protocols. The high loading capacity of such a bead matrix is the key to the high enrichment factors reported [43]. Solid phase extraction (SPE) principles have, therefore, been adapted to the robotised protocols of proteomics sample handling. The use of ZipTips™ (Millipore, Bedford, MA, USA), i.e. disposable pipette tips that have the end of the tip loaded with a solid phase, has thus been widely accepted as a way of proteomic sample clean-up and concentration prior to MS analysis. The basic work using micro-capillary columns, or nano-capillaries for simultaneous sample clean-up and sample enrichment, was performed by Annan, Kussmann, and others [44,45].

Gobom et al. [46] presented a simple, cheap and highly efficient method for sample clean-up and enrichment by using in-gel loader tips containing reversed phase packing materials from where the protein samples were directly eluted onto a MALDI target plate with a displacing eluent containing the matrix. This methodology is also compatible with static nano-electrospray MS/MS. For this approach, the capillaries are mounted into the electrospray needle and the protein sample is directly eluted into the needle tip. Later developments utilising nano-capillary extraction for phosphoprotein speciation were carried out by Stensballe and Jensen [47].

As sample volumes shrink and amounts of analytes become very low, sample handling protocols have to become more integrated to avoid undesired exposure of the sample to surfaces where the analyte may be lost/adsorbed before identification. When miniaturising proteomic sample processing into a chip format it is evident that SPE protocols also have to follow the same route. The benefit is in the totally integrated sample handling that may be accomplished. Prior to development of the lab-on-a-chip concept, capillaries packed with SPE beads were reported as miniaturised devices for sample enrichment [48–50]. Later, several groups reported bead-based SPE on chips. Oleschuk et al. [51] described a glass chip with a weir as the bead trapping structure. Similar approaches using immuno-activated beads have also been reported [52,53]. Gyros AB, Uppsala, Sweden is marketing a polymer-fabricated microfluidic device for proteomic sample preparation incorporating an SPE microarray for sample enrichment and clean-up prior to MALDI-TOF MS readout [54]. Recently, a CE-chip with integrated solid phase enrichment and an electrospray interface was described by Li et al. [55], which had a peptide mapping throughput of 12 samples per hour and a detection level of 5 nM, i.e. 25 fmol on-chip. Antibody-activated beads were also used to detect target peptides at a level of 20 ng/ml.

The use of a weir to trap beads in a flow channel is a common way to create a solid phase packed bed. However, this design may not always be optimal from a microfluidic standpoint because the weir actually is a constriction of the flow channel, and when the channel widens again after the weir, stagnant fluidic zones are obtained that introduce undesired dispersion to the system. Fig. 8 shows a cross-section of the flow profile in a weir type of bead trap. This data was derived using microfluidic modelling of the weir microstructure in ANSYS R5.7.1 FLOTRAN (Ansys, Canonsburg, PA, USA). The dark zones indicate stagnant

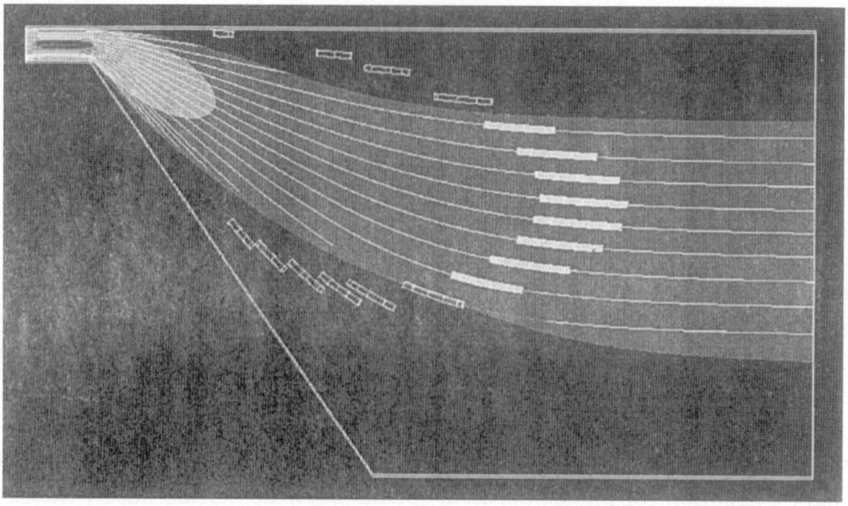

Fig. 8. Microfluidic modelling of the flow profile of a weir style bead trap. Dark zones indicate low flow velocities that cause sample dispersion. The rectangles indicate the magnitude of the flow velocity vector and dispersion of the sample plug.

areas which give rise to sample dispersion as the enriched sample is eluted from the solid phase bed. In order to investigate the influence of dispersion due to the weir, an optional bead trap design composed of thin standing walls in the flow channel has been studied [56]. Fig. 9 shows a light microscope photograph of the standing wall bead trap with the packed bed (left) and a fused silica capillary coupled to the chip outlet (right). The insert photograph is a scanning electron micrograph of the microfabricated bead trap.

MALDI-TOF MS data verified the modelling results described earlier. After 60 s, peptides were no longer eluted from the standing wall SPE chip, whereas after 120 s peptides were still detected when eluted from the weir SPE chip. The performance of the standing wall SPE chip was further investigated by loading 10 μl of a 10 nM peptide mixture (angiotensin I, ACTH clip 1-17, ACTH clip 18-39 and ACTH clip 7-38) onto the packed bed. The MALDI mass spectrum from the pure peptide mixture in Fig. 10a shows no peptides, whereas in Fig. 10b all four peptides were unambiguously identified when eluted from the SPE chip. A more detailed investigation of the standing wall microextraction chip has been published [56].

As discussed earlier, miniaturised techniques enabling the integration of SPE both with ESI MS and MALDI MS have a bright future. Totally integrated sample handling is of key importance to maximise system sensitivity, and optimised microfluidic design will have a strong impact on the performance of miniaturised proteomic sample processing.

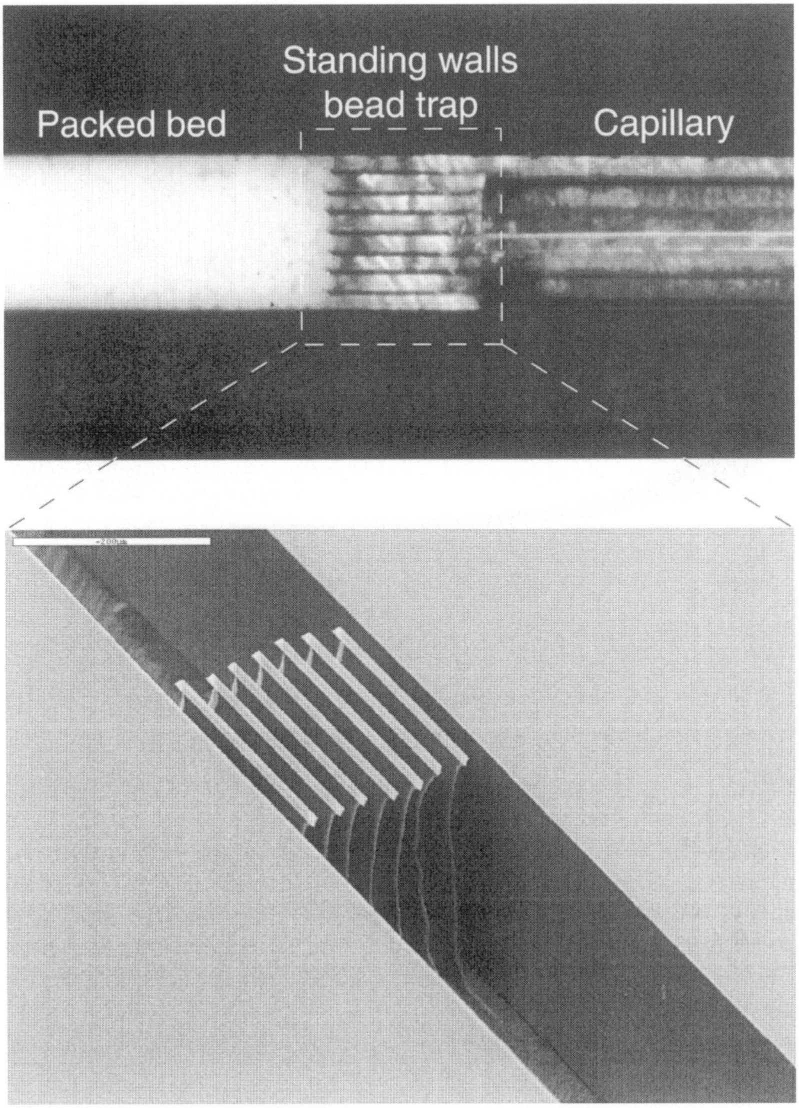

Fig. 9. The standing wall solid phase microextration chip imaged in a light microscope with the standing walls (centre), the packed bed (left) and a fused silica capillary coupled to the chip outlet (right). The insert photo is a scanning electron micrograph of the bead trap. The flow channel is 220 μm wide.

5.3. Microdispensing to interface MALDI

As dimensions are reduced, surface tension becomes a critical parameter when handling liquids. Reproducibility is not easily guaranteed when processing

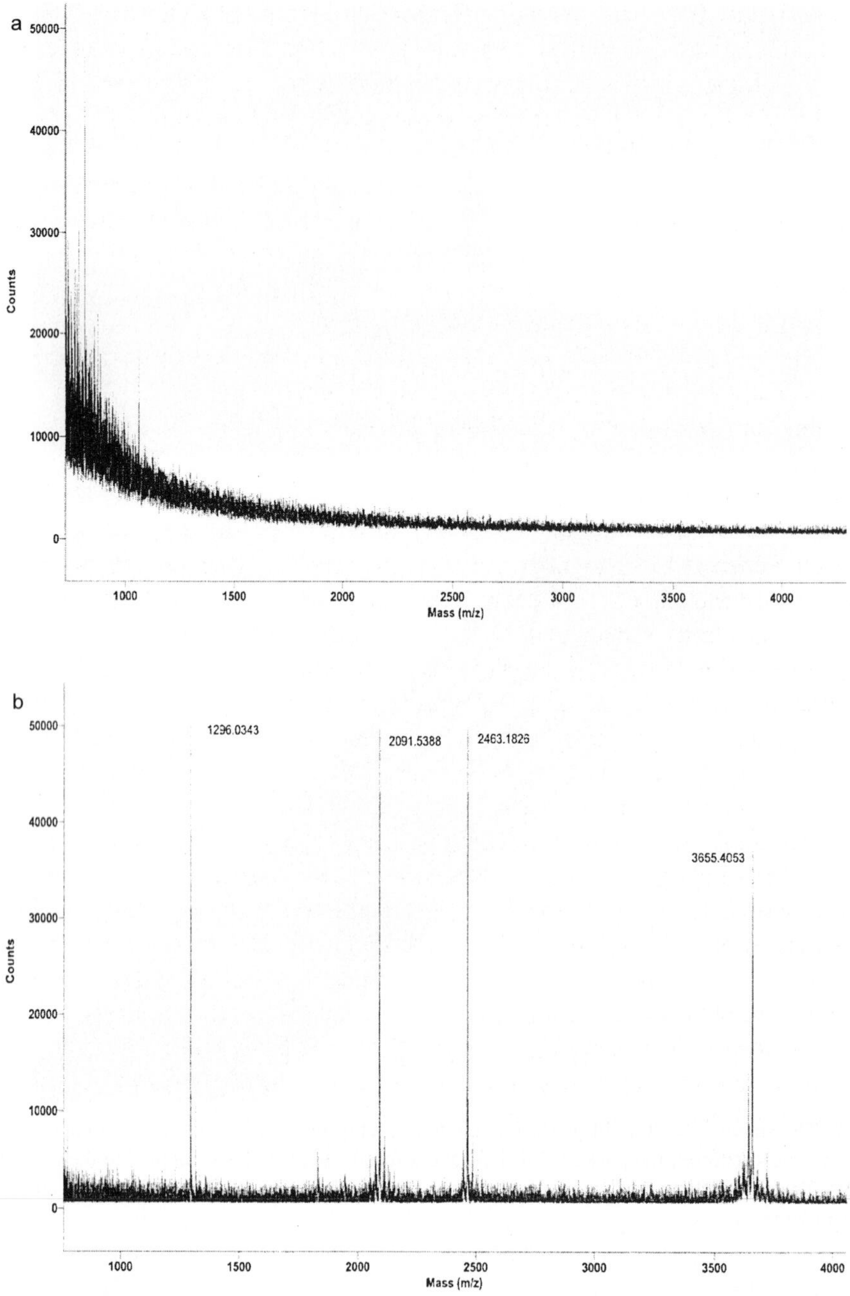

Fig. 10. MALDI mass spectrum of a 10 μl, 10 nM, peptide mixture (angiotensin I, ACTH clip 1-17, ACTH clip 18-39 and ACTH clip 7-38) (a) before SPE microchip enrichment; and (b) after enrichment and elution from the SPE microchip.

volumes below 0.5–1 μl. When performing on-chip solid phase enrichment, the eluted sample may be localised in a volume of a 100 nl or less. In order to handle such a volume optimally, pipetting is no longer an alternative, and thus reproducible sample deposition from a chemical microsystem onto a target plate for MALDI is not a trivial task.

Sample deposition on the MALDI target from fused silica capillaries is one option that has been explored. In 1995 Chiu [57] demonstrated a fraction collection system that spotted the effluent from a capillary electrophoresis system on a rotating target. The same group also demonstrated the potential of using this fraction collection technique as an off-line interface to MALDI-TOF MS. Later Stevenson et al. [58] reported on a similar approach where a capillary was mounted in a modified pen plotter to translate the capillary in the *x*- and *y*-directions and to move it up and down for sample deposition on the target, which was dried and transferred to the MALDI mass spectrometer vacuum chamber. Preisler et al. [59] demonstrated a very promising online interface to MALDI MS by means of a capillary that continuously deposited the effluent mixed with MALDI matrix on a moving tape. As the sample dried on the tape it was continuously moved into a vacuum chamber where the laser desorption of the sample took place. Subsequently, an eight capillary array format was described [60]. A related technique to interface capillary separations with MALDI MS was the rotating ball inlet reported by Orsnes et al. [61,62].

An alternative approach to interface continuous flow systems to MALDI-TOF MS was recently launched by Waters Inc. [63] where the capillary effluent is nebulised/sprayed by a heated gas sheet flow surrounding the capillary. The capillary end is positioned in the vicinity of the target plate to enable continuous tracks of chromatograms to be deposited. Sensitivity levels of 200 amol/μl have been reported. Likewise, Hensel [64] reported electrospraying samples directly onto a target plate for subsequent MALDI-TOF MS. The electrosprayed samples displayed a very fine microcrystalline surface and highly reproducible mass spectra were obtained as compared to conventionally air-dried sample spots.

Piezo electric dispensing has emerged as a promising alternative to the above described sample deposition methods using capillaries. Little et al. [65] demonstrated deposition of DNA samples in arrays of $800 \times 800\ \mu m^2$ wells for subsequent MALDI-TOF analysis by means of piezoelectric dispensing. Shortly thereafter Önnerfjord et al. [66] reported the first proteomic sample processing using piezoelectric dispensing and MALDI-TOF MS. In order to overcome the limitations of commercial piezo dispensers, which have only one inlet for loading and one outlet through the dispensing nozzle, the flow-through microdispenser was developed in the authors' laboratory [67,68]. The dispenser featured a chip integrated flow-through channel incorporating a microfabricated nozzle for online sample ejection by means of a piezoelectric element coupled to the dispenser chip. Fig. 11 shows schematic cross-sections of the flow-through dispenser design.

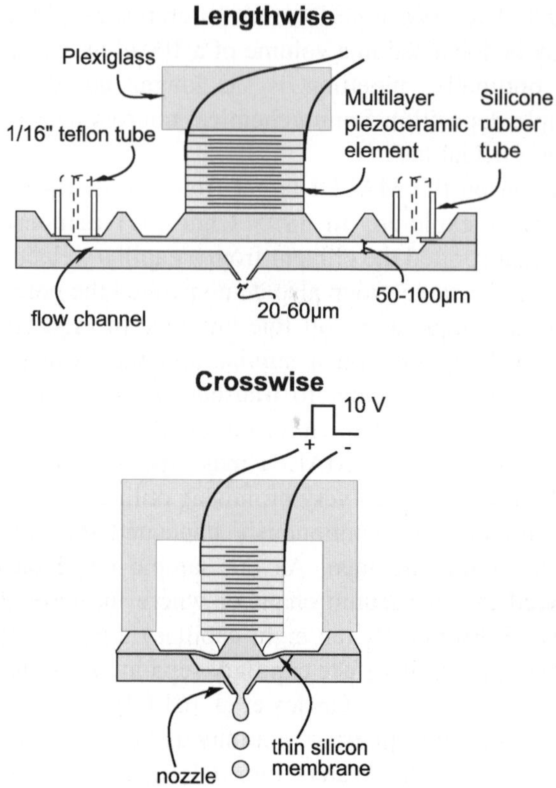

Fig. 11. Schematic cross-sections of the flow-through piezoelectric microdispenser.

As the piezoelectric element is actuated, the thicker silicon push-bar, which is in situ fabricated on a thin (7 μm) silicon membrane, is depressed into the underlying flow channel. The generated pressure pulse subsequently ejects a droplet through the microfabricated nozzle in the opposing channel wall.

A key feature of the microdispenser is the microfabricated nozzle (Fig. 12a) which ensures stable droplet formation and directivity. Fig. 12b shows a stroboscopic video image of high-speed droplet ejection at 5 kHz. The fine nozzle front surface area is not prone to contamination due to surface tension effects. Possible contaminants that may disturb the droplet formation tend to be collected at the side of the nozzle or the planar surface surrounding the nozzle, as illustrated in Fig. 13, rather than at the nozzle front surface, thereby avoiding any physical contact with the droplet formation process.

The flow-through microdispenser currently has an internal volume of 200 nl from inlet to nozzle. The nozzle opening is typically 40 μm providing droplets of approximately the same diameter, i.e. a droplet volume of 35 pl.

a

b

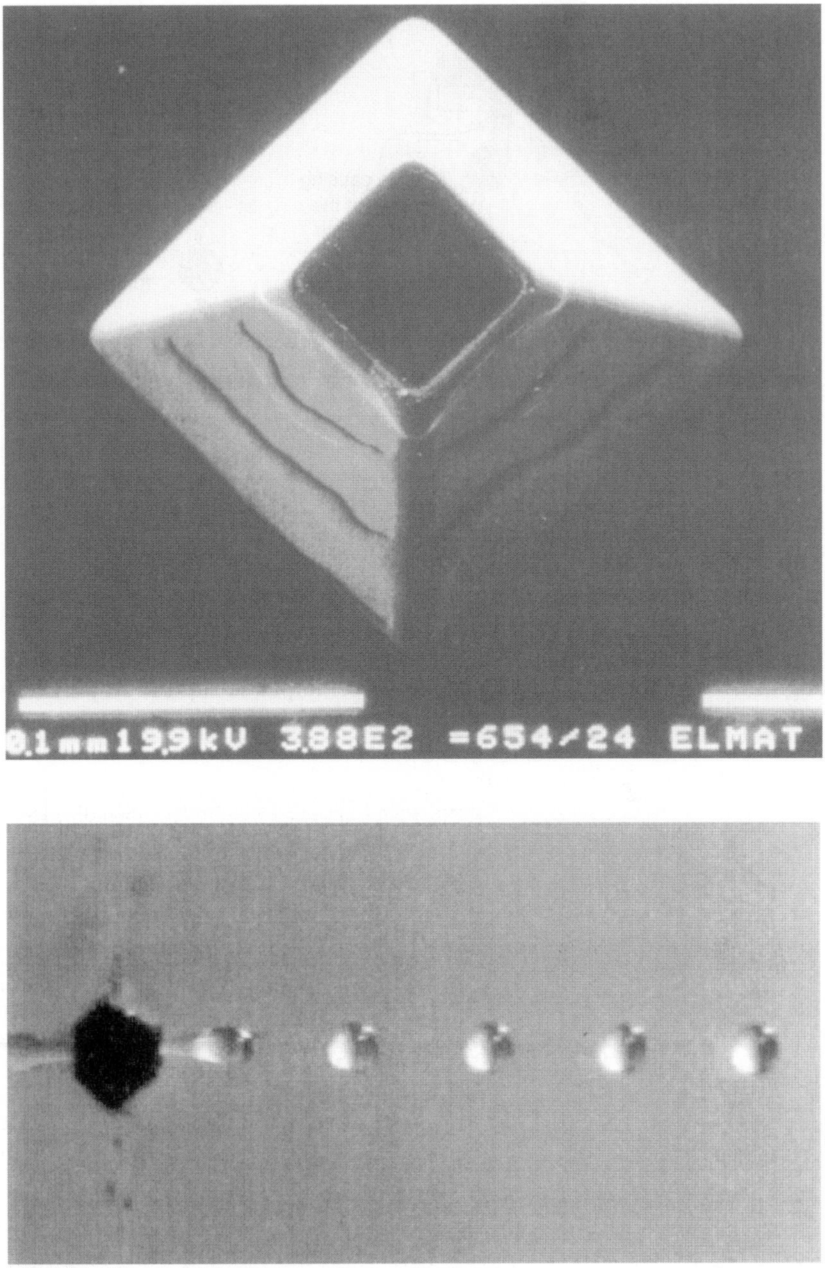

Fig. 12. Scanning electron micrograph of the droplet ejecting nozzle (a) and 5 kHz droplet ejection sequence recorded by a stroboscopic microscope set-up (b).

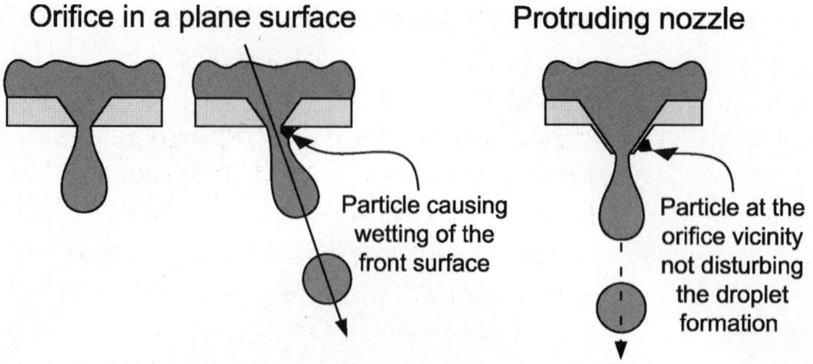

Fig. 13. Schematic drawing showing the improved droplet directionality due to reduced tendency of wetting or contamination of the nozzle front surface area as compared to a nozzle with an orifice in a planar surface.

The design of the dispenser enables online coupling of piezoelectric dispensing to any chemical process performed in a flow-through format. Chromatography systems and flow injection analysis (FIA) systems can easily be linked online to the dispenser for sample transfer to a MALDI target plate, serving as a generic interface to MALDI-TOF MS. Capillary liquid chromatography has successfully been interfaced to MALDI-TOF MS by means of the flow through microdispenser. Fig. 14 shows the principal set-up used by Milotis et al.

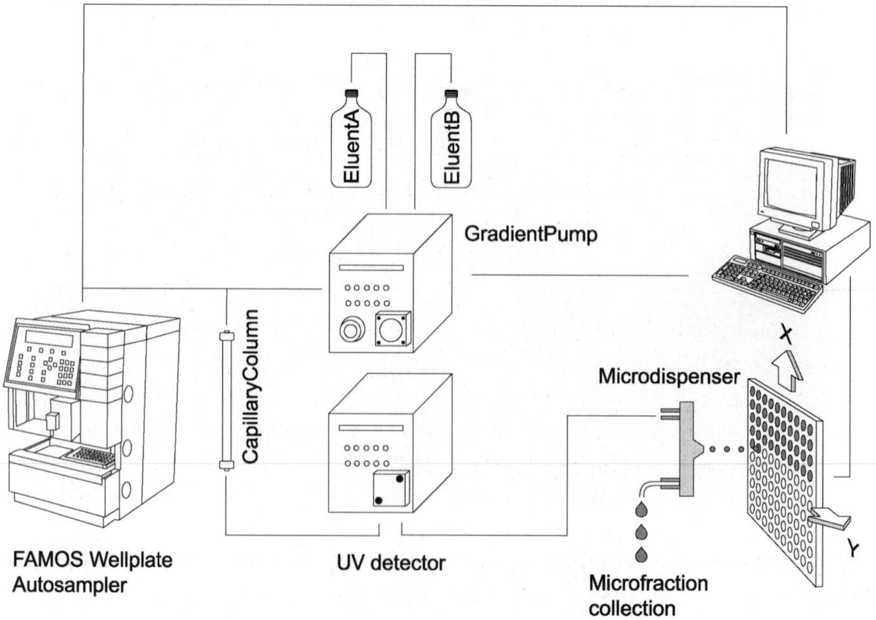

Fig. 14. A typical set-up for interfacing microliquid chromatography to MALDI-TOF MS using the flow-through microdispenser.

[69] to interface μLC to MALDI-TOF MS by means of flow-through microdispensing.

The dispenser platform enables reproducible generation of high-density arrays for MALDI-TOF MS by translating the target plate with a motorised precision *x-/y*-stage. This topic is further discussed in Section 5.4.

As discussed earlier, the evaporation time of a 40 μm water droplet from a glass or a silicon surface is of the order of a second. This feature becomes extremely favourable as samples are transferred to a target plate by the deposition of multiple droplets that are allowed to dry on the same spot area. If the dispense rate is adapted to the evaporation rate, a small spot size is maintained, and consequently the surface density of the analyte in the sample spot is rapidly increased. Because the MALDI MS readout basically is a process that records surface density of the analyte, the repeated deposition of sample on a fixed spot area is a direct enrichment process. The dramatic impact of on-spot enrichment is demonstrated in Fig. 15, which compares a mass spectrum from a 50 nM peptide mixture deposited on a standard MALDI target plate by pipetting 50 fmol (1 μl) with a mass spectrum of the same peptide mixture deposited by microdispensing 500 amol (10 nl) of the solution on a small spot. Further investigation of this enrichment protocol shows that the enrichment factor primarily scales linearly with the number of droplets deposited [26].

5.4. Nanovial MALDI target arrays

5.4.1. Aspects of microdispensing for MALDI MS

When using microdispensing as the sample transfer methodology to the MALDI target plate, it is clear that dense array formats are easily obtained. A standard MALDI target, which is typically $4.5 \times 4.5\ cm^2$, holds 100 spots when loaded manually. In contrast, array densities of 3000–5000 spots with microdispensing are easily obtained on the same plate. When dispensing samples on a planar surface, the spot size grows until the evaporation from the spot equals the supply rate from the dispenser. In order to make the spot size reasonably independent of the supply rate, nanovials can be etched in a silicon chip. Each nanovial serves as a sample reservoir with a volume in the nanoliter to picoliter range. Fig. 16 shows a scanning electron micrograph of such a silicon MALDI target nanovial array.

As a sample is dispensed into such a nanovial, it is confined to the spot area predefined by the nanovial geometry. The confinement of the sample volume to a predefined nanovial also means that a suitable sample volume can now be deposited at any dispense rate without giving rise to varying spot sizes and thus varying enrichment factors. Fig. 17 illustrates the effect on spot size (d) when

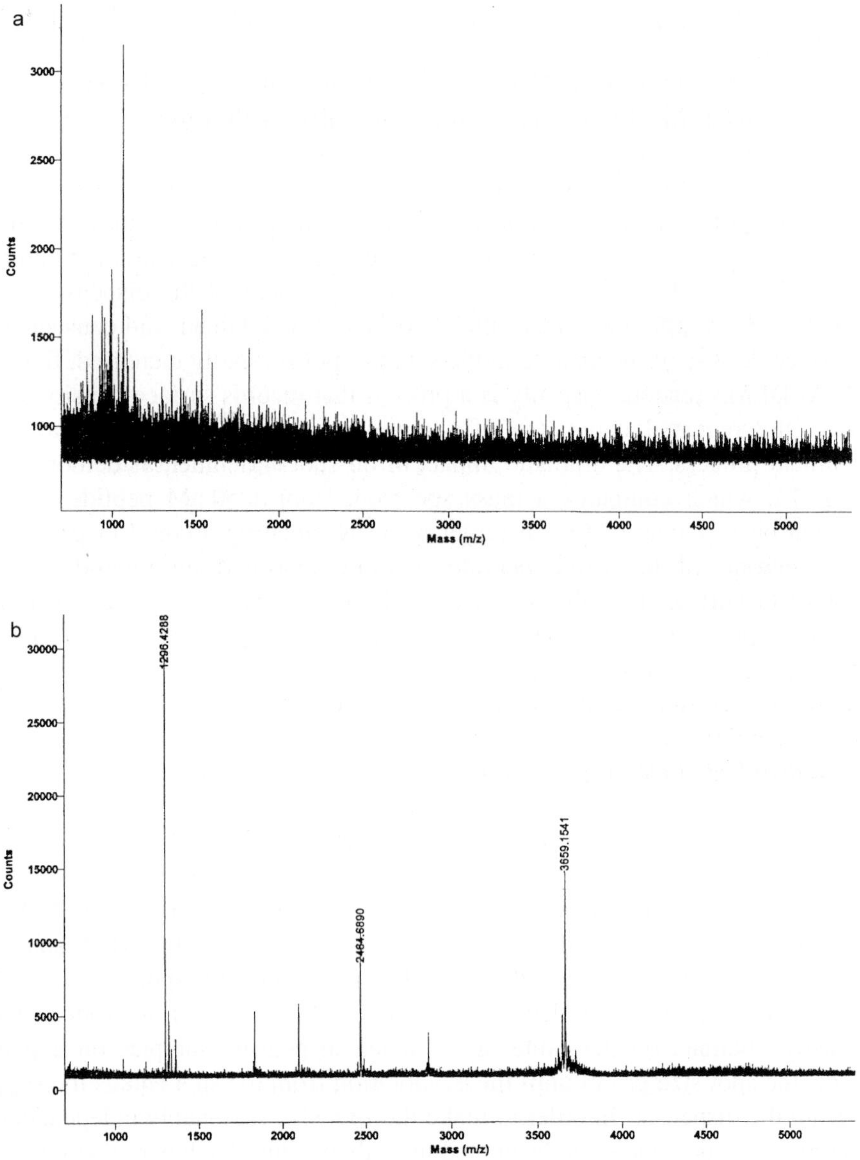

Fig. 15. A comparison of mass spectra obtained from a 50 nM standard peptide mixture, demonstrating the enrichment effect that can be obtained by microdispensing the proteomic sample onto the MALDI target rather than pipetting. (a) 50 fmol (1 μl) of the peptide mixture is deposited on the target plate and (b) 500 amol (10 nl) of the same solution is deposited by microdispensing.

Fig. 16. A silicon microfabricated MALDI target nanovial array. The vial size is $100 \times 100\ \mu m^2$.

a sample is deposited at different dispense rates on a planar surface and in a nanovial configuration. Fig. 18 compares mass spectrometer signal intensities derived from a peptide sample that was dispensed onto a planar target at different dispense rates. Each spot was supplied with an equal total sample volume. As

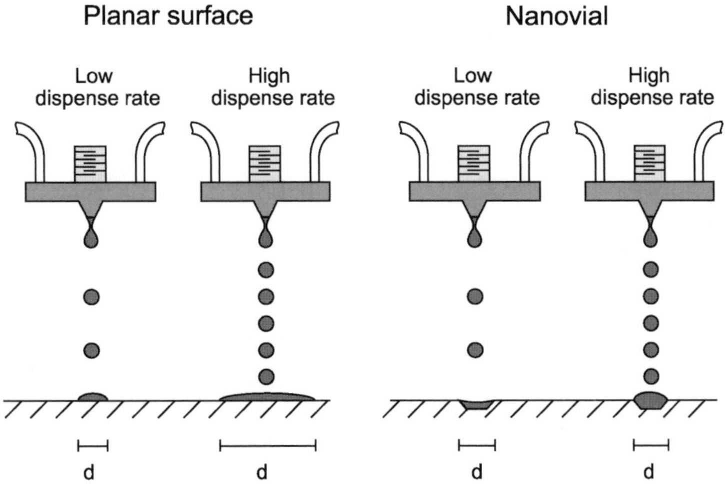

Fig. 17. A schematic drawing of the effect on spot size (d) as a fix volume of a sample is deposited at different dispense rates (a) on a planar surface and (b) in a nanovial.

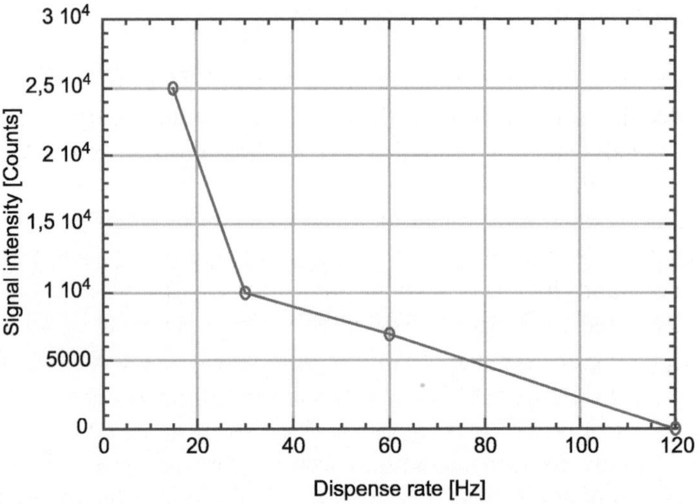

Fig. 18. The MS signal of a model peptide is reduced as the dispense rate is increased due to the large spot size that is obtained at higher dispense rates. Equal sample volumes were deposited for each dispense rate.

the dispense rate is increased, the MS signal intensity is reduced due to the fact that the spot size grows and thus a lower surface density of the analyte is obtained [70]. Fig. 19 shows the spot sizes obtained when dispensing varying number of droplets of deionised water on a glass surface (dispense rate of 50 Hz) [66].

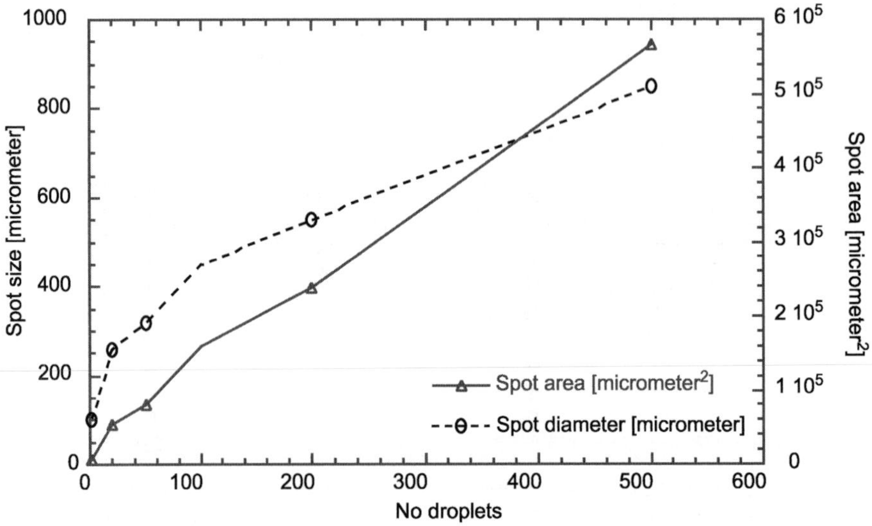

Fig. 19. Spot size and spot area versus number of deposited droplets. Deionised water was dispensed at 50 Hz.

5.4.2. Silicon nanovial MALDI target arrays

An important aspect of nanovial design for MALDI-TOF MS is the depth of the nanovial. As the sample is crystallised along the boundary surfaces of the nanovial, the analyte has a spatial distribution along the ion flight path in the mass spectrometer, which corresponds to the depth of the vial. If vials that are too deep are utilised, this distribution will significantly influence the mass resolution obtained in the MS readout. Fig. 20 shows a comparison of mass resolution as nanovials (deep and shallow) of different sizes were investigated and compared to a sample deposited on a flat silicon surface. Deep nanovials were obtained by anisotropic etching of silicon to create nanovials in the shape of inverted pyramids and shallow nanovials were obtained by interrupting the anisotropic etching after an etch depth of only 20 μm was obtained. As seen in Fig. 20, mass resolution deteriorates rapidly for deep nanovials as the size, and thus the vial depth, is increased. A nanovial depth well below 50 μm did not significantly reduce mass resolution on the MALDI instrument used in these studies, which was a Perseptive DE Pro (Applied Biosystems, Boston, Ma, USA).

An optional way of confining the sample to a predefined surface has been developed by Bruker Daltonik (Bremen, Germany), which launched hydrophobic anchor point MALDI targets. These targets are covered by a thin Teflon film and

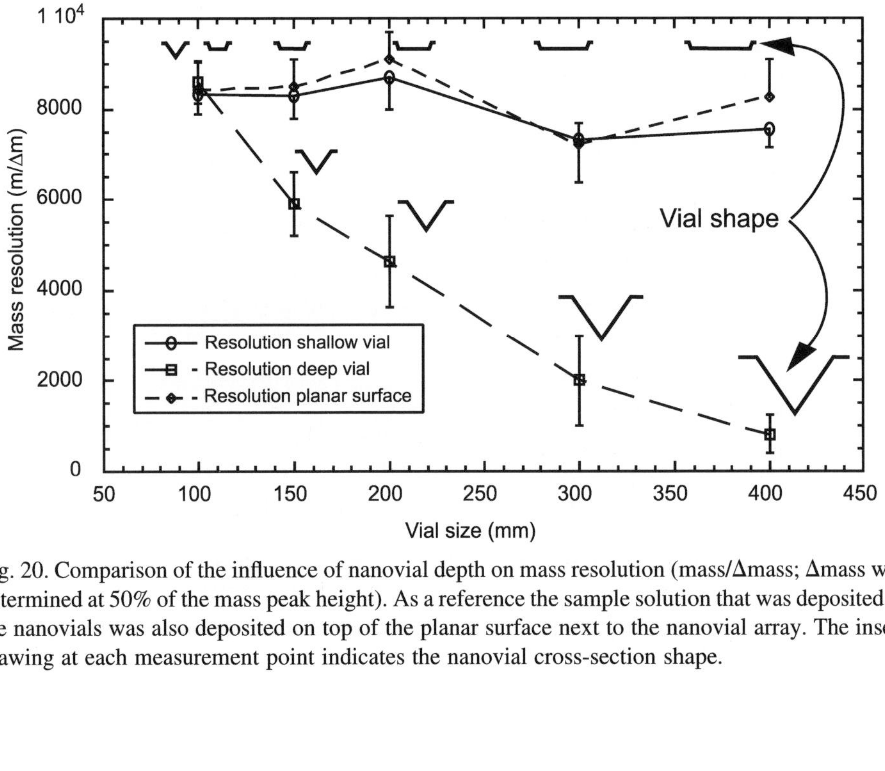

Fig. 20. Comparison of the influence of nanovial depth on mass resolution (mass/Δmass; Δmass was determined at 50% of the mass peak height). As a reference the sample solution that was deposited in the nanovials was also deposited on top of the planar surface next to the nanovial array. The insert drawing at each measurement point indicates the nanovial cross-section shape.

each MALDI spot is made of a small opening (diameter 600 μm) in the polymer film that exposes the underlying metal plate. As a sample is deposited, it spreads beyond the metal spot, but during drying of the droplet, the analyte is enriched on the metal spot due to surrounding hydrophobic Teflon surface. This sample enrichment in a well-defined surface area outlined by the hydrophilic zone is similar to that described previously for repetitive dispensing into a nanovial configuration. Single femtomole peptide detection was reported for neurotensin clips 1-11 and 1-13 with these targets [71].

5.4.3. Polymer nanovial MALDI target arrays

To obtain a disposable and low-cost alternative to silicon as the base material of the target plate, a polymer nanovial array may optionally be obtained by microfabrication technologies. When changing to a polymer-based material, a silicon master structure is commonly fabricated to maintain the high fidelity of the structures. The silicon master structure is subsequently electroplated with metal, and then the silicon is removed by etching to leave an inverse metal structure that can be used as the master template in a polymer injection moulding process analogous to the method used to fabricate music compact disks. Fig. 21 shows a typical processing scheme for fabricating polymer nanovial MALDI-target arrays.

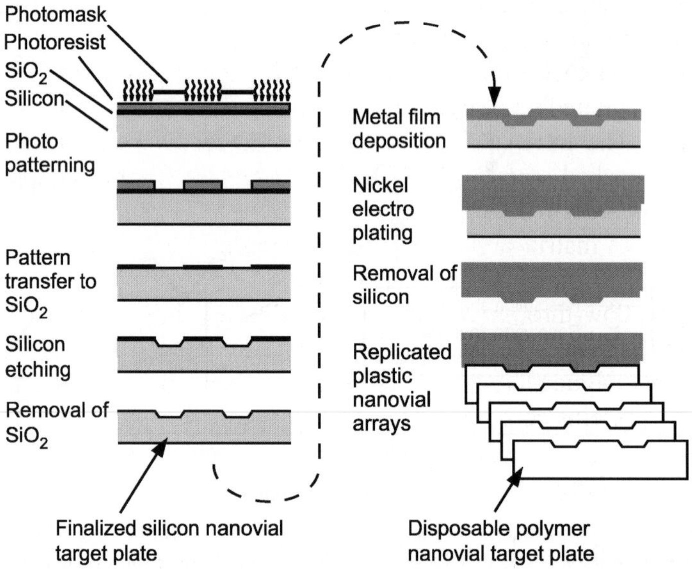

Fig. 21. Processing scheme for fabricating the silicon master, the metal inverse, and finally the injection moulded polymer nanovial array.

Because strong solvents are commonly used in the sample handling process, e.g. acetonitrile, when extracting samples from a reverse phase bed, either the selected polymer base has to be inert to the solvent used or optionally a metal film, e.g. gold, can be deposited on top of the target plate.

Polymer nanovial target plates have proven to be an attractive alternative to their more costly silicon counterparts. In initial studies, nanovial arrays were fabricated by cold embossing a metal stamping tool, shaped as an inverted nanovial array, into polymer sheets that were cut to fit into a reloadable MALDI target tray [72]. The target plates were gold covered by evaporation prior to use. Fig. 22a shows the cold embossing stamping tool and Fig. 22b shows a photograph of the obtained polymer target plate that holds 1200 spots (diameter 300–500 μm) arranged in 10×10 subarrays. Although the machining quality of the cold embossed nanovial arrays cannot match the quality obtained by injection moulding with a metal master originating from a silicon original, cold embossing is a cheap and rapid prototyping alternative. The initial studies of the cold embossed polymer nanovials showed that the nanovial microstructure quality was sufficient for good MALDI-TOF MS readout.

Disposable polymer targets with nanovial diameters of 400 μm were evaluated with respect to reproducibility of the MS signal intensity across the entire target plate. A standard 10 nM peptide mixture of angiotensin I, II, III and bradykinin was used and 300 amol of analyte was dispensed in each nanovial. The masses showed a relative standard deviation in signal intensity (peak height) of 30% ($N = 100$) across the complete target plate.

These target plates were also used in proteomic studies. A TGF-β stimulated human fibroblast cell (HFL-1) supernatant was separated by 2D gel electrophoresis. Selected spots that differed compared with the corresponding 2D gel from a non-TGF-β stimulated cell fraction were excised and submitted to a tryptic digestion step. The extracted tryptic peptides were enriched on custom-made microcolumns packed with POROS 20 R2 (Perseptive Biosystems Framingham, MA, USA) [46]. After washing the microcolumns, the bound peptides were eluted with a mixture of matrix and acetonitrile (10 mg/ml CHCA in 60% ACN/0.5% TFA) directly into the nanovials on the polymer target plate by injecting the sample into the flow-through microdispenser. Detailed information on the sample handling steps have been reported [73]. Fig. 23 shows the obtained mass spectra of three of the unambiguously identified proteins: kinesin, moesin and transkeltolase. None of these proteins could be identified before the SPE and in-vial enrichment.

5.4.4. Matrix handling

An optimal crystalline sample/matrix surface has to be accomplished in order to obtain high quality MALDI MS data. Cohen et al. [74] have investigated the influence of crystallisation rate and matrix/solvent composition with respect to a

b

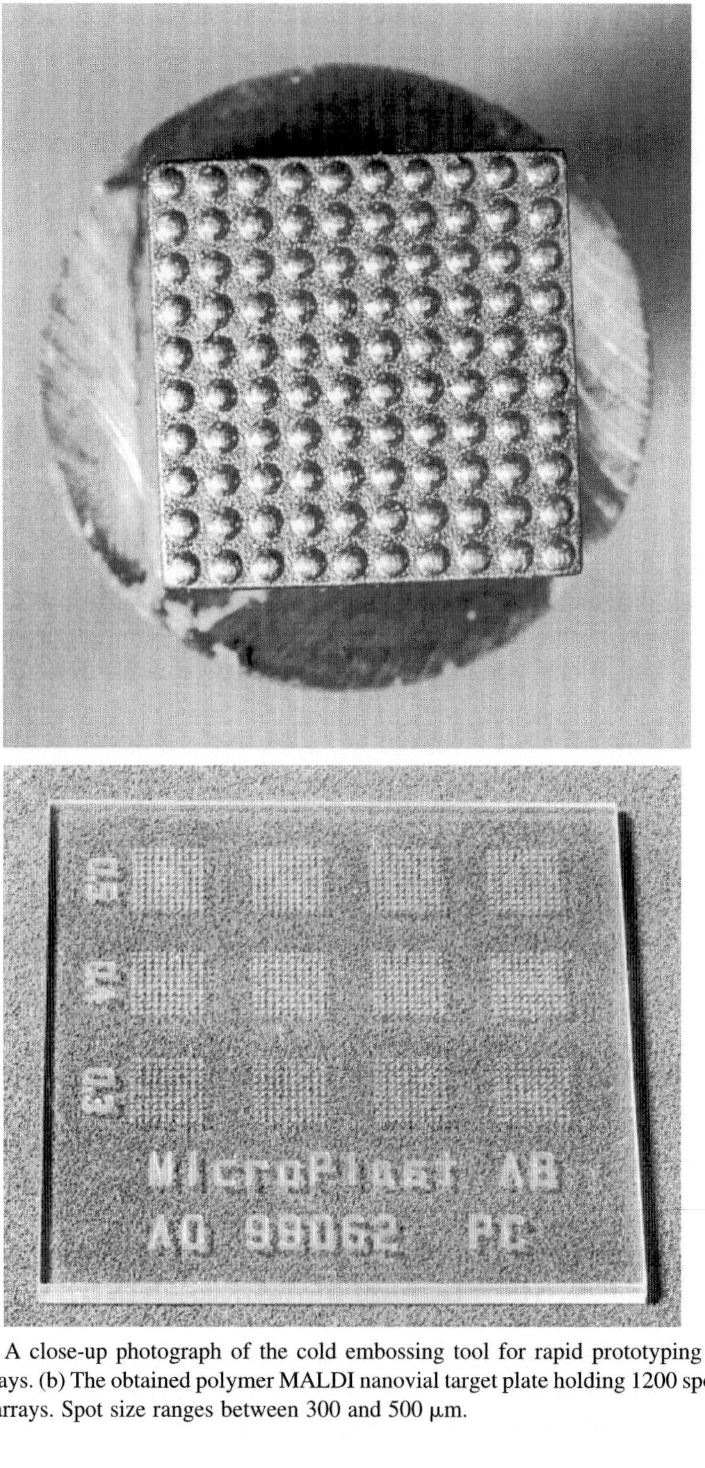

Fig. 22. (a) A close-up photograph of the cold embossing tool for rapid prototyping of polymer nanovial arrays. (b) The obtained polymer MALDI nanovial target plate holding 1200 spots arranged in 10 × 10 arrays. Spot size ranges between 300 and 500 μm.

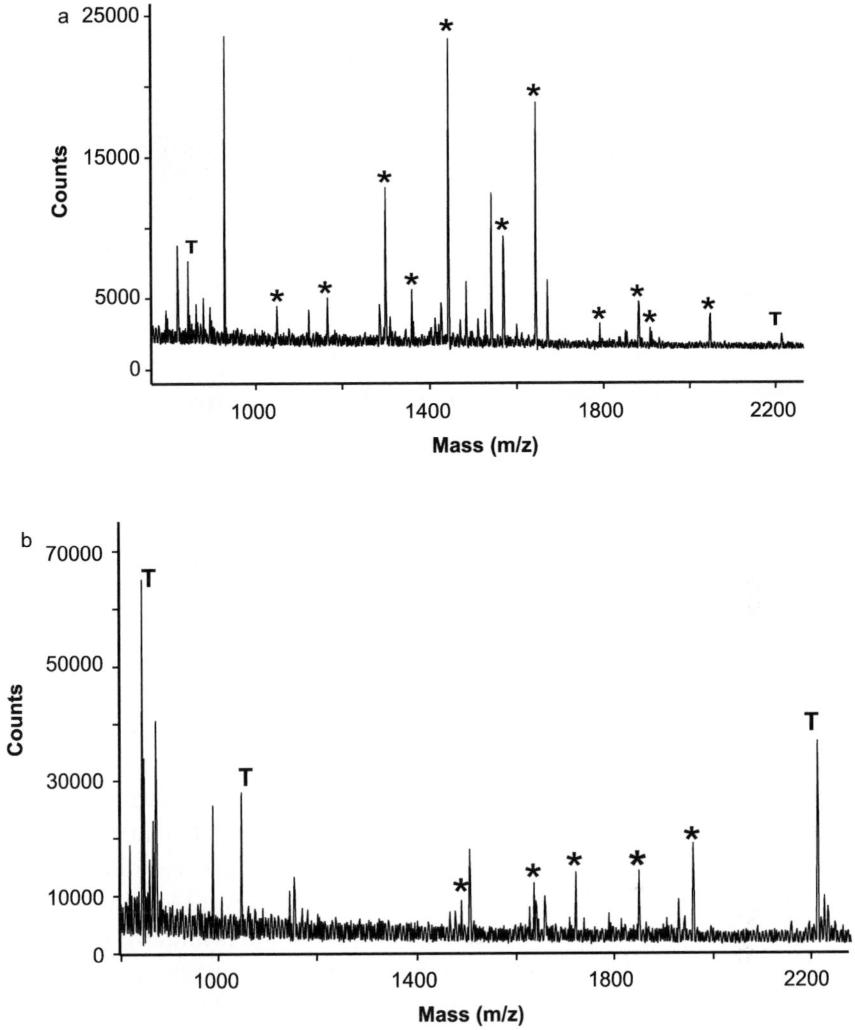

Fig. 23. Proteomic samples identified using the microdispenser assisted in vial enrichment. A gold-coated disposable polymeric nanovial MALDI-target plate was used. The identified proteins were (a) kinesin, (b) moesin, and (c) transkeltolase. $*$ indicates identified peptides, and 'T' indicates tryptic autodigest peaks.

MALDI data quality. It is also known that sample spots that display fine homogeneous crystalline surfaces provide more reproducible and higher quality mass spectra [75]. The reasons for this can again be partially traced back to surface area phenomena, i.e. a highly differentiated crystal surface with extremely fine sample/matrix crystals provides more surface area for laser desorption ionisation and thus more ions are produced per laser shot. Another important aspect on

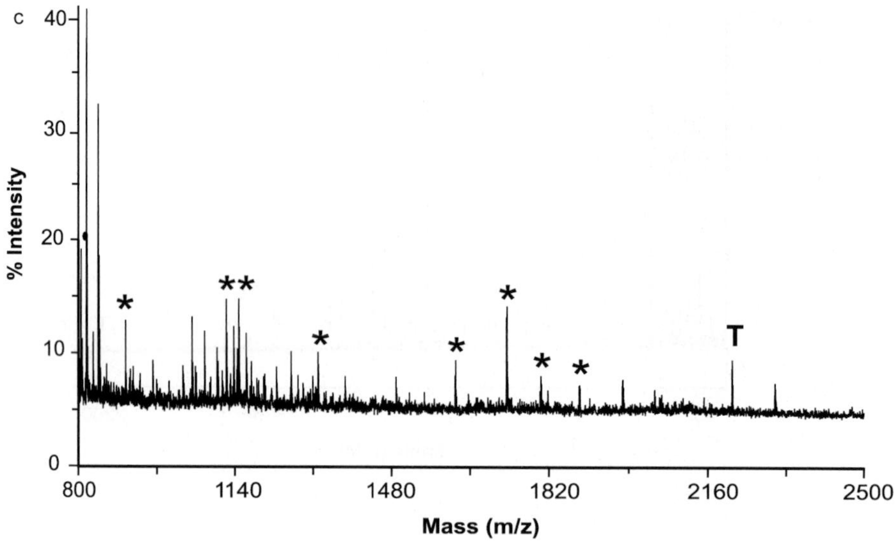

Fig. 23 *continued.*

surface homogeneity is the fact that the formation of a fine homogeneous crystalline surface avoids the need for sweet spot searching [76], thus improving the MS throughput.

A technique that provides fine and homogeneous crystalline surfaces is the fast evaporation technique proposed by Vorm et al. [77], in which matrix is dissolved in a highly volatile solvent, applied to the MALDI target and allowed to dry rapidly. An alternative to this methodology is the seed-layer technique described by Westman et al. [78] where a very dilute matrix solution is initially applied to the target plate and allowed to dry, followed by loading the sample/matrix mixture. In both methods the initially applied matrix provided a thin homogeneous layer of very fine matrix crystals that served as predefined nucleation sites for the subsequent deposition of a sample/matrix mixture. Later, Önnerfjord et al. [75] also reported homogeneous sample preparations using the seed-layer preparation combined with microdispensing and high-density arraying for MALDI-TOF MS. Fig. 24 shows a close-up of the microcrystalline matrix surface (αCHCA) that was obtained with the seed-layer preparation and microdispensing.

The time required for a sample to dry on a spot influences the crystal morphology and slow evaporation generally yields large crystals. A faster evaporation is, as mentioned previously, known to provide finer crystals. Fast evaporation is also an aspect that can easily be beneficially utilised when preparing sample spots in a microformat for MALDI due to the short evaporation times of

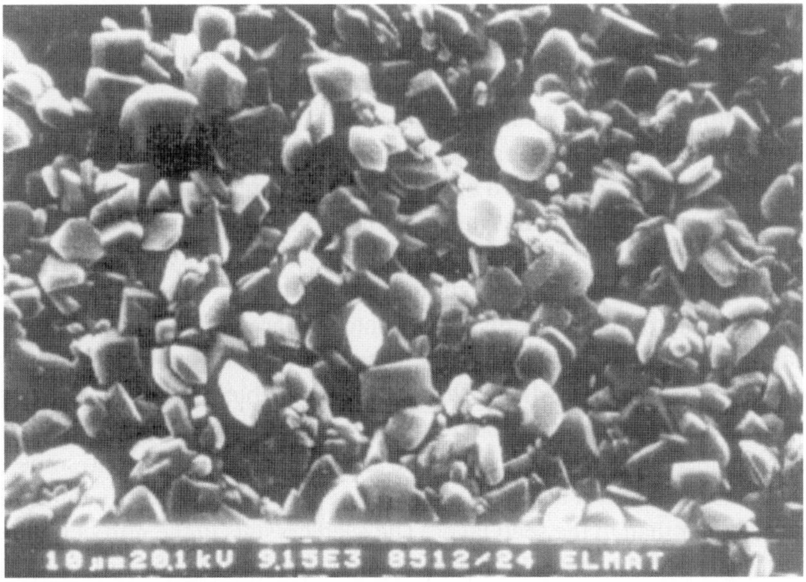

Fig. 24. A scanning electron micrograph of a seed-layer preparation. The matrix used was α̃CHCA. Note the homogenous surface coverage and crystallite size.

picoliter to nanoliter volumes. Samples derived from a chemical microsystem, which inherently has small sample volumes, are thus very well suited to microarray for MALDI-TOF analysis.

5.4.4.1. Dispensing. An inherent problem with handling the MALDI matrix in a piezoelectric microdispenser is the fact that it is prone to crystallisation in the nozzle opening, inhibiting any further dispensing. Microdispensing of the sample matrix mixture is therefore preferably done at lower matrix concentrations than commonly used when, for example, making a dried droplet preparation. Önnerfjord et al. [66] investigated the influence of matrix solvent ratio on the microdispenser operation and found that acetonitrile concentrations of 30% or lower did not interfere with dispensing. It was also shown that microdispensing 100 droplets of a sample (insulin 2 μM)/matrix mixture onto each spot of a seed-layer preparation provided nice homogeneous MALDI spots with an intensity variation less than 30%.

The matrix can, of course, also be dispensed after the sample has been deposited. The same acetonitrile concentration conditions ($<30\%$) for stable matrix dispensing are then still valid.

5.4.4.2. Airbrush matrix/nitrocellulose precoated targets. An alternative to dispensing the matrix onto the MALDI target is to use target plates that are already precoated with a thin film of a matrix/nitrocellulose mixture. The samples can then be directly dispensed onto the target plate where the sample dries and forms a fine crystalline layer. This technique has been thoroughly investigated by Miliotis et al. [79,80].

The thin film is achieved by airbrushing a matrix/nitrocellulose/acetone mixture onto a target plate. By adjusting the spray and the distance to the target plate, basically all acetone is evaporated before the solution reaches the target. The dried target is thereby supplied with a uniform thin film of submicron matrix crystals (not visible in microscope) embedded in a nitrocellulose layer that has a strong affinity to the subsequently deposited analyte.

When the analyte solution is dispensed onto the target, the surface immediately recrystallises and the resulting sample spot has a very fine crystalline texture. Fig. 25 shows an airbush prepared MALDI target plate capable of holding 2500 sample spots. The inserts show the reproducibility of the spot array and the fine crystal structure obtained with this technique. The airbrush preparation can to some extent be regarded as a seed-layer technique, but the seed crystals in the nitrocellulose film are much smaller than in the seed layer. The seed crystals in the nitrocellulose film were neither resolvable by light microscopy nor by scanning electron microscopy.

The very fine sample preparations obtained with the airbrush technique are also illustrated by the sensitivity of the mass spectrometry readout. A standard peptide mixture of angiotensin I, II, III and bradykinin was deposited on the thin film. Fig. 26 shows a mass spectrum of a 0.25 nM peptide mixture where a total of 25 amol was deposited. As these data show, low attomole levels of detection can be accomplished with these protocols, and a most attractive feature is that samples can be at very low concentrations, i.e. in the subnanomolar range.

It is important to adapt the dispense rate to the evaporation and crystallisation rates such that complete dissolution of the seed crystals in the nitrocellulose film is avoided. Otherwise the recrystallised sample tends to form spots with few or no crystals in the centre, and instead the crystals form in a ring around the edge of the spot.

5.4.4.3. Compound microdispensing. An optional dispensing technique is currently being developed in order to overcome the problem with crystallisation in the nozzle when dispensing matrix directly onto the target [81]. By covering the matrix-dispensing nozzle with a solvent and ejecting the matrix through the solvent layer no matrix crystallisation at the nozzle is observed. This has been achieved by sandwiching a third silicon plate to the original flow-through microdispenser. The third plate provides a second flow channel and nozzle which serves as the flow path for the liquid that protects the inner nozzle from evaporation. Fig. 27a shows a schematic cross-section of the compound microdispenser and Fig. 27b its principle of operation.

The compound dispenser has successfully been demonstrated to dispense matrix from the inner compartment through a water/acetonitrile (50/50) filled

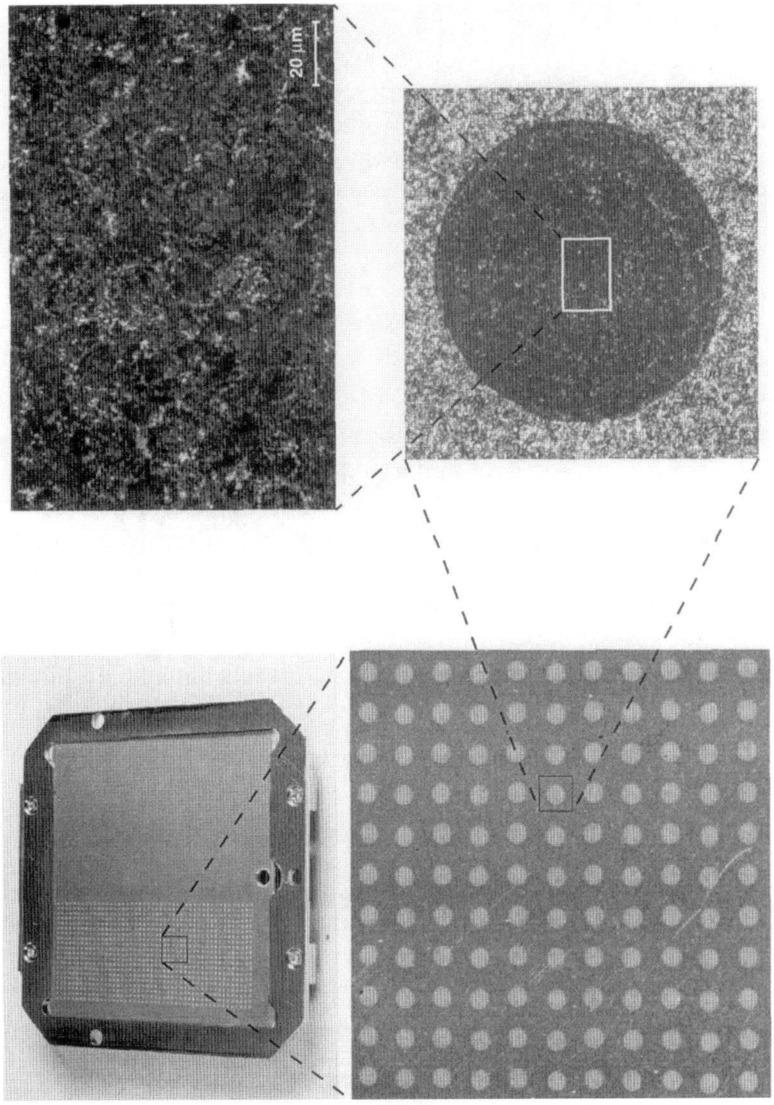

Fig. 25. A matrix/nitrocellulose airbrush precoated MALDI target plate with over 900 sample spots deposited in a high-density array. The target plate size is 45 × 47 mm^2 and the spot size is approximately 400 µm in diameter.

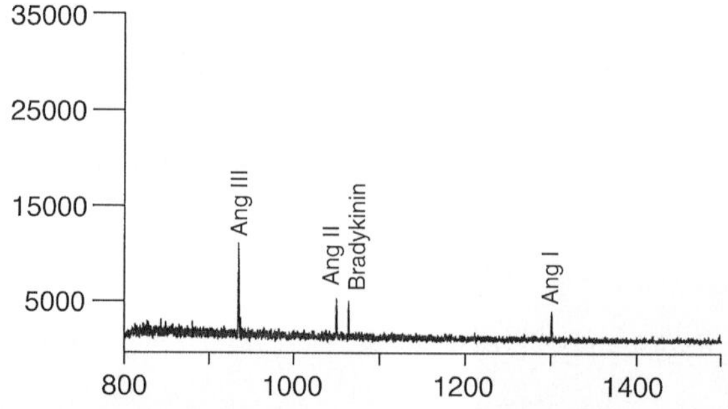

Fig. 26. Mass spectra of a 0.25 nM peptide mixture (25 amol dispensed: angiotensin I, II, III and bradykinin) dispensed at 15 Hz on a matrix/nitrocellulose airbrush precoated MALDI target plate.

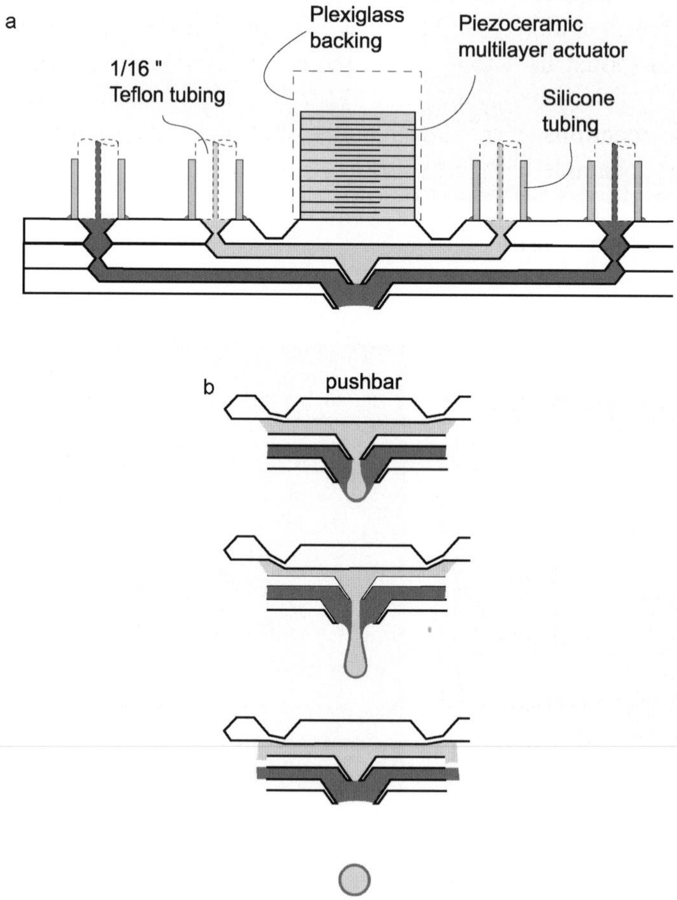

Fig. 27. (a) Schematic cross-section of the compound microdispenser. (b) Schematic sequence demonstrating the operating principle of the compound microdispenser.

outer compartment without observing any matrix deposition on the dispenser nozzle. Further, as seen in the mass spectra shown in Fig. 28, liquid is ejected from both fluid compartments as demonstrated by dispensing a sample volume where angiotensin I was in the inner compartment fluid and angiotensin II was in the outer compartment.

5.4.5. On-target high-speed biochemistry

New protocols can be developed that benefit from the increased kinetics that can be obtained in the picoliter to nanoliter volumes readily achieved by piezoelectric dispensing. Nanovials naturally become the format in which such reactions can be carried out. Early work by Jespersen et al. [82] showed the possibility of analysing low amounts of peptides and tryptic digests contained in nanovials by MALDI-TOF MS. Litborn et al. [83–85] reported work on performing chemical reactions in a nanovial format for subsequent analysis by capillary electrophoresis. Also, recent work by Ericsson et al. [86] and Ekström et al. [73] investigated the benefits of running reactions in a nanovial format combined with piezoelectric microdispenser sample handling and MALDI MS readout.

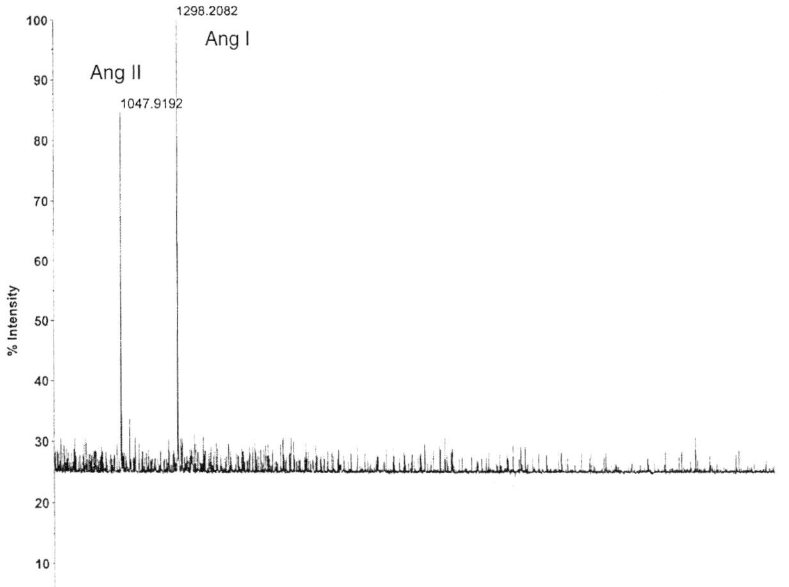

Fig. 28. Mass spectra showing that both the peptide in the inner (angiotensin I) and the outer (angiotensin II) compartments of the dispenser are dispensed in the ejected volume.

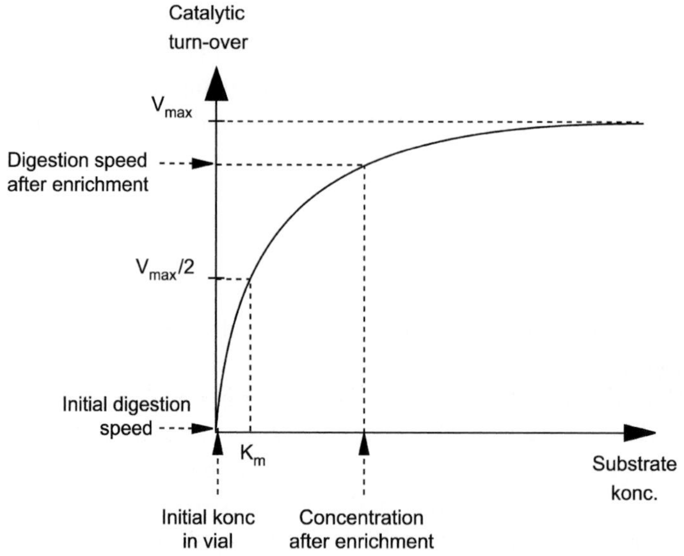

Fig. 29. Schematic illustration of the transition from a non-digestive region to a high-speed digestion region of the Michaelis–Menten kinetics for a proteolytic process by continuously supplying substrate to an enzyme-coated nanovial.

Very high protein digestion rates can be obtained by using a continuous supply of substrate (a protein) to a nanovial that has been precoated/loaded with proteolytic enzymes. The substrate supply is set to match the evaporation rate, and even though the initial concentration of the substrate is very dilute, it will reach a level where the turnover in the enzymatic reaction accelerates. Fig. 29 illustrates the transition from a non-digestive region on the Michaelis–Menten kinetics to a high-speed digestion region of the kinetic curve by enriching an initially very dilute protein sample.

Protocols for performing high-speed nanodigestions have been thoroughly investigated by Ericsson et al. [86]. Fig. 30 shows the mass spectra of a high-speed (10 min) digestion of a 10 nM hemoglobin solution. In total, 600 nl (6 fmol) of protein solution, providing a 300-fold concentration increase, was added to the nanovial. NH_4HCO_3 was used as the buffer.

At higher substrate concentrations the digestion rate rapidly increases. Fig. 31 shows the mass spectrum result from tryptic digestion of a 500 nM lysozyme sample with a total of 50 fmol in-vial. The digestion was only allowed to run for 75 s, and a sequence coverage of 62% was obtained.

Nanovial sample processing of proteomic samples can be readily accomplished with the microdispenser platform for detailed biological studies such as determination of phosphorylation sites in proteins. Ekström et al. [73] investigated

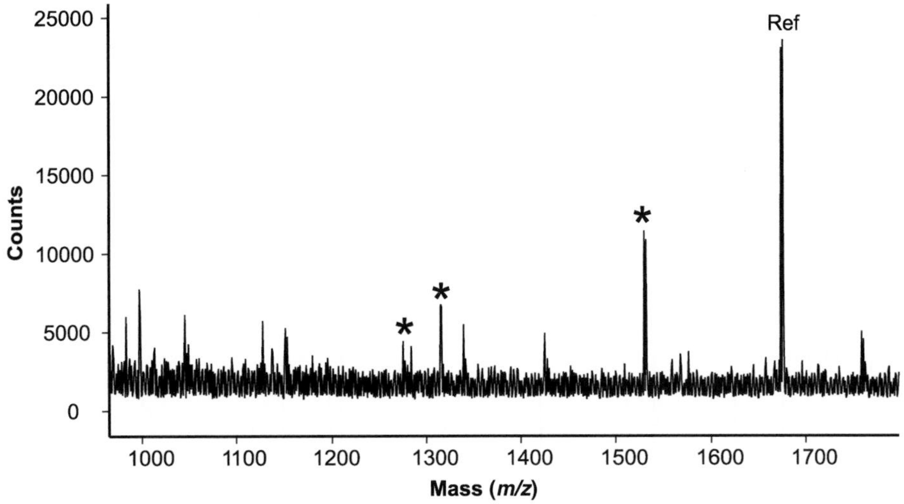

Fig. 30. Mass spectra showing a high-speed tryptic digest of a dilute (10 nM) hemoglobin sample. Six femtomole of the sample was continuously microdispensed for 10 min into an enzyme coated nanovial. * indicates identified peptides and 'Ref' is the internal mass standard (neurotension) used.

nanoscaled protocols for dephosphorylation of peptides by precoating nanovials with alkaline phosphatase. The identification of phosphopeptides in β-casein was demonstrated using 10 fmol of tryptically digested β-casein which was dispensed over 60 s into a 300 μm nanovial precoated with enzyme. The peptide found in the

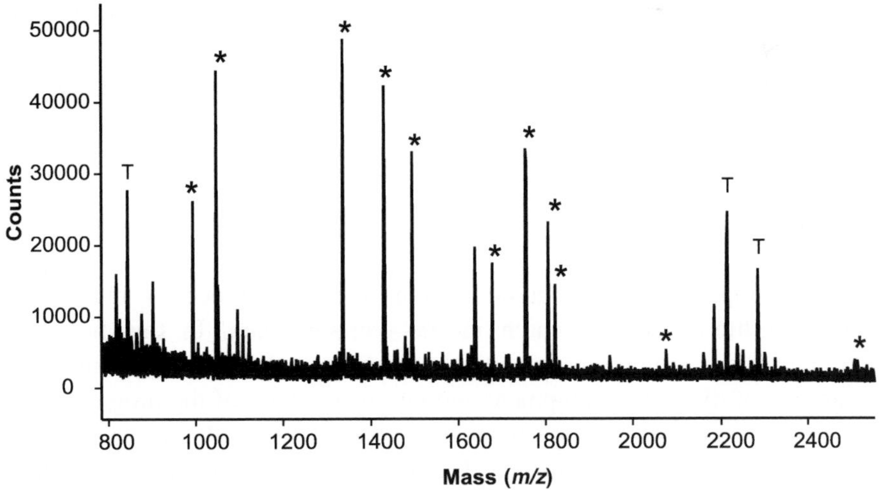

Fig. 31. Mass spectra showing a 75 s tryptic digestion of a 500 nM lysozyme sample with a total of 50 fmol in-vial. * indicates identified lysozyme peptides and 'T' indicates tryptic autodigest peaks.

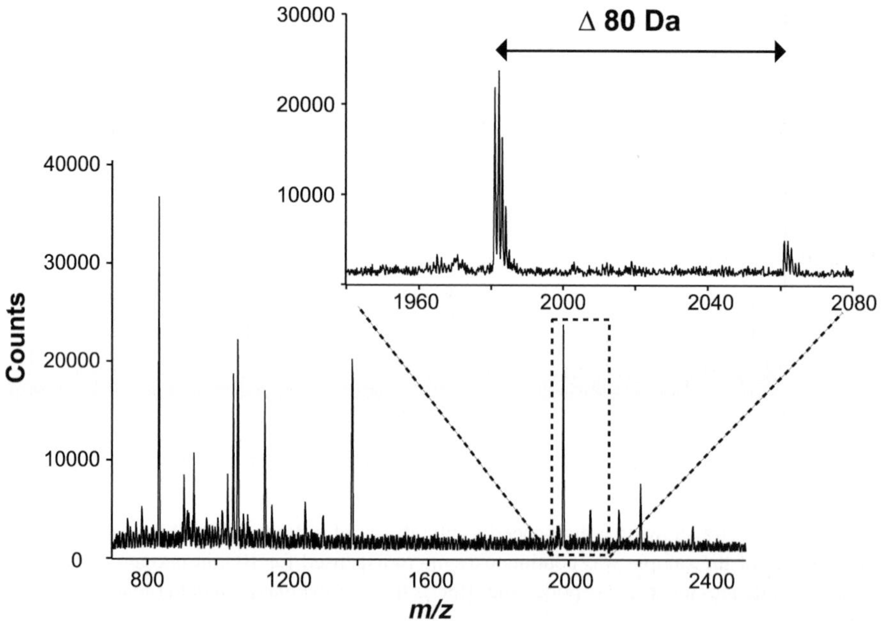

Fig. 32. A tryptic digest of β-casein (10 fmol) was submitted to an in-vial dephosphorylation step by microdispensing alkaline phosphatase into the nanovial for 60 s, verifying that the original peptide at $m/z = 2061.3$ Da was a phosphorylated peptide that appeared at $m/z = 1981.3$ Da ($\Delta m/z = -80$ Da) after dephosphorylation.

original mass spectrum of the digested β-casein at $m/z = 2061.3$ was a suspected phosphopeptide. The corresponding mass spectra obtained after in-vial dephosporylation confirmed this by the appearance of a peptide peak shifted 80 Da lower in the mass spectrum at $m/z = 1981$. Because complete dephosphorylation was not obtained, the original peptide peak was also visible (Fig. 32).

6. Summary

Proteome mapping and protein interaction studies will occupy much of the attention in life science research in the coming years. In this field, mass spectrometry is an important analytical tool for the deciphering of underlying mechanisms of disease development and progress. Much of the progress will be linked to the development of new bioanalytical technologies that will pave way for improved bioanalytical performance. The developments described in this chapter illustrate the critical role micro- and nanotechnology may play in the global proteomics quest. It is also apparent that new miniaturised analytical technologies are needed due to the ever-increasing demands for sensitivity and the ability to process constantly decreasing sample volumes.

The current developments within micro- and nanotechnology predict a dramatic future. The rapid merge of biology and technology research into a dynamic, interdisciplinary, and very exciting field promises an era of even more stunning progress in life science research and more comprehensive understanding of human biology and disease. Mircofluidics is one important part of the micro/nanotechnology developments that so far has impacted this field strongly and further advances are expected. As nanotechnology continues to mature, the importance of surface technology and control of surface properties will rise, and its direct impact on life science will become even more crucial for new biotechnical and biomedical developments.

Acknowledgements

The authors express their gratitude for financial support. Especially thanks are directed to the Carl Trygger Foundation, the Crafoord Foundation, the Royal Physiographic Society in Lund, the ELFA Foundation, the Swedish Foundation for Strategic Research, the Swedish National Science Council, NUTEK/Vinnova, Astra Zeneca AB, Microplast AB, SWEGENE and the Lund Institute of Technology.

References

1. Venter, J.C., Adams, M.D., Myers, E.W., et al., The sequence of the human genome. *Science*, **291**, 1304–1351 (2001).
2. Lander, E.S., Linton, L.M. and Birren, B., Initial sequencing and analysis of the human genome. *Nature*, **409**, 860–921 (2001).
3. Mural, R.J., Adams, M.D., Myers, E.W., et al., A comparison of whole-genome shotgun derived mouse chromosome 16 and the human genome. *Science*, **296**, 1661–1671 (2002).
4. Wilkins, M.R., Williams, K.L., Apelsand, R.D. and Hochstrasser, D.F. (Eds.), *Proteome Research: New Frontiers in Functional Genomics*. Springer, Berlin, 1997.
5. Madou, M. (Ed.), *Fundamentals of Microfabrication*. CRC Press, Boca Raton, FL, 1997.
6. Gad-el-Hak, M. (Ed.), *The MEMS Handbook*. CRC Press, Boca Raton, FL, 2002.
7. Terry, S.C., *A gas chromatography system fabricated on a silicon wafer using integrated circuit technology*. PhD Dissertation, Stanford University, 1975.
8. Terry, S.C., Jerman, J.H. and Angell, J.B., Gas-chromatographic air analyzer fabricated on a silicon wafer. *IEEE Trans. Electron. Devices*, **26**, 1880–1886 (1979).
9. Petersen, K.E., Silicon as a mechanical material. *Proc. IEEE*, **70(5)**, 420–457 (1982).
10. Le, H.P., Progress and trends in ink-jet printing technology. *J. Imaging Sci. Technol.*, **42(1)**, 49–62 (1998).
11. Manz, A., Graber, N. and Widmer, H.M., Miniaturized total chemical-analysis systems — a new concept for chemical sensing. *Sensors Actuat. B*, **1**, 244–248 (1990).

12. Micro total analysis systems, In: Van den Berg, A. and Bergveld, P. (Eds.), *Proceedings of the μTAS '94 Workshop*. Kluwer, Dordrecht, 1995.
13. Kuschel, M., Lab-on-a-chip technology — applications for life sciences. *Pharm. Technol. Eur.*, **May** (2001).
14. Anderson, R.C., McCall, G. and Lipshutz, R.J., Polynucleotide arrays for genetic sequence analysis, In: Manz, A. and Becker, H. (Eds.), *Microsystem Technology in Chemistry and Life Sciences*. Springer, Berlin, 1998, pp. 117–130.
15. Zhu, H., Klemic, J.F., Chang, S., Bertone, P., Casamayor, A., Klemic, K.G., Smith, D., Gerstein, M., Reed, M.A. and Snyder, M., Analysis of yeast protein kinases using protein chips. *Nat. Genet.*, **26**, 283–289 (2000).
16. Zhu, H., Bilgin, M., Bangham, R., Hall, D., Casamayor, A., et al., Global analysis of protein activities using proteome chips. *Science*, **293**, 2101–2105 (2001).
17. Lottspeich, F., Proteome analysis: a pathway to the functional analysis of proteins. *Angew. Chem. Int. Ed.*, **38**, 2477–2492 (1999).
18. Aebersold, R. and Goodlett, D.R., Mass spectrometry in proteomics. *Chem. Rev.*, **101**, 269–295 (2001).
19. Huang, L., Baldwin, M.A., Maltby, D.A., Medzihradszky, K.F., Baker, P.R., Allen, N., Rexach, M., Edmondson, R.D., Campbell, J., Juhasz, P., Martin, S.A., Vestal, M.L. and Burlingame, A.L., The identification of protein–protein interactions of the nuclear pore complex of *Saccharomyces cerevisiae* using high throughput matrix-assisted laser desorption ionization time-of-flight tandem mass spectrometry. *Mol. Cell. Proteom.*, **1**, 434–450 (2002).
20. Gygi, S.P., Corthals, G.L., Zhang, Y., Rochon, Y. and Aebersold, R., Evaluation of two-dimensional gel electrophoresis-based proteome analysis technology. *Proc. Natl Acad. Sci. USA*, **97**, 9390–9395 (2000).
21. Washburn, M.P., Wolters, D. and Yates, J.R., Large-scale analysis of the yeast proteome by multidimensional protein identification technology. *Nat. Biotechnol.*, **19**, 242–247 (2001).
22. Licklider, L., Wang, X.-Q., Desai, A., Tai, Y.-C. and Lee, T.D., A micromachined chip-based electrospray source for mass spectrometry. *Anal. Chem.*, **72**, 367–375 (2000).
23. Wachs, T. and Henion, J., Electrospray device for coupling microscale separations and other miniaturized devices with electrospray mass spectrometry. *Anal. Chem.*, **73**, 632–638 (2001).
24. Deng, Y., Henion, J., Li, J., Thibault, P., Wang, C. and Harrison, D.J., Chip-based capillary electrophoresis/mass spectrometry determination of carnitines in human urine. *Anal. Chem.*, **73**, 639–646 (2001).
25. Schuerenberg, M., Luebbert, C., Eickhoff, H., Kalkum, M., Lehrach, H. and Nordhoff, E., Prestructured MALDI-MS sample supports. *Anal. Chem.*, **72**, 3436–3442 (2000).
26. Ekström, S., Ericsson, D., Önnerfjord, P., Bengtsson, M., Nilsson, J., Marko-Varga, G. and Laurell, T., Signal amplification using "Spot-on-a-chip" technology for the identification of proteins via MALDI-TOF MS. *Anal. Chem.*, **73**, 214–219 (2001).
27. Cussler, E.L., *Diffusion Mass Transfer in Fluid Systems*, Cambridge University Press, Cambridge, 1997.
28. Manz, A., Harrison, D.J., Verpoorte, E. and Widmer, H.M., Planar Chips Technology for Miniaturization of Separation Systems: A Developing Perspective in Chemical Monitoring, In: Brown, P.R. and Grushka, E. (Eds.), Advances in Chromatography, **Vol. 33**. Marcel Dekker, New York, 1993.
29. Gygi, S.P., Rist, B., Gerber, S.A., Turecek, F., Gelb, M.H. and Aebersold, R., Quantitative analysis of complex protein mixtures using isotope-coded affinity tags. *Nat. Biotechnol.*, **17**, 994–999 (1999).
30. Canham, L. (Ed.), *Properties of Porous Silicon*. INSPEC, London, 1997.

31. Herino, R., Bomchil, G., Barla, K., Bertrand, C. and Ginoux, J.L., Porosity and pore size distributions of porous silicon layers. *J. Electrochem. Soc.: Solid-State Sci. Technol.*, **134**, 1994–2000 (1987).
32. Drott, J., Rosengren, L., Lindström, K. and Laurell, T., Pore morphology influence on catalytic turn-over for enzyme activated porous silicon matrices. *Thin Solid Films*, **330**, 161–166 (1998).
33. Clerc, P.-A., Dellman, L., Grétillat, F., Grétillat, M.-A., Indermühle, P.-F., Jeanneret, S., Luginbuhl, P., Marxer, C., Pfeffer, T.L., Racine, G.-A., Roth, S., Staufer, U., Stebler, C., Thiébaud, P. and De Rooij, N.F., Advanced deep reactive ion etching: a versatile tool for microelectromechnical systems. *J. Micromech. Microengng*, **8**, 272–278 (1998).
34. McAuley, S.A., Ashraf, H., Atabo, L., Chambers, A., Hall, S., Hopkins, J. and Nicholls, G., Silicon micromachining using a high-density plasma source. *J. Phys. D: Appl. Phys.*, **34**, 2769–2774 (2001).
35. McDonald, J.C., Duffy, D.C., Anderson, J.R., Chiu, D.T., Wu, H.K., Schueller, O.J.A. and Whitesides, G.M., Fabrication of microfluidic systems in poly(dimethylsiloxane). *Electrophoresis*, **21**, 27–40 (2000).
36. Becker, H. and Gartner, C., Polymer microfabrication methods for microfluidic analytical applications. *Electrophoresis*, **21**, 12–26 (2000).
37. Laurell, T., Biocatalytic porous silicon microreactors, In: Baltes, H., Fedder, G.K. and Korvink, J.G. (Eds.), *Sensors Update: Sensor Technology — Applications — Markets*. Wiley-VCH, Weinheim, 2002, pp. 3–32.
38. Ekström, S., Önnerfjord, P., Nilsson, J., Bengtsson, M., Laurell, T. and Marko-Varga, G., Integrated microanalytical technology enabling rapid and automated protein identification. *Anal. Chem.*, **72**, 286–293 (2000).
39. Bengtsson, M., Ekström, S., Marko-Varga, G. and Laurell, T., Improved performance in silicon enzyme microreactors obtained by homogeneous porous silicon carrier matrix. *Talanta*, **56**, 341–353 (2002).
40. Xue, Q.F., Dunayevskiy, Y.M., Foret, F. and Karger, B.L., Integrated multichannel microchip electrospray ionization mass spectrometry: analysis of peptides from on-chip tryptic digestion of melittin. *Rapid Commun. Mass Spectrom.*, **11**, 1253–1256 (1997).
41. Wang, C., Oleschuk, R., Ouchen, F., Li, J.J., Thibault, P. and Harrison, D.J., Integration of immobilized trypsin bead beds for protein digestion within a microfluidic chip incorporating capillary electrophoresis separations and an electrospray mass spectrometry interface. *Rapid Commun. Mass Spectrom.*, **14**, 1377–1382 (2000).
42. Ekstrom, S., Malmström, J., Wallman, L., Löfgren, M., Nilsson, J., Laurell, T. and Marko-Varga, G., On-chip microextraction for proteomic sample preparation of in-gel digests. *Proteomics*, **2**, 413–421 (2002).
43. Malmström, J., Larsen, K., Hansson, L., Löfdahl, C.-G., Jensen, O.N., Marko Varga, G. and Westergren-Thorson, G., Proteoglycan and proteome profiling of central human pulmonary fibrotic tissue utilizing miniaturized sample preparation: a feasibility study. *Proteomics*, **2**, 394–404 (2002).
44. Annan, R.S., Mculty, D.E. and Carr, S.A., *Proceedings of the 44th ASMS Conference on Mass Spectrometry and Allied Topics*, Portland, OR, 1996, p. 702.
45. Kussmann, M., Nordhoff, E., Rahbek-Nielsen, H., Haebel, S., Rossel-Larsen, M., Jakobsen, L., Gobom, J., Mirgorodskaya, E., Kroll-Kristensen, A., Palm, L. and Roepstorff, P., Matrix-assisted laser desorption/ionization mass spectrometry sample preparation techniques designed for various peptide and protein analytes. *J. Mass Spectrom.*, **32**, 593–601 (1997).
46. Gobom, J., Nordhoff, E., Mirgorodskaya, E., Ekman, R. and Roepstorff, P., Sample purification and preparation technique based on nano-scale reversed-phase columns for the

sensitive analysis of complex peptide mixtures by matrix-assisted laser desorption/ionization mass spectrometry. *J. Mass Spectrom.*, **34**, 105–116 (1999).

47. Stensballe, A. and Jensen, O.N., Simplified sample preparation method for protein identification by matrix-assisted laser desorption/ionization mass spectrometry: in-gel digestion on the probe surface. *Proteomics*, **1**, 955–966 (2001).
48. McGinley, M.D., Davis, M.T., Robinson, J.H., Spahr, C.S., Bures, E.J., Beierle, J., Mort, J. and Patterson, S.D., A simplified device for protein identification by microcapillary gradient liquid chromatography–tandem mass spectrometry. *Electrophoresis*, **21**, 1678–1684 (2000).
49. Gatlin, C.L., Kleeman, G.R., Hays, L.G., Link, A.J. and Yates, J.R., *Anal. Biochem.*, **263**, 93–101 (1998).
50. Figeys, D., Zhang, Y. and Aebersold, R., Optimization of solid phase microextraction capillary zone electrophoresis mass spectrometry for high sensitivity protein identification. *Electrophoresis*, **19**, 2338–2347 (1998).
51. Oleschuk, R.D., Shultz-Lockyear, L.L., Ning, Y.B. and Harrison, D.J., Trapping of bead-based reagents within microfluidic systems: on-chip solid-phase extraction and electrochromatography. *Anal. Chem.*, **72**, 585–590 (2000).
52. Sato, K., Tokeshi, M., Odake, T., Kimura, H., Ooi, T., Nakao, M. and Kitamori, T., Integration of an immunosorbent assay system: analysis of secretory human immunoglobulin A on polystyrene beads in a microchip. *Anal. Chem.*, **72**, 1144–1147 (2000).
53. Sato, K., Tokeshi, M., Kimura, H. and Kitamori, T., Determination of carcinoembryonic antigen in human sera by integrated bead bed immunoasay in a microchip for cancer diagnosis. *Anal. Chem.*, **73**, 1213–1218 (2001).
54. Palm, A., Wallenborg, S.R., Gustafsson, M., Hedström, A., Togan-Tekin, E. and Andersson, P., Integrated sample preparation and MALDI MS on a disc, In: Ramsey, J.M. and Van den Berg, A. (Eds.), *Micro Total Analysis Systems 2001*, Proceedings of the μTAS 2001 Symposium, Kluwer, Dordrecht, 2001, pp. 216–218.
55. Li, J., LeRiche, T., Tremblay, T.-L., Wang, C., Bonneil, E., Harrison, D.J. and Thibault, P., Application of microfluidic devices to proteomics research. *Mol. Cell. Proteom.*, **1**, 157–168 (2002).
56. Bergkvist, J., Ekstrom, S., Wallman, L., Löfgren, M., Marko-Varga, G., Nilsson, J. and Laurell, T., Improved chip design for integrated solid-phase microextraction in on-line proteomic sample preparation. *Proteomics*, **2**, 422–429 (2002).
57. Chiu, R.W., Walker, K.L., Hagen, J.J., Monnig, C.A. and Wilkins, C.L., Coaxial capillary and conductive capillary interfaces for collection of fractions isolated by capillary electrophoresis. *Anal. Chem.*, **67**, 4190–4196 (1995).
58. Stevenson, T.I. and Loo, J.A., A simple off-line sample spotter for coupling HPLC with MALDI MS. *LC–GC*, **16**, 54–58 (1998).
59. Preisler, J., Foret, F. and Karger, B.L., On-line MALDI-TOF MS using a continuous vacuum deposition interface. *Anal. Chem.*, **70**, 5278–5287 (1998).
60. Preisler, J., Hu, P., Rejtar, T., Moskovets, E. and Karger, B.L., Capillary array electrophoresis-MALDI mass spectrometry using a vacuum deposition interface. *Anal. Chem.*, **74**, 17–25 (2002).
61. Orsnes, H., Graf, T. and Degn, H., Stopped-flow mass spectrometry with rotating ball inlet: application to the ketone-sulfite reaction. *Anal. Chem.*, **70**, 4751–4754 (1998).
62. Orsnes, H., Graf, T., Degn, H. and Murray, K.K., A rotating ball inlet for on-line MALDI mass spectrometry. *Anal. Chem.*, **72**, 251–254 (2000).
63. Wall, D.B., Finch, J.W. and Cohen, S.A., A MALDI LC/MS interface for continuous sample deposition from reverse-phase HPLC. *Genet. Engng News*, 22 (2002).

64. Hensel, R.R., King, R.C. and Owens, K.G., Electrospray sample preparation for improved quantitation in matrix-assisted laser desorption/ionization time-of-flight mass spectrometry. *Rapid Commun. Mass Spectrom.*, **11**, 1785–1793 (1997).
65. Little, D.P., Cornish, T.J., Odonnell, M.J., Braun, A., Cotter, R.J. and Koster, H., MALDI on a chip: analysis of arrays of low femtomole to subfemtomole quantities of synthetic oligonucleotides and DNA diagnostic products dispensed by a piezoelectric pipet. *Anal. Chem.*, **69**, 4540–4546 (1997).
66. Önnerfjord, P., Nilsson, J., Wallman, L., Laurell, T. and Marko-Varga, G., Picoliter sample preparation in MALDI-TOF MS using a micromachined silicon flow-through dispenser. *Anal. Chem.*, **70**, 4755–4760 (1998).
67. Wallman, L., Drott, J., Nilsson, J. and Laurell, T., A micromachined flow-through cell for continuous pico-volume sampling in an analytical flow. *Digest of Technical Papers. The Eighth International Conference on Solid-State Sensors and Actuators and Eurosensors IX.* Stockholm, Sweden, June 25–29, 1995.
68. Laurell, T., Wallman, L. and Nilsson, J., Design and development of a silicon microfabricated flow-through cell for on-line picoliter sample handling. *J. Micromech. Microengeng*, **9**, 369–376 (1999).
69. Milotis, T., Kjellström, S., Nilsson, J., Laurell, T., Edholm, L.-E. and Marko-Varga, G., Capillary liquid chromatography interfaced to matrix-assisted laser desorption/ionization time-of-flight mass spectromatry using an on-line coupled piezoelectric flow-through microdispenser. *J. Mass Spectrom.*, **35**, 369–377 (2000).
70. Milotis, T., Kjellström, S., Nilsson, J., Laurell, T., Edholm, L.-E. and Marko-Varga, G., Ready-made matrix-assisted laser desorption/ionization target plates coated with thin matrix layer for automated sample deposition in high-density array format. *Rapid Commun. Mass Spectrom.*, **15**, 1–10 (2001).
71. Johnson, T., Bergquist, J., Ekman, R., Nordhoff, E., Schürenberg, M., Klöppel, K.-D., Müller, M., Lehrach, H. and Gobom, J., A CE-MALDI interface based on the use of prestructured sample supports. *Anal. Chem.*, **73**, 1670–1675 (2001).
72. Marko-Varga, G., Ekström, S., Helldin, G., Nilsson, J. and Laurell, T., Disposable polymeric high-density nanovial arrays for MALDI-TOF mass spectrometry — a novel concept. Part 1 — microstructure development and manufacturing. *Electrophoresis*, **22**, 3978–3983 (2001).
73. Ekström, S., Nilsson, J., Helldin, G., Laurell, T. and Marko-Varga, G., Disposable polymeric high density nanovial arrays for MALDI-TOF Mass Spectrometry — a novel concept: Part II — biological applications. *Electrophoresis*, **22**, 3984–3992 (2001).
74. Cohen, S.L. and Chait, B.T., Influence of matrix solution conditions on the MALDI-MS analysis of peptides and proteins. *Anal. Chem.*, **68**, 31–37 (1996).
75. Önnerfjord, P., Ekström, S., Bergkvist, J., Nilsson, J., Laurell, T. and Marko-Varga, G., Homogenous sample preparation for automated high throughput analysis with MALDI-TOF-MS. *Rapid Commun. Mass Spectrom.*, **13**, 315–322 (1999).
76. Strupat, K., Karas, M. and Hillenkamp, F., 2,5-Dihydroxibenzoic acid — a new matrix for laser desorption ionization mass-spectrometry. *Int. J. Mass Spectrom. Ion Processes*, **111**, 89–102 (1991).
77. Vorm, O., Roepstorff, P. and Mann, M., Improved resolution and very high-sensitivity in MALDI-TOF of matrix surfaces made by fast evaporation. *Anal. Chem.*, **66**, 3281–3287 (1994).
78. Westman, A., Nilsson, C.L. and Ekman, R., Matrix-assisted laser desorption/ionization time-of-flight mass spectrometry analysis of proteins in human cerebrospinal fluid. *Rapid Commun. Mass Spectrom.*, **12**, 1092–1098 (1998).

79. Miliotis, T., Kjellström, S., Önnerjord, P., Nilsson, J., Laurell, T., Edholm, L.-E. and Marko-Varga, G., Protein identification platform utilising microdispensing technology interfaced to MALDI-TOF MS analysis. *J. Chromatogr. A*, **886**, 99–110 (2000).
80. Miliotis, T., Kjellström, S., Nilsson, J., Laurell, T., Edholm, L.-E. and Marko-Varga, G., MALDI target plates pre-coated with a thin layer of matrix and nitrocellulose for automated sample deposition in high density array formats. *Rapid Commun. Mass Spectrom.*, **16**, 117–126 (2002).
81. Nilsson, J., Bergkvist, J., Ekström, S., Wallman, L. and Laurell, T., Compound microdispensing, In: Ramsey, J.M. and Van den Berg, A. (Eds.), *Micro Total Analysis Systems 2001*, Proceedings of the μTAS 2001 Symposium, Kluwer, Dordrecht, 2001, pp. 75–77.
82. Jespersen, S., Niessen, W.M.A., Tjaden, U.R., van der Greef, J., Litborn, E., Lindberg, U. and Roeraade, J., In: Burlingame, A.L. (Ed.), *Mass Spectrometry in the Biological Sciences*. Humama Press, Totowa, NJ, 1996, pp. 2221–2226.
83. Litborn, E., Emmer, Å. and Roeraade, J., Liquid lid for biochemical reactions in chip-based nanovials. *J. Chromatogr. B*, **745**, 137–147 (2000).
84. Litborn, E., Emmer, Å. and Roeraade, J., Chip-based nanovials for tryptic digest and capillary electrophoresis. *Anal. Chim. Acta*, **401**, 11–19 (1999).
85. Litborn, E., Emmer, Å. and Roeraade, J., Parallel reactions in open chip-based nanovials with continuous compensation for solvent evaporation. *Electrophoresis*, **21**, 91–99 (2000).
86. Ericsson, D., Ekström, S., Bergqvist, J., Nilsson, J., Marko-Varga, G. and Laurell, T., Dispenser aided on-target (in-vial) enzymatic digestion of proteins for MALDI-TOF analysis. *Proteomics*, **1**, 1072–1081 (2001).

Index